Agronomy and Economy of Black Pepper and Cardamom

Agronomy and Economy of Black Pepper and Cardamom

The "King" and "Queen" of Spices

K.P. Prabhakaran Nair

Formerly Professor, National Science Foundation, The Royal Society, Belgium & Senior Fellow, Alexander von Humboldt Foundation, The Federal Republic of Germany

ELSEVIER

AMSTERDAM • BOSTON • HEIDELBERG • LONDON • NEW YORK • OXFORD
PARIS • SAN DIEGO • SAN FRANCISCO • SINGAPORE • SYDNEY • TOKYO

Elsevier
32 Jamestown Road London NW1 7BY
225 Wyman Street, Waltham, MA 02451, USA

First edition 2011

British Library Cataloguing in Publication Data
A catalogue record for this book is available from the British Library

Library of Congress Cataloging-in-Publication Data
A catalog record for this book is available from the Library of Congress

ISBN: 978-0-323-16533-4

This book has been manufactured using Print on Demand technology. Each copy is produced to order and is limited to black ink. The online version of this book will show color figures where appropriate.

Contents

Dedication

With much love I dedicate this book to Prahlad, our grandson, who brought sunshine into our family, and to Black and Charlie, our canine fleet, who are always a source of joy to me.

Preface

This book is about world's two most important spice crops: Black pepper, known as the "King" of spices and Cardamom, known as the "Queen" of spices.

Black pepper has a checkered history, dating back to the times of Queen Sheeba and King Solomon (B.C. 1015–B.C. 66), which has influenced the destiny of nations and their people, spread across the world, both economically and culturally. Today, Black pepper commands the leading position among the different spices as the spice of immense commercial importance in world trade, and it is finding its way onto the dining table of millions around the world, even on the European and North American continents and in Japan. Pepper use ranges from a simple dietary constituent to that of immense pharmacological benefits. Though beset with many problems, both agronomic and economic, it is a safe bet that pepper will emerge as the world's most sought-after spice; its global demand is predicted to escalate colossally to about 280,000 metric tons by the year 2020, which will further climb to 360,000 metric tons by 2050.

This book's exhaustive review details the various constraints in enhancing pepper productivity and charts contours of a new course of action. Among the primary constraints in pepper production is the absence of an ideal ideotype that combines many positive traits to boost production potential while, at the same time, resisting the ravages of nature—such as the dreaded disease foot rot, caused by the *Phytophthora* fungi. The book also includes an exclusive chapter on a revolutionary soil management concept, developed by the author over three decades of research in Europe, Africa, and Asia, now globally known as the "nutrient buffer power concept," which has challenged the conventional wisdom of industrial agriculture, euphemistically known as the "green revolution." The book specifically details the work of the author on pepper nutrition with zinc, the most important micronutrient, which is suspected to be an intermediary in causing the *Phytophthora* onslaught in the quick wilt disease.

Cardamom, popularly known as the "Queen" of spices has an interesting history dating back to the Vedic period (ca. B.C. 3000) and is among the ingredients poured into the sacrificial fire during the Hindu marriage. Today, cardamom commands a leading position among the spices of immense commercial importance and is finding its way into the dietary habits of millions around the world—even on the European and North American continents and in Japan, hitherto unaccustomed to its use. Cardamom use ranges from a simple dietary constituent to that of immense pharmacological benefits. Although beset with many problems, both agronomic and economic, it is certain that next to black pepper, cardamom will emerge as an important commercial spice in the world trade. India was the world leader in cardamom production, but, starting in 1970, the country began to slide in both production and productivity. Guatemala took over as the leading cardamom producer, like what Vietnam could achieve in growing pepper. However, Guatemalan cardamom is of inferior quality, the same being the case with Vietnam pepper.

Among the primary constraints of production, it is the absence of an ideal ideotype, as in the case of black pepper, that is the biggest constraint in enhancing productivity. An ideal ideotype must combine superior productivity traits with strong resistance capacity to the dreaded Katte disease, found widespread in Karnataka State of southern India. Fertility management of cardamom is still rooted in textbook knowledge, but the book contains an exclusive chapter on the "nutrient buffer power concept," based on the author's research, with specific reference to cardamom nutrition with the major plant nutrient potassium, which is required in abundance for optimum cardamom production. The chapter details the author's research with reference to cardamom production in the two most important cardamom-growing states of the southern Indian region: Kerala and Karnataka.

1 The Agronomy and Economy of Black Pepper (*Piper nigrum* L.)—The "King of Spices"

1.1 Introduction

Known as the "King" of spices, black pepper (*Piper nigrum*), a perennial crop of the tropics, is economically the most important and the most widely used spice crop of the world. The history of spices is very much entwined with the history of mankind. But, within the family of spices, black pepper predominates. In ancient Egypt, when the mummified body of the Pharaoh was laid to rest in the Pyramids, it was black pepper (along with gold and silver) that was kept adjacent to the body, in the belief that even in the afterlife this very important spice would be of use. The ancient scriptures, Bible, Koran, and the Vedas mention the use of spices. According to the Bible, it was during the royal visit of Queen Sheeba to King Solomon (BC 1015–BC 66) that a caravan load of spices, primarily pepper, was presented by the former to the latter. Nearly 3000 years before the birth of Christ, both Babylonians and Assyrians were trading in spices, primarily black pepper, with the people of the Malabar coast in the state of Kerala on the Indian subcontinent. Also, the ancient Indian medical texts, such as *Ashtangahridaya* and *Samhitas*, mention the use of pepper in rare and unique medical formulations. That spices, and in particular, pepper, had such a lasting impact on the economic prosperity of places is revealed by the fact that cities like Alexandria, Genoa, and Venice can trace their economic prosperity back to the vigorous trade in spices (Rosengarten, 1973). Parry (1969) observed that due to the increased demand and consumption of pepper in England and Europe, a guild known as "Pepperers"—the wealthiest of the merchants—was established in London. The high price of pepper made it the exclusive commodity in use by the rich for culinary purposes. But, it is interesting to note that with the arrival of pepper, the Western kitchen transformed when "dishes took on a fullness of flavor previously unknown, beverages glowed with a redolent tang, and life experienced a new sense of warmth and satisfaction" (Parry, 1969). King Solomon of Israel and the Phoenician King, Hiram of Tyre, obtained their spices from the coast of Malabar (Rosengarten, 1973). Though the Jews and Arabs were well into the spice trade by this time, it was the latter who had a domineering role, which was later usurped by the Romans. However, the Arabs were privy to the knowledge of the country of origin of pepper

Agronomy and Economy of Black Pepper and Cardamom. DOI: 10.1016/B978-0-12-391865-9.00001-3 and even secretly kept the sea route knowledge to themselves. But, the onslaught of the Roman Empire changed all that during the first century A.D. when Rome captured Egypt. It was around A.D. 40, during the reign of Roman Emperor Claudius that the enterprising Greek merchant mariner Hippalus, who—after discovering the full power and velocity of wind movement of the Indian ocean, a secret guarded by the Arabs—let it become known that a round trip to Egypt *via* India could be completed in less than 1 year (Rosengarten, 1973). A consequence of the great economic scale of this discovery was that when the sea route came to be known, the dependence on the overland trade routes declined substantially; several wealthy cities on the overland trade route went into economic penury (Ummer, 1989). Hippalus and many others reached the coast of Malabar and returned to Rome with loads of pepper and other spices. The Roman dominance on the spice trade completely eclipsed the Arabs. Interestingly, among all the spices, it was pepper that the Romans fancied most, despite its "obnoxious pungency" as noted by Pliny the Elder in his Natural History compiled between A.D. 23 and A.D. 79. It is also interesting to note that just about the time Arabs were getting involved in the spice trade with the Malabar coast, the Chinese also entered the fray. As early as the first century A.D., a royal messenger reached the Malabar coast in search of spices, and the trade relationship between China and the Malabar coast began to flourish as recorded by travelers, such as Sulaiman, who reached the Malabar coast in A.D. 851 (Ummer, 1989). It was the Chinese who played an important part in spreading pepper to southeast Asia and far-east Asia, and it was during the voyages of Zheng He (1405–1433) that China imported as much pepper as the total quantity imported into Europe during the first half of the tenth century (T'ien, 1981). The Chinese trade was an imperial monopoly. In the fifteenth century, soldiers in China were partly paid in pepper, as were government officials, as observed by T'ien, (1981). The fabulous Chinese ships that carried pepper were greatly praised by the venerable explorer Marco Polo (Mahindru, 1982).

The colonization of the Indian subcontinent had much to do with the pepper trade. It was the lure of the spice trade that led Vasco de Gama, the great Portuguese explorer, to discover the sea route to India, and he landed on the Malabar coast at the Kappad beach near the presently named city of Kozhikode (Calicut) on 20 May 1498. Vasco de Gama and his men returned to Portugal immensely rich, but in the process also put an end to the Arab trade. The Portuguese King, Dom Mannuel, was so greatly impressed with the pepper bounty that Gama brought back home that he sent a naval contingent with a fleet of 13 battle ships to India under the command of Pedro Alwarez Cabral in A.D. 1500; he went further to declare sovereignty over India, along with other countries such as Ethiopia and Arabia (Rosengarten, 1973). The Portuguese domination of Kerala through pepper production and trade was so complete that this tiny state of India turned out to be the cradle of world pepper. The Portuguese rulers were ruthless, and their only aim was to make the most out of the pepper trade. However, the scene changed in the first quarter of the seventeenth century with the arrival of the Dutch. The Dutch were temporarily successful in elbowing out the Portuguese until the British came on the scene in A.D. 1600. In hindsight, what is most astonishing is how a trade war in pepper between the Dutch and the British led to the establishment of the British Empire on the Indian subcontinent.

The historians Collins and Lapieree, in their monumental work *Freedom at Midnight* (1976), make a remarkable observation as follows.

Sometimes history's most grandiose accomplishments have the most banal of origins. Great Britain was set on the road to the great colonial adventure for five miserable shillings. They represented the increase in price of a pound of pepper proclaimed by the Dutch privateers who then controlled the spice trade. Incensed at what they considered a wholly unwarranted gesture, 24 merchants of the city of London gathered on the afternoon of 24 September 1599 A.D. in a decrepit building on Leaden Hall Street. Their purpose was to find a modest trading firm with an initial capital of £72,000 subscribed by 125 shareholders. Only the simplest of concerns, that is, profit, inspired their enterprise, which expanded and transformed, would ultimately become the most noteworthy creation of the age of imperialism—the British Raj.

The British East India Company came into being on 31 December 1600 (with the stamp of approval by Queen Elizabeth I). Just 36 years later (1636), when the 500-ton ship *Hector* landed in the Surat port, north of Bombay, the Company laid the long and tortuous road to the subjugation of the vast millions of Indians through the pepper trade. In more than one sense, pepper was the cause of India losing its sovereignty.

It was only in the last quarter of the eighteenth century that the Americans entered the pepper trade. The first sponsored trip to the East Indies was organized by Capt. Jonathan Carnes in 1795. Though the American pepper trade flourished until 1810, it later declined, coinciding with the American Civil War in 1861. Compared with the Portuguese, Dutch, and British, the impact of Americans on the pepper trade was only marginal. The Americans traveled to Sumatra to fetch pepper. By 1933, pepper was introduced to Brazil, and in 1938, it reached the Republic of Malagasy. By 1954, pepper was introduced to the African continent.

Within the Indian Republic, the tiny state of Kerala can pride itself as the home of pepper, in particular, the coastal region of Malabar, in the state of Kerala, which accounts for 95% of the country's area and production (Anon., 1997). Besides Kerala, two other states in the southern region—namely, Tamil Nadu and Karnataka—contribute the remainder. The first research station of pepper in the world was established in Kerala in a small town named Panniyur on the Malabar coast during 1952 and 1953. In addition, the first hybrid pepper—Panniyur-1—in the world was released by this station in 1966 (Fig. 1.1). Following research at Panniyur, pepper research began in Sarawak (Malaysia) in 1955. Following the success of the "green revolution" in India, several co-ordinated research projects catering to the specific needs of individual crops were established by the Indian Council of Agricultural Research, headquartered in New Delhi. Thus, the All India Co-ordinated Research Project on Spices, with an intense focus on pepper, was established in 1971. More than 30 years later, the project has included other spice crops as well, yet pepper continues to receive the most attention against the background of the current liberalization and globalization process and the World Trade Organization (WTO)–mandated changes. The future of pepper is most crucial to the economy of India vis-à-vis the economy of the state of Kerala, where it is the economic mainstay.

Figure 1.1 Panniyur-1, the first pepper hybrid developed in the world at the Panniyur Pepper Research Station, in Kerala State, India.

Many other tropical countries have made concerted efforts to grow pepper in view of its global economic importance, the most important being Indonesia, where the International Pepper Community (IPC) is headquartered in Jakarta. Indonesia is second to India as a pepper-growing country. It was either through the Polynesian seafarers or through the Babylonian–Chinese sea route linking the Malabar coast and southeast and far-east Asia that pepper reached Indonesia. Indonesia, known as the Dutch East Indies during the pre–World War II period, was the largest pepper producer. It was during the Japanese occupation during the war that many plantations were abandoned and production declined sharply. During the pre-war period, Indonesia had close to 30,000 ha of pepper (Lawrence, 1981). Malaysia and Sri Lanka are the other two major pepper producers. It was the European settlers who introduced pepper to Malaysia, while in Sri Lanka, pepper was grown as a mixed crop with other crops like cocoa. The foreign occupation of Sri Lanka helped pepper cultivation expand in the country. In the southeast Asian countries such as Thailand, Vietnam, Cambodia, South Korea, and parts of south China, pepper cultivation took hold in the post-war years. Within the southeast Asian countries, Vietnam is beginning to emerge as a major pepper grower. Within the South Pacific islands, Fiji is the most important pepper-growing country. In South America, Brazil is the leader, followed by Mexico, Guatemala, Honduras, Saint Lucia, Costa Rica, and Puerto Rico. Within the African continent, Madagascar leads the pack followed by Malawi, Zimbabwe, Benin, Kenya, Nigeria, Cameroon, Congo, and Ethiopia.

Within Asia, where pepper production is concentrated, Malaysia takes third place next to Indonesia, with an Agricultural Research Station in Kutching, Sarawak. Indonesia, which takes second place, has the Research Institute for Spice and Medicinal Plants in Bogor. Though India tops the list among the producers in acreage, with a total area of 191,426 ha and a production of 56,200 metric tons, its productivity is the lowest in the world with just 294 kg ha^{-1}, while Thailand, with a total

Table 1.1 Acreage, Production, and Productivity of Major Pepper-Growing Countries in the Last Decade of the 20th Century

Country	Area (ha)	Production (metric tons)	Productivity (kg ha^{-1})
Brazil	26,500	23,400	883
China	13,170	11,045	839
India	191,426	56,200	294
Indonesia	110,580	45,240	409
Madagascar	4,228	2,160	511
Malaysia	8,960	16,920	1,888
Mexico	1,294	1,112	859
Sri Lanka	12,080	5,058	419
Thailand	2,808	10,091	3,594
Vietnam	15,700	17,266	1,100

area of only 2808 ha, tops the list with a productivity of 3594 kg ha^{-1} (Table 1.1). The International Trade Center (ITC) in Geneva, Switzerland, estimates the current trade in spices at 400,000–450,000 metric tons with a total value of US $1.5–$2 billion annually. With an annual growth rate of 3.6% in quantity and 8.4% in value in spices, pepper contributes 34% of total trade in spices. Within the industrialized west, Denmark tops the list in pepper consumption, followed closely by Germany and Belgium. The USA is a sizable consumer, while Canada and Switzerland tail the list.

Post-Doha (World Trade Organization (WTO) held an important meeting in Doha, Quattar, on global trade regulations, including agriculture) negotiations, agriculture will increasingly play a crucial role in the world economy. Among the spices, pepper will play a major role. By 2010, projected world consumption will reach 230,000 metric tons, which scales up to 280,000 metric tons by 2020, which means an annual increase of 5000 metric tons. Present production is close to 200,000 metric tons. For the next two decades, close to 100,000 metric tons will be needed to balance the projected demand and consumption. Worldwide, especially in the industrialized countries, there is a growing demand for premium organically grown pepper. The potential for the organic food market is close to US $8 billion now in the USA, which is followed by Germany and Japan, each with a market share of close to US $2.5 billion. A substantial part of this market will be for pepper. This would imply that future production strategies would need to increasingly focus on clean pepper production, which has to withstand both biotic and abiotic stresses without recourse to high input chemical technology—the hallmark of the so-called green revolution.

There are areas of pepper production that simultaneously pose great challenges while opening up new avenues. One of the most daunting in the former category is the evolution of a pepper variety that is totally resistant to "foot rot," caused by the fungus *Phytophthora*, which has wiped out many pepper plantations. Also, pepper nutrition is still far from being thoroughly understood. The fact that it is a perennial crop adds to the lack of thorough understanding. Despite the complexity of soil science and emergent soil management practices, the basic concept of soil as a medium of plant growth can be expected to persist for an indefinite length of time (Nair, 1996). But it is becoming clearer that the earlier views on soil as merely a "supportive medium" for plant growth is giving place to new ones on managerial concepts of this supportive medium. This is amply illustrated by the shift in focus from the green revolution phase of the 1960s to mid-1970s—where application of increasing quantities of soil inputs, such as fertilizers and pesticides, was emphasized—to the sustainable agriculture phase of the early 1980s continuing to the present; sustainable agriculture places more reliance on biological processes by adopting genotypes to adverse soil conditions, enhancing soil biological activity, and optimizing nutrient cycling to minimize external inputs (such as fertilizers) and maximize their efficiency of use. In fact, the paradigm of the earlier phase has given way to an emergent new paradigm (Sanchez, 1994), and this is clearly reflected in the dialogue of the world leaders during the Earth Summit in 1991 in Rio de Janeiro, Brazil, where Agenda 21 incorporated six chapters on soil management (Keating, 1993). This review on pepper, while discussing its overall production profile in the world, will place a special emphasis on the second paradigm inasmuch as prescriptive soil management for pepper production is concerned with regard to understanding the soil nutrient bioavailability and its efficient management in the pepper production. On account of the paucity of published literature on this aspect, the focus will only be with regard to specific nutrients, such as zinc, which is becoming increasingly important in pepper production.

1.2 The Pepper Plant—Its Botany and Chemistry

1.2.1 Pepper Botany

Black pepper (*P. nigrum*) is a perennial plant and derives its name—"Piper"—perhaps from the Greek name for black pepper, *Piperi* (Rosengarten, 1973); most of the European names for black pepper were derived from the ancient Indian language Sanskrit, such as *Pippali*, the name for long pepper (*P. longum*). It was the great botanist Linnaeus (1753) who established the genus *Piper* in his work *Species Plantarum*. In this monumental work, Linnaeus recognized 17 species in the Piper family, all of which were included in the same genus. In 1974, Ruiz and Pavon introduced the second genus in the family, namely, Peperomia (Trelease and Yuncker, 1950). The family name Piperaceae was used for the first time in 1815 by L. C. Rich in Humboldt, Bonpland, and Kunth's *Nova Genera et Species Plantarum* (Yuncker, 1958). All the species known in the family Piperaceae during the early years of systematic classification were included in the classical monographic study *Systema Piperacearum* published in 1843 by F. A. W. Miquel.

It was Rheede who made the earliest record of the description of Piper in the Indian subcontinent in 1678. In his *Hortus Indicus Malabaricus*, the earliest printed document of plants on the Malabar coast, Rheede described five types of wild pepper that included both black and long pepper. Linnaeus (1753) included 17 species from India in his monumental work *Species Plantarum.* The first major study of the *Piper* spp. from the Indian subcontinent was that of Hooker (1886) in his book *Flora of British India.* However, the most authoritative floristic study of the Western Ghats of southern India was that of Gamble in 1925 in his book *Flora of Presidency of Madras,* in which the following 13 species with their taxonomic keys were described: *P. argyrophyllum, P. attenuatum, P. barberi, P. brachystachyum, P. galeatum, P. hapnium, P. hookeri, P. hymenophyllum, P. longum, P. nigrum, P. schmidtii, P. trichostachyon,* and *P. wightii.* Following the publication of Gamble's text, no new additions to the list were made until 1981, when a new species, *P. bababudani*—from the Bababudin hills of the state of Karnataka (the neighboring state of Kerala)—was added (Rahiman, 1981); however, this was not published. Ravindran et al. (1987) reported a new species, *P. silentvalleyensis*—the only bisexual wild species from the world-renowned Silent Valley in the Western Ghats. Other new species are *P. pseudonigrum* (Velayudhan and Amalraj, 1992) and *P. sugandhi* (Nirmal Babu et al., 1993a).

Black pepper cultivars could possibly have originated from wild ones through the process of domestication and selection. More than a hundred cultivars are known, but most have vanished due to the onset of devastating diseases like foot rot and replacement by hybrids. Human migration has contributed to the spread of these cultivars. Pepper distribution is extensive in moist evergreen forests and, to a lesser extent, in semievergreen and moist deciduous forests of the Western Ghats of southern India, growing from almost sea level to a height of 1500 m above mean sea level. Population structure of any species is determined mainly by the breeding system of the species; by the pollen, fruit, and seed dispersal mechanism; and the presence or absence of isolation mechanisms. In *Piper*, male, female, and hermaphrodite forms exist. The cultivated *P. nigrum* is monoecious, having hermaphrodite flowers, while the wild ones are mostly dioecious. Human selection has played a major role in the directional evolution of hermaphroditism in the cultivated pepper. Predominantly self-pollinated, pepper pollen dispersal is aided by rain or dewdrops and also by geitonogamy—the gravitational descent of pollen. Though flowers are protogynous, in the absence of an active pollen transfer mechanism, protogyny is ineffective in outbreeding. Active and efficient pollen and seed dispersal mechanisms ensure gene flow within and between population segments, leading to the establishment of intergrading populations. Absence of such a mechanism in *Piper* ensures effective isolation barriers among individuals and population units. Segregation in the seedling progenies leads to variations in such units, as well as mutations and chance crossing followed by segregation. Any such variation stemming in population is fixed immediately because of the vegetative mode of propagation, and such a unit may, over the course of time, diverge from other similar units. Quite often, different types of *P. nigrum* or different species climb up a single tree, which enhances the chances of outcrossing, which results in hybrid seedlings. Progenies from such chance crosses grow and later climb up the same or nearby trees and chance outcrossing with the parental vine or its clonal or other seedling progenies, which result in further back crossing or hybrid progenies—all of which lead to substantial variation within the population. These forces acting together would have most likely contributed to the evolution of many present-day pepper cultivars (Ravindran et al., 1990).

Though pepper has originated in the evergreen forests of the Western Ghats in southern India, with extensive occurrence of wild pepper in the less disturbed forest areas, no study has so far been carried out to find out the origin of *P. nigrum*. The basic chromosome number of *Piper* is $x = 13$ and that of *P. nigrum* $2n = 52$, a tetraploid (Mathew, 1958, 1972; Jose and Sharma, 1984). On the basis of morphological and biosystematic studies, Ravindran (1991) reported that three species—namely, *P. wightii, P. galeatum*, and *P. trichostachyon*—could be the putative parents of *P. nigrum*. These three are woody climbers with similar texture and leaf morphology. Their spikes and fruits are more similar to that of *P. nigrum* than those of other species. Of the three, *P. wightii* and *P. galeatum* are the most probable ancestors of *P. nigrum*.

Pepper is a shade-loving plant that requires a constant moisture supply during dry spells because it has a high evapotranspiration coefficient (Raj, 1978). However, even under favorable soil moisture conditions, when exposed to direct sunlight the plant develops certain physiological disorders (Vijayakumar et al., 1984). Pepper is highly sensitive to the growth-light regime (GLR), and plant parts exposed to high GLR produce more fruits per unit surface area than those exposed to low GLR (Montaya et al., 1961). Where grown in permanently shaded situations, productivity is poor. Under such conditions, plants respond positively to radiation, and a positive correlation has been shown to exist between productivity and radiation. Except in the case of "bush pepper," normal pepper vines are grown on artificial support. These supports are called "standards." Ramadasan (1987) reports three types of pepper canopies. Canopy shaping is largely based on the shading provided by live support and that of the adjoining shading trees. When dead supports such as brick pillars or reinforced concrete poles are used, as in Thailand, the canopy does not taper at the top. Trees that grow adjacent to dead supports lead to tapering of the canopy at the base due to partial shading. Such tapering at the base is not observed when there are no competing trees. In Sarawak, the wood of Bornean ironwood (*Eusideroxylon zwageri*) is used for support, and generally, no shade tree is grown. In such dead supports, the top of the canopy unfolds like an umbrella.

The pepper plant is endowed with two advantages that many others do not have. First, it can be vegetatively propagated; second, it is viable for sexual reproduction. Both offer a large scope for the exploitation of hybrid vigor as well as selection breeding. Published literature show that clonal selection, hybridization, open pollinated progeny selection, mutation, and polyploidy have been successfully experimented with to improve pepper yield and impart desirable plant traits. In recent times, biotechnological methods are also being employed to impart pathogen resistance.

Quality is the most important criterion in breeding as far as spices are concerned, and pepper is no exception. The quality of pepper is decided mainly by the amounts of piperine and oleoresin (essential oil). An alkaloid, piperine contributes to the pungency of the berry, and oleoresin content enhances the flavor. The unique pepper flavor is influenced by the large number of chemical compounds present in the essential oil, and a number of genes control this unique character.

Conservation of pepper genetic resources is the key to sound pepper breeding. Among the 18 biodiversity hot spots on this planet, the Western Ghats of southern India is one of the most unique, with its different forest types—such as tropical wet evergreen, tropical moist and dry deciduous, montane subtropical and temperate ones. It is on the western side of the Western Ghats that the population and diversity of *Piper* is in abundance. It is important to realize that during the past century, the Western Ghats have been ravaged the most (leading to ecological breakdown) due to clandestine timber felling and uncontrolled human encroachment. The Western Ghats is home to spices that are endemic to the region. Of the 17 species of *Piper* found here, 11 are endemic to the region (Ravindran and Peter, 1995). More than a hundred cultivars were under cultivation in the region, but most of them have vanished from the pepper growers' gardens, probably due to foot rot (*Phytophthora*) and slow decline as well as replacement of old and unproductive cultivars by the few high-yielding elite ones. It takes millennia for nature to nurture plant diversity, but human activity, productive or otherwise, can change all that in a very short span of time.

As part of the conservation efforts, some pepper-growing countries, notably India, have made a concerted effort to collect and catalogue pepper germplasm in a systematic manner. Mention should be made of the efforts in this direction undertaken at the Indian Institute of Spices Research at Kozhikode, Kerala State. The Institute has established a National Conservatory for in situ conservation. Conservation is carried out in four stages (Ravindran and Nirmal Babu, 1994) as follows: (1) conservation as a nursery gene bank by trailing each accession in split-bamboo pieces of about 2–3 ft long serially, which will be maintained and continuously multiplied; (2) conservation in the clonal repository, where 10 rooted cuttings from each accession are maintained; (3) conservation in the field, as a field gene bank, by planting accessions for pilot yield performance and characterization; and (4) conservation *in vitro* and in cryogen banks. With the establishment of the All India Co-ordinated Research Project on Spices in 1971, collection and conservation of pepper germplasm on a smaller scale has also been made by the Panniyur Pepper Research Station (the first pepper research station established in the world) in the state of Kerala; the Pepper Research Station at Sirsi, in the state of Karnataka; the Horticultural Research Station, Chintapalli, in the state of Andhra Pradesh; and the Horticultural Research Station, Yercaud, in the state of Tamil Nadu. Among these centers, the research station in Yercaud has the largest collection, but it is the Indian Institute of Spices Research that can pride itself as the one holding the largest collection in the world. On a much smaller scale, collections and conservation have also been made, notably in Indonesia and Malaysia, with the former holding 40 cultivars and 7 *Piper* species and Sarwak (Malaysia) holding 18 cultivars of *P. nigrum*, 18 identified *Piper* spp., and 98 unidentified accessions (De Waard, 1984). Despite the fact that they are not closely related to the cultivated black pepper, the vast number of *Piper* spp. found in central and southern America, northeastern India, Indonesia, and Malaysia have not been collected and systematically catalogued. The vast germplasms in these regions could form a very important reservoir of gene pool to develop varieties with in-built disease and pest resistance (Table 1.2).

Table 1.2 Some Improved Pepper Varieties from Major Pepper-Growing Countries

Country of Origin and Name of Variety	Average Yield ($kg\,ha^{-1}$)	Remarks
India		
Panniyur-1	1242	Well suited to most regions; performs rather poorly at higher elevations and under shade
Panniyur-2	2750	Well suited to most regions; shade tolerant
Panniyur-3	1953	Well suited to most regions; rather late in maturity
Panniyur-4	1277	Consistently good yielder; suited to most regions
Panniyur-5	1098	Most important characteristic is tolerance to nursery diseases and shade
Subhakara	2352	Suited to most regions; berries of good quality
Sreekara	2677	Suited to most regions; berries of good quality
Panchami	2828	Suited to most regions; late in maturity
Pournami	2333	Good tolerance to root knot nematode
PLD-2	2475	Suited to most regions; berries of high quality
Indonesia		
Natar1	–[a]	High yield; particularly tolerant to foot rot and nematodes
Natar 2	–	High yield; particularly tolerant to foot rot and nematodes
Malaysia		
Semongok Perak	–	High yield
Semongok Emas	–	Uniquely high yield; tolerant to foot rot, black berry disease, and pepper weevil
Madagascar		
Sel. IV.1	–	High yield
Sel. IV.2	–	High yield
Sri Lanka		
PW 14	–	Claimed to be resistant to *Radopholus similis*

[a]Average yield unknown.

A new avenue for pepper improvement is through biotechnological research, which of late has made great strides in many areas of biology. The biotechnological approaches permit the researchers to manipulate plant tissues and cells *in vitro*. A better understanding of genetics is provided by tissue culture techniques and investigations at the molecular, cellular, and organismal levels. The possibilities of cell culture and plant tissue culture for breeding and plant propagation purposes have led to large-scale commercial exploitation. Technical perfection by manipulating plant cells at the molecular level through recombinant DNA technology has opened up enormous possibilities in the creation of transgenic plants, and this is an area that is assuming industrial proportions. Inasmuch as pepper is concerned, one area where recombinant DNA technology could be successfully exploited is the evolution of pepper varieties tolerant to foot rot caused by *Phytophthora capsici*. None of the varieties in cultivation is tolerant to this disease. Within the *Piper* spp., a distant relative of *P. nigrum*—namely, *P. colubrinum* from South America—is resistant to *Phytophthora* and, hence, biotechnological approaches that will enable transfer of resistance-governing genes from *P. colubrinum* to *P. nigrum* will be the only mode to contain the devastation from this dreaded disease.

Another area of interest is micropropagation, where *in vitro* culture methods for pepper cloning is done. Micropropagation was first introduced by Broome and Zimmerman (1978). For micropropagation, several plant parts—such as shoot tips, nodal segments, and apical meristems—from both juvenile and mature plants have been used (Agarwal, 1988; Joseph et al., 1996; Lissamma et al., 1996; Mathews and Rao, 1984; Nazeem et al., 1993; Nirmal Babu et al., 1993a, b; Philip et al., 1992). However, a serious limitation to micropropagation is the incidence of bacterial contamination (Kelker and Krishnamoorthy, 1996), though contamination is minimal in seedling explants. The authors (Kelker and Krishnamoorthy, 1996) suggest the incorporation of antibiotics in the culture media to arrest the endogenous bacterial contamination in the *in vitro* pepper cultures. Field establishment of micropropagated pepper is feasible, though endogenous bacterial contamination and phenolic exudates from the cut surface could adversely affect field establishment (Fitchet, 1988a, 1988b; Raj Mohan, 1985). The authors suggest treating the explants with fungicides prior to routine sterilization followed by frequent transfer to fresh medium to keep in check the problem of contamination. Madhusudhanan and Rahiman (1996) suggest the use of activated charcoal, at the rate of $200\,mg\,l^{-1}$, which could reduce browning of the explants and culture medium. Micropropagation has been in use for quite some time (Chua, 1981; Fitchet, 1988a, 1988b). Culture media commonly used in micropropagation are those of Murashige and Skoog (1962), Schenk and Hildebrandt (1972), and McCown and Amos (1979). Indole acetic acid (IAA), naphthalene acetic acid (NAA), and 2,4-dichlorophenoxy acetic acid (2,4-D) are some of the commonly used growth regulators and additives used in the culture formulation.

One other area of pepper improvement is the *Agrobacterium* mediated gene transfer. Reference to the technique is made by Sasikumar and Veluthambi (1994) and Sim et al. (1995). The technique can also be helpful in transferring disease resistance genes from *P. colubrinum* to *P. nigrum* in the light of the foot rot disease. Work on transgenic pepper to evolve delayed ripening ensuring uniform maturity—which will ensure reduction in labor costs for harvest—in Sarawak (Malaysia) has led to promising results.

Conservation of pepper germplasm assumes high practical importance in improving the crop for enhanced productivity and acquisition of many desirable traits, the most important being tolerance to diseases such as *Phytophthora*-caused foot rot. Normal conservation of crop germplasm is through seed banks. Because pepper is propagated only vegetatively (seeds are heterozygous and unpredictable in behavior) *in vitro* storage of germplasm is a viable alternative. *In vitro* conservation of pepper germplasm and its related species—such as *P. barberi*, *P. colubrinum*, *P. beetle*, and *P. longum*—by defining protocols have been made by Geetha et al. (1995) and Nirmal Babu et al. (1996) by maintaining cultures at reduced temperatures in the presence of osmotic inhibitors at low nutrient levels and also by reducing evaporative losses to a minimum using closed containers. When the conserved materials were transferred to a multiplication medium after storage, all species exhibited normal multiplication rates. When the normal-sized plantlets were transplanted into field soil, an establishment rate exceeding 80% was obtained. Normal pepper plants, morphologically similar to mother plants, were observed to develop.

There is a steep decline in viability in pepper seeds associated with moisture depletion. Seeds cryopreserved in liquid nitrogen at − 196 °C with an initial moisture percentage of 12 have a survival rate of only 45%, while at 6% moisture it reduced to just 10.5% (Chaudhary and Chandel, 1994), which clearly shows how seed moisture is crucial to seed survival. There is yet another manner in which pepper plants can be propagated: by producing synthetic seeds, which consist of somatic embryos or shoot buds encapsulated in a protective coating that is biodegradable. The system facilitates not only low-cost propagation but also germplasm conservation and exchange. Encapsulation of disease-free bud employing tissue culture techniques helps production of disease-free plantlets. Pepper shoot buds 0.5 cm long were used for production of synthetic seeds, which could be stored up to 9 months in sterile water. The method is described by Sajina et al. (1996).

1.2.2 Pepper Chemistry

It is the unique pungency and aroma of pepper that has both intrigued and fascinated pepper chemists. As already mentioned, the essential oil in the berry—oleoresin—contributes to the aroma, while the alkaloid piperine imparts the unique pungency. Oleoresin has very great commercial value and is extracted from the dry powdered berries. Dramatically put, it is the chemistry of pepper emanating from oleoresin that imparts that rare and unique blend of both aroma and flavor that has been the cause of wars and enslavement (as much as love and tragedy), which has been detailed in the introductory part of this review. The earliest investigations on pepper chemistry go back more than half a century (Guenther, 1950), and subsequent research has enriched it (Govindarajan, 1977; Lawrence, 1981; Purseglove et al., 1981; Wealth of India, 1969).

The various compounds occurring in *Piper* spp. are listed by Parmar et al. (1977). It was in the early nineteenth century that the first report on essential oil in pepper was made by Dumas and later by Subeiran and Capitaine (Guenther, 1950). These researchers concluded that pepper oil is almost free of oxygenated constituents. By treating a fraction of the oil boiling at 176 °C with acid and alcohol, Eberhardt obtained terpin hydrate and the presence of 1-phellendrene, caryophyllene, and tentatively dipentene were reported by Schimmel and company and by Schreiner and Kremers (Guenther, 1950). The steam distillation of essential oil obtained from dry powdered Malabar pepper berries showed the presence of α-pinene, β-pinene, 1-α-phellendrene, DL-limonene, piperonal, dihydrocarveol (a compound melting at 161 °C), β-caryophyllene, and a piperidine complex (Hasselstrom et al., 1957). Additionally, the presence of cryptone, epoxydihydrocaryophyllene, and, possibly, citronellol and an azulene were also reported. The presence of α- and β-pinenes, limonene, and caryophyllene in the hydrocarbon part of black pepper oil was confirmed by infrared spectroscopy (Jennings and Wrolstad, 1961).

It is important to mention that a renewed thrust was given to the study of chemical compounds in pepper after the advent of gas chromatography. Modern researchers used thin-layer chromatography, gas chromatography, column chromatography, vacuum distillation, and the like to separate the constituents and employed ultraviolet, infrared, nuclear magnetic resonance, and mass spectroscopy for identification.

As many as 135 compounds—consisting of monoterpenoids, sesquiterpenoids, aliphatic, aromatic, and those of miscellaneous nature—have been reported (Debrauwere and Verzele, 1975a, 1975b, 1976; Gopalakrishnan et al., 1993; Ikeda et al., 1962; Lawrence, 1981; Muller and Jennings, 1967; Muller et al., 1968; Nigam and Handa, 1964; Richard and Jennings, 1971; Russel and Else, 1973; Sharma et al., 1962; Wrolstad and Jennings, 1965).

Different researchers have reported wide variations in the chemical composition of essential oils in pepper, and these variations originate for several reasons, such as varietal differences, their geographic origin, variations in the maturity of raw material, procedural differences in oil extraction, and nonresolution of constituents in early gas chromatographic analysis employing packed columns. To a certain extent, composition of essential oils will depend on the method of preparation; for example, steam distillation will produce oils containing about 70–80% monoterpene hydrocarbons, 20–30% sesquiterpene hydrocarbons, and less than 4% oxygenated constituents. Vacuum-distilled oils will contain less monoterpene hydrocarbons and more sesquiterpene hydrocarbons and oxygenated constituents. It is the result of incomplete distillation and the poor recovery of the high boiling sesquiterpene hydrocarbons and oxygenated constituents that lead to their low contents in steam-distilled oils compared with vacuum-distilled oils. Seventeen pepper cultivars from the state of Kerala were analyzed for their essential oil contents by Lewis et al. (1969) and Richard et al. (1971) and comparable results were obtained. Gopalakrishnan et al. (1993) analyzed the four Panniyur genotypes (Panniyur-1, Panniyur-2, Panniyur-3, and Panniyur-4) developed at the Pepper Research Station in Panniyur using a combination of gas chromatography and mass spectroscopy and Kovats indices on a methyl silicone capillary column and found, on the whole, that Panniyur-4 contained the most chemical constituents as compared with the other three genotypes. Significant differences could also be found in 12 samples from Lampong and 16 samples from Sarawak, both in Malaysia, using the procedure of Richard et al. (1971) by Rusel and Else (1973).

An interesting aspect of research refers to sensory evaluation of essential oils in pepper. There are but scanty references with regard to the relationship between odor characteristics and oil composition. The characteristic odor of pepper oil has been attributed to the meager amounts of oxygenated constituents (Hasselstrom et al., 1957). Arctander (Purseglove et al., 1981) describes the odor of pepper oil as fresh, dry–woody, warm–spicy, and identical to that of black pepper corn. Pangburn et al. (1970), after a systematic and thorough sensory evaluation of pepper oil obtained from pepper along the Malabar coast and using column chromatographic fractions and mixtures of fractions of the oil, reported that the early fraction was considered pepperlike and floral; the late fraction pepperlike, fresh, and woody; and the middle fraction remaining in between. Direct sniffing at the eluting port of the gas chromatographic columns helped distinguish the distinct odor of black pepper and ascribe it to the three areas of the late fractions. Distinctive odor analysis has been developed by Harper et al. and Sydow et al. (Govindarajan, 1977). Using a descriptive odor profile based on a four-point category scale and subjecting the oils to a ranking list, Gopalakrishnan et al. (1993) have described the odor evaluation of the four Panniyur genotypes described earlier in this review.

Besides aroma (odor), the other aspect that has been of interest to the researchers is the pungency emanating from the alkaloid piperene. From the early nineteenth century onward, the pungency of pepper has been investigated—starting with the first report of Oersted, who isolated piperine in 1819 (Guenther, 1950). Piperene is a yellow crystalline substance that was subsequently identified as the *trans* form of piperoyl piperidine. Subsequent researchers showed that piperine was not the only substance imparting pungency to pepper, and Bucheim (Govindarajan, 1977) obtained a dark oily resin, which he called “chavicine,” subsequent to the removal of piperine from the oleoresin. Chavicine was supposedly far more pungent than piperine, leaving a much sharper bite on the tongue than the crystalline piperine. This was shown to be incorrect by subsequent researchers who demonstrated the extreme pungency of piperine in solution. The debate, whether it was chavicine or other possible isomers, such as isopiperine or isochavicine, which are more pungent than piperine, continued for almost a century. However, recent investigations reveal that it is piperine that is the most pungent; chavicine is a mixture of piperine and many other minor alkaloids.

The question of why pepper blackens has interested many researchers. Lewis et al. (1976) attributed blackening of pepper berries to enzymatic oxidation of polyphenolic substrates present in the skin of green pepper. The blackening of pepper is a chemical reaction akin to browning in fruits and vegetables and can have an enzymatic or nonenzymatic origin. Formation of coloring pigments because of enzymatic browning is triggered by the enzyme polyphenol oxidase. The authors Lewis et al. (1976) developed a process in which green pepper was blanched to arrest the enzyme reaction with subsequent drying in a cross-flow drier to obtain dehydrated green pepper. Pruthi et al. (1976) reported that green pepper could also be preserved in brine containing either acetic or citric acid.

In addition to the previously described chemical constituents, starch is a predominant constituent of pepper; its content may vary between 35% and 40% in terms of weight. In addition to starch, pepper also contains protein and fat. Reports on both starch and protein are practically nonexistent. As for fat, pepper contains 1.9–9% fat. Bedi et al. (1971) and Salzer (1975) have determined the fatty acid composition of pepper and reported that of the different fatty acids, such as palmitic, oleic, linoleic, and linolenic, it is the linoleic acid that is most abundant (25–35%) and linolenic is the least abundant (8–19%).

The survey of pepper chemistry shows that though some of the major constituents have been researched during the last few decades, in the future, one may still come across constituents hitherto unknown. The research so far has mostly been confined to the varieties in cultivation in different regions worldwide, and yet there are far more cultivars that remain totally out of the chemists' laboratory. It is important to research these, because one may come across unique chemical compositions and flavor profiles in the different cultivars. As with disease resistance, unique chemical constitution is also of immense commercial importance. Only when pepper breeders, biotechnologists, and pepper chemists join hands in an intense search for unique traits will the research on pepper chemistry have any meaningful outcome, because it is only through these intense efforts that unique pepper lines could be identified and isolated for future breeding and crop improvement programs.

1.3 Pepper Agronomy

Agriculture systems differ from natural systems in one fundamental aspect: while there is a net outflow of nutrients by crop harvests from soils in the first, there is no such thing in the second (Sanchez, 1994). This is because nutrient losses due to physical effects of soil and water erosion are continually replenished by weathering of primary minerals or atmospheric deposition. Hence, the crucial element of sustainability is the nutrient factor. Of all the factors, however, the nutrient factor is the least resilient (Fresco and Kroonenberg, 1992). The thrust of high input technology—the hallmark of the green revolution, in retrospect—or the moderation by low input technology—the foundation stone of sustainable agriculture, in prospect—both dwell on this least-resilient nutrient factor. If the pool of nutrients in the soil, both native and added, could be considered as the "capital," efficient nutrient management might be analogous to raising the "interest" accrued from this capital in such a way that there is no great danger of erosion of this capital. Hence, sound prescriptive soil management should aim at understanding the actual link between the capital and the interest so that meaningful practices can be prescribed (Nair, 1996). This is all the more important for a perennial crop like pepper whose life span can be upward of 25 years. An effort in this direction was initiated recently by Nair (2002). But, before we dwell on it, it is only fair that an objective review is made of what has been done so far on the agronomy of black pepper.

1.3.1 The Pepper Soils

Pepper easily grows in a variety of soils. De Waard (1969) reported that in Malaysia, pepper mostly grows on soils that are developed from slate or sandstone and even in soils of alluvial origin low in fertility. Pepper is seen to grow in Indonesia in all types of soils—ranging from fertile and friable volcanic soils to sticky clayey soils. In Sri Lanka, pepper growers prefer red clay loam or sandy loam. In India, pepper is seen

Table 1.3 Type, Order, and Suborder of Major Pepper-Growing Soils and Their Statewise Distribution in India

Soil Type	Order	Suborder	Indian State
Forest loam	Mollisols	Udolls	Kerala, Karnataka, and Tamil Nadu
		Ustolls	
Laterite	Alfisols	Ustalfs	Kerala, Karnataka, and Tamil Nadu
	Oxisols		
Alluvium	Entisols	Ustorthcut	Kerala and Karnataka
Red loam	Alfisols	Ustalfs	Kerala and Karnataka
		Ustults	

to grow in a variety of soils, but, in its native Kerala State, it is grown mostly in red laterites. Pepper prefers a well-drained soil having adequate water-holding capacity and that is rich in humus and essential plant nutrients. In India, the major pepper-growing soils are oxisols (6%), alfisols (70%), mollisols (10%), and entisols (4%). The statewise distribution is outlined in (Table 1.3) (Sadanandan, 1994).

Forest loam is acidic in reaction, with a pH range of 5.0–5.5, and is confined mainly to the Western Ghats Highlands. The soils are rich in organic carbon, well drained, brown to black in color, and very well suited for pepper. Soils, in general, are quite fertile, rich in nitrogen and potassium, medium in phosphorus. Laterites are, again, acidic, with a pH range of 5.0–6.2, low in fertility, and invariably run into problems with phosphorus fertilization because of high amounts of soluble iron and aluminum, which render applied phosphorus immobile. Excessive soluble aluminum leads to aluminum toxicity. Of late, micronutrient deficiencies—in particular, zinc—have been reported, and recently Nair (2002) demonstrated the lacunae in zinc nutrition of pepper based on classical textbook knowledge of zinc fertilization in these problematic soils. Alluvium is confined to the riverbanks and their tributaries. The soils are only moderately fertile and respond well to management. The soils are acidic in nature, with pH ranging from 5.0 to 6.5. Red loams are highly porous and friable, low in fertility, and, like laterites, contain high amounts of soluble aluminum and iron, which lead to problems with phosphorus fertilization. Except in forest loam, where pepper is grown as a monocrop because the soil is highly suitable for the crop, in all the others it is grown both singly or in association with other crops, such as coconut (*Cocus nucifera*), arecanut (*Areca catachu*), and the like. Unlike the large exclusive pepper plantations, these soils support what is generally known as "homestead farming" in Kerala—small, less than 5 ha, self-supporting family farming—where even milch animals are reared in association.

1.3.2 Nutrition of Black Pepper

The fundamental difference between agricultural systems and natural systems is that while there is a net outflow of plant nutrients by crop harvests from soils in the former, there is no such thing in the latter. This is because nutrient losses due to physical effects of soil and water erosion are continually replenished by weathering of primary minerals or atmospheric deposition in natural systems. Hence, the crucial element of sustainability of crop production is the nutrient factor. As discussed earlier, the nutrient factor is the least resilient to management. For production sustainability, the nutrient factor is the most crucial. It is becoming increasingly clear that it is not the quantum per se of a specific nutrient in the soil that is crucial, as far as plant requirements are concerned, but its bioavailability. In the final analysis, it is the plant and plant alone that will decide whether a nutrient inherently present in the soil or externally added to it will ultimately be "available" to the plant or not (Nair, 2002). Formidable effort has gone into the task of defining this availability. It is the conviction of this author that if sound and accountable prescriptive fertilizer recommendations have to be made to sustain crop production in the decades to come, one must have a clear understanding of the dynamics of nutrient availability from which accurate predictions can be made and sustainable and accountable field practices devised. These considerations are particularly important to devise accurate fertilizer recommendations for pepper because, unlike most other annual or biennial crops, pepper is perennial, and the utilization pattern of applied or native nutrients over several years, often running into decades, could be uniquely different. This part of the review will, at first, dwell on the available pool of information relating to the general and prevalent mode of pepper nutrition, and then will go on to the next stage, where it will encompass the newer vision, which is now known, the world over, as "the nutrient buffer power concept."

1.3.3 Evolution of Pepper Manuring

Pepper has been in cultivation for many decades in India and elsewhere. In earlier days, farmers used only meager quantities of organic manures, such as leaf litter, animal manure (principally cow dung, which is a widespread practice of homestead farming in Kerala even now), or slashed stems and leaves of live support, such as *Erythrina indica*. Chemical manuring, through factory-manufactured fertilizers, is a relatively recent phenomenon. It was the pepper boom—and the escalating prices of pepper in the world market, especially in the post-World War II period—that encouraged pepper farmers to use artificial manure. In addition, the spectacular initial yield increases in wheat and rice in the North Indian regions—the green revolution effect owing to liberal doses of chemical fertilizers—prompted affluent pepper farmers to use factory-produced fertilizers. Though there is now a blacklash against the use for chemical fertilizers—in terms of degraded soils, drying aquifers, and vanishing biodiversity, all of which are attributed to the indiscriminate use of chemical fertilizers—chemical fertilizer use in a systematic manner is still confined only to affluent owners of pepper plantations. Organic manuring is still the mode of pepper nutrition in homestead farming. Significantly, even pepper planters of fairly large farms are gradually switching to organic farming, because "organic pepper" raised by the use of only organic manures fetches a much better price, and a steady and premium market is developing, especially in European countries such as Germany, where the spices are much fancied lately.

1.3.4 Response of Pepper to Mineral Nutrients

The major emphasis on pepper fertilization is still confined to nitrogen, phosphorus, and potassium. There is an extensive body of literature on nitrogen fertilization in pepper globally (De Waard, 1964; Mohanakumaran and Cheeran, 1981; Pillai and Sasikumaran, 1976; Pillai et al., 1979, 1987; Sadanandan, 1994; Sim, 1971). These routine studies have dealt with partitioning of nitrogen in various plant parts (Adzemi et al., 1993), response functions (Pillai et al., 1987; Sadanandan, 1990), greenhouse cultivation (Murni and Faodji, 1990), and foliar nitrogen application (Anon., 1995). Adzemi et al. (1993) found the maximum nitrogen concentration in leaves (2.30%), while the branches transported the most (47.6 mg plant^{-1} year^{-1}). In Kerala State, for Panniyur-1, the first hybrid variety released in the world from the Pepper Research Station, Panniyur, 50 kg nitrogen with 100 kg P_2O_5 and 150 kg ha^{-1} K_2O were found to be the optimum rates for laterite soil (Pillai et al., 1987). In greenhouse conditions, 1.1 g plant^{-1} nitrogen as urea and 2.0 g plant^{-1} K_2O produced the maximum of dry matter (Murni and Faodji, 1990). Between Panniyur-1 and Karimunda, another improved variety from the Panniyur Research Station, 292 kg ha^{-1} year^{-1} and 183 kg ha^{-1} year^{-1} of nitrogen, respectively, were found to be the optimum rates (Sadanandan, 1990). In Sarawak (Malaysia), foliar application of 0.7% urea at weekly intervals, totaling nine sprays, were found to increase pepper yield by 22% (Anon., 1995) (Table 1.4).

Comparatively, research on phosphorus in pepper is less, viewed against that of nitrogen. Unlike nitrogen, pepper stems contain the most phosphorus (3.96 g vine^{-1}) because of large transport and a total of 22.8 kg ha^{-1} of phosphorus is removed from the soil (Adzemi et al., 1993). Availability of phosphorus in laterite soil is a universal problem, because of the presence of excessive soluble aluminum and iron, which render the applied phosphorus fertilizer immobile, and some researchers have focused on alternate sources of phosphorus as a possible remedy. Phosphate rocks (PRs) are ideally suited for plantation crops and where indigenously available could be profitably utilized. In northern India, a popular PR is Mussoorie rock phosphate (MRP). Sadanandan (1994) reported that MRP is superior to ordinary superphosphate. Both cumulative yield increase and relative agronomic effectiveness were superior in MRP as compared to ordinary superphosphate.

Pepper is a prolific user of potassium, and even with 2% content of potassium in the pepper plant, potassium deficiency will manifest (De Waard, 1969). Field trials conducted using Panniyur-1 as the test crop showed that as much as 200 g of K_2O vine^{-1} is required (Pillai et al., 1987). Response functions to potassium fertilizer application indicated that pepper needs a very high dose of potassium (270 kg K_2O ha^{-1}) for high yield (Sadanandan, 1990). Pepper plants remove as much as 203.2 kg ha^{-1} of potassium and white pepper stores most of it (42.0 g vine^{-1}) as reported by Adzemi et al. (1993). Most pepper growers adopt varying nitrogen-to-phosphorus-to-potassium (N:P:K) ratios, while Pillai et al. (1979) suggest an optimum ratio of 5:5:10 for nitrogen, phosphorus, and potassium to obtain the maximum yield.

As far as the secondary nutrients (calcium, magnesium, and sulfur) are concerned, the only published work refers to calcium indirectly, through the effect of liming.

Table 1.4 Nutrient Deficiency Symptoms in Pepper

Nutrient	Deficiency Symptoms	Cited by
Nitrogen	General chlorosis and stunted growth. Yellowing of older leaves followed by younger ones. Bottom leaf tips and margins become brown and necrotic.	De Waard (1969); Nybe and Nair (1986)
Phosphorus	Bronzing of older leaves accompanied by necrosis of leaf tips and margins. Stunted growth.	De Waard (1969); Nybe and Nair (1986)
Potassium	Browning and necrosis of older leaf tips and margins. Symptoms later spread to younger leaves.	De Waard (1969); Nybe and Nair (1986)
Calcium	Young leaves develop tiny, brown necrotic pinhead spots that later spread to older leaves. Interveinal chlorosis can also be seen.	De Waard (1969); Nybe and Nair (1987a)
Magnesium	Pale yellow discoloration of leaf margins and tips, followed by necrosis and defoliation. Major veins remain green, and laterals turn yellow.	De Waard (1969); Nybe and Nair (1987a)
Sulfur	Late stage chlorosis in younger leaves, turning to bright yellow color in interveinal areas. Premature leaf fall and die-back of growing tip.	Chin et al. (1993); Nybe and Nair (1987a)
Zinc	Stunted growth, small leaves, interveinal chlorosis. Leaf margins pucker.	Chin et al. (1993); Nybe and Nair (1987c)
Manganese	Interveinal chlorosis, with major veins remaining green. Chlorotic leaves turn yellow or white later, and necrotic mature leaves. Symptoms can often be confused with that of magnesium deficiency.	Chin et al. (1993); Nybe and Nair (1987b)
Iron	Interveinal chlorosis in younger leaves, youngest leaves becoming totally chlorotic	Chin et al. (1993); Nybe and Nair (1987b)
Copper	Interveinal chlorosis in younger leaves; necrosis on leaf tips and margins.	Chin et al. (1993); Nybe and Nair (1987b)
Boron	Stunted growth, necrosis, and interveinal chlorosis. Necrotic lesions are seen on main vein.	Chin et al. (1993); Nybe and Nair (1987c)

Pillai et al. (1979), however, did not observe any positive response to lime application in a laterite soil growing Panniyur-1. Adzemi et al. (1993) report the highest accumulation of both calcium and magnesium in stems (11.5 and 5.98 g vine^{-1}, respectively), with a total accumulation of 54.5 and 36.4 kg ha^{-1}, respectively.

Among the micronutrients, zinc is the most important followed by molybdenum and boron (De Waard, 1969). Zinc deficiency is beginning to be seen in many tropical countries in a very significant manner and is adversely affecting crop yields. However, to date, corrective measures still bank on routine soil testing with DTPA as the most commonly used extractant. That such an approach does not lead to satisfactory results uniformly is shown through the work of this author (Nair, 2002), which will be discussed later in this section. A combination of zinc, molybdenum, and boron, along with nitrogen, phosphorus, and potassium, was shown to result in the highest yield of two locally developed varieties—namely, Subhakara and Sreekara at the Indian Institute of Spices Research experimental farm (IISR, 1977), where zinc, boron, and molybdenum were used at a ratio of 5:2:1. Spraying a solution of $ZnSO_4$ (0.5%) reduced spike (inflorescence which eventually develops into pepper berries) shedding by 48.4% (Geetha and Nair, 1990).

Organic manuring is a very commonly adopted practice in pepper production in India and parts of Asia. This can either be through the use of fresh vegetative matter or through the use of "burnt earth" (Bergman, 1940; Harden and White, 1934). In the former category, freshly chopped materials (such as leaves, stems, and the like) from a number of trees are used. The trees generally used are *E. indica*, *Garuga pinnata*, and *Grevillea robusta* (Sivakumar and Wahid, 1994). Organic manures are widely used in pepper production in Sarawak (Malaysia), which include soybean cake, guano, prawn, and fish refuse (Purseglove et al., 1981). Holes are dug and about 85 g of manure are placed about 20 cm away from the main stem of the vine. This is done once in a 2-month period. Some farmers dig trenches all around the vines and put large quantities of manure once in 2 months. Of late, sterilized animal meat and bone meal admixture, fortified with potassium, are gaining popularity as an organic manure.

Another method of organic manuring is through the use of "burnt earth." For this, forests are first cleared; then, soil from vacant patches is spread on top of the heaped vegetation, the heap is set on fire, and the fire is allowed to slowly burn for 2–3 weeks. At the end of this period, both the wood ash and burnt soil are thoroughly mixed and applied at a rate of 18 kg $vine^{-1} year^{-1}$. The application of burnt ash has many beneficial effects in improving the physical, chemical, and biological properties of soil. Compared to the acidic soil, having a pH of 4–5, the burnt earth has a pH of 7–8 and acts as a good soil ameliorant in correcting pH. There is extensive literature on this method (Bergman, 1940; De Waard, 1969; Harden and White, 1934; Huitema, 1941). Owing to environmental hazards, especially because of large-scale burning and the emission of smoke, the Government of Sarawak banned the practice in 1940. Figures 1.2, 1.3, and 1.4 represent potassium, magnesium, and zinc deficiency, respectively.

1.4 The Role of the Nutrient Buffer Power Concept in Pepper Nutrition

Historically, soil testing has been used to quantify availability of essential plant nutrients to field-grown crops. However, contemporary soil tests are based on philosophies and procedures developed several decades ago without significant changes in their general approach. For a soil test to be accurate, one needs to clearly understand the physico-chemico-physiologic processes at the soil–root interface, and an

Figure 1.2 Potassium deficiency symptom.

Figure 1.3 Magnesium deficiency symptom.

understanding of soils and plant root systems as polycationic systems is as essential. It is this knowledge that leads to sound prescriptive soil management practices in nutrient availability vis-à-vis fertilizer application, because, of all the factors that govern sustainability of crop production, the nutrient factor is the most important, and it is also the least resilient to management. This section of the review will focus on the buffering of plant nutrients, with specific reference to zinc, and discuss experimental results that relate to pepper nutrition.

1.4.1 The Buffer Power and Its Effects on Nutrient Availability

Before being able to understand the significance of the *nutrient buffer power concept* on plant nutrient availability, certain basic concepts must be addressed, and the following review starts with this rationale.

Figure 1.4 Zinc deficiency symptom.

1.4.2 Basic Concepts

In any nutrient management approach that is sound and reproducible, one must start with a basic understanding of the chemical environment of plant roots. When we consider this, the first term that we come across is the *soil solution* because the plant root is bathed in it and it is most affected by its chemical properties. The Soil Science Society of America (1965) defines soil solution as "the aqueous liquid phase of the soil and its solutes consisting of ions dissociated from the surfaces of the soil particles and of other soluble materials." Adams (1971) has given a simple definition: "The soil solution is the aqueous component of a soil at field moisture content." Perhaps it is important to emphasize here that much contemporary soil testing has considered a soil extract as synonymous with the soil solution. Because soil extraction is supposed to simulate plant root extraction, it is pertinent to consider the chemical environment of the root, though briefly, from this angle. It is worth noting that the chemical environment of roots in natural soil systems is so obviously complex that both soil scientists and plant physiologists have been unable to provide a precise definition. If this complex chemical system is to be accurately quantified, thermodynamic principles will need to be used to evaluate experimental data. Even then, the limitations are obvious, as in the case of potassium, where the thermodynamic investigations are quite often inapplicable under field conditions. Thus, although a quasi-equilibrium in potassium exchange can be achieved in the laboratory, these conditions are seldom, if ever, attained under field conditions (Sparks, 1987). Agricultural soils are, for the most part, in a state of disequilibrium owing to both fertilizer input and nutrient uptake by plant root. It appears that a universal and accurate definition of a root's chemical environment awaits the proper application of thermodynamics for the root's ambient solution (Adams, 1974) or even kinetics, as in the case of potassium (Sparks, 1987), where thermodynamics have been found to be inadequate.

Soil extractions with different extractants provide a second approach in defining the root's chemical environment. This approach has been particularly successful in understanding cases like phosphorus insolubility, soil acidity, and potassium fixation. However, this approach also fails to precisely define the root's chemical environment. Though this approach also suffers deficiencies—such as the extractants removing arbitrary and undetermined amounts of solid-phase electrolytes and iron (or the extractants causing precipitation of salts or ions from the soil solution) and the soil–plant interrelationship defined in terms of solid-phase component of the soil, even though the solid phase is essentially inert except as it maintains thermodynamic equilibria with the solution phase (Adams, 1974)—the latter part could be researched more to understand how the solid phase–solution phase equilibria can be interpreted to give a newer meaning to quantifying nutrient availability. It is in this context that the role of the plant nutrient *buffer power* assumes crucial importance.

The close, almost linear relationship in a low concentration range of $<0.5\,mM$ for $NO_3^- - N$, $NH_4^+ - N$, K^+, $H_2PO_4^-$, and HPO_4^{2-}, which has been established by numerous solution culture experiments, can be quantitively described by the equation:

$$U = 2\pi r \alpha C_r \tag{1.1}$$

where U is the uptake of a 1-m root segment, r is the root radius, C_r is the concentration of the ion at the root surface, and α is the root-absorbing power (Mengel, 1985). The metabolic rate of the root determines its absorbing power. A high root-absorbing power would imply that a relatively high proportion of nutrient ions coming in contact with the root surface is absorbed and vice versa. The nutrient ion concentration at the root surface (C_r) depends on α because a high root-absorbing power tends to decrease C_r; it also depends on the rate of movement from bulk soil toward the root (Mengel, 1985). Diffusion and/or mass flow control this movement. But it is now established that nearly 95% of this movement for nutrients such as phosphorus, potassium, and zinc (among heavy metals) and, possibly, NH_4^+ is by way of diffusion. When root uptake of an ion species is less than its movement toward it, accumulation of the ion species on the root surface is bound to occur, as has been shown to be the case with Ca^{2+}, where mass flow contributes to this accumulation (Barber, 1995). The diffusive path for ions such as phosphorus and potassium, which plant roots take up at high rates but which are in low concentration in the soil solution near the root, is the concentration gradient. In a sense, the effective diffusion coefficient that quantifies the diffusive path and the buffer power are analogous because the diffusive flux across the root surface is integrally related to the nutrient buffer power. This has been shown to be true in the case of phosphorus, where a highly significant positive correlation between the two was found to exist in 33 soil samples obtained from experimental sites in the USA and Canada (Kovar and Barber, 1988). However, in a routine laboratory setup, it is far easier to measure the buffer power than the effective diffusion coefficient, and this review will further focus on the question of how buffer power can be quantified without recourse to cumbersome analytical techniques and how its integration into routine soil test data will considerably improve predictability of nutrient uptake.

1.4.3 Measuring the Nutrient Buffer Power and Its Importance in Affecting Nutrient Concentration on Root Surfaces

The ability to predict the mobility of dissolved chemicals, such as fertilizers, in the soil is of considerable value in managing fertilizer applications. Soil testing in its essence aims to achieve this. While modeling transport and retention of ions from thermodynamic (Selim, 1992), kinetic (Sparks, 1989), and mechanistic (Barber, 1984) angles could be very informative, the importance of translating this information into practically feasible procedures in crop production calls for an understanding not only of the basic concepts but of their intelligent application as well. In a dynamic state of plant growth, the concentration of any nutrient on the root surface is nearly impossible to measure because both the nutrient in the plant tissue and the root-absorbing power, which directly affects it, change quickly due to root metabolic processes. The inability of even mechanical mathematical models to accurately predict the nutrient influx rate has recently been highlighted (Lu and Miller, 1994). Hence, if an effective soil-testing procedure is to be devised for a nutrient, which is an alternative to defining the plant root's chemical environment, one must resolve the problems of quantifying the nutrient concentration on the root surface indirectly, even if it is impossible to resolve it directly, for the reasons mentioned earlier. Using Fick's first law,

$$F = -D\left(\frac{\mathrm{d}C}{\mathrm{d}x}\right) \tag{1.2}$$

where F is the flux, $\mathrm{d}C/\mathrm{d}x$ is the concentration gradient across a particular section, and D is the diffusion coefficient. Nye (1979) has suggested that the formula can be applied to both ions and molecules. The negative sign for D implies net movement from a high to a low concentration. Although, for molecules in simple systems, like dilute solutions D may be nearly constant over a range of concentrations, for ions in complex systems like soils and clays, D will usually depend on the concentration of the ion and on that of other ions as well (Nye 1979). Nye has further suggested that though Fick's first law may be derived from thermodynamic principles in ideal systems, in a complex medium such as the soil, Equation (1.2) may be regarded as giving an operational definition of the diffusion coefficient. Thus, Nye (1979) defines the diffusion coefficient as

$$D = D_1\theta f_1\left(\frac{\mathrm{d}C_1}{\mathrm{d}C}\right) + D_E \tag{1.3}$$

where D_l is the diffusion coefficient of the solute in free solution, θ is the fraction of the soil volume occupied by solution and gives the cross-section for diffusion, f_l is an impedance factor, C_l is the concentration of solute in the soil solution, and D_E is an excess term which is zero when the ions or molecules on the solid have no surface mobility, but represents their extra contribution to the diffusion coefficient when they are mobile. D_E can generally be neglected because only in rare instances will it play any role in diffusion of plant nutrient ions in soil (Mengel, 1985). From the point of view of nutrient availability, dC_l/dC, which represents the concentration gradient, assumes crucial importance as we shall see next.

The term dC_l/dC, where C_l is the concentration of the nutrient ion in the soil solution and C is the concentration of the same ion species in the entire soil mass, assumes considerable significance in lending a practical meaning to nutrient availability. If we ascribe the term "capacity" or "quantity" to C and "intensity" to C_l, we have in this term an integral relationship between two parameters that may crucially affect nutrient availability. Because the concentration gradient of the depletion profile of the nutrient in the zone of nutrient uptake depends on the concentration of the ion species in the entire soil mass (represented by "capacity" or "quantity") in relation to the rate at which this is lowered on the plant root surface by uptake (represented by "intensity"), it could be argued that a quantitative relationship between the two should represent the rate at which nutrient depletion and/or replenishment in the rooting zone should occur (Nair, 1984a). This relationship has been functionally quantified by Nair and Mengel (1984) for phosphorus in eight widely differing central European soils (Table 1.5) and the term dC_l/dC has been referred to as the nutrient buffer power. Nair and Mengel (1984) used electroultrafiltration to quantify C_l, while using an incubation and extraction technique to quantify C. For phosphorus, C was found to closely approximate isotopically exchangeable phosphorus (Keerthisinghe and Mengel, 1979), but in the experiments conducted by Nair and Mengel (1984), it was estimated by the extraction of incubated soil with an extractant that was a mixture consisting of 0.1 M calcium lactate + 0.1 M calcium acetate + 0.3 M acetic acid at pH 4.1. The extractant exchanges adsorbed phosphate and dissolves calcium phosphates except apatites; the method known as the "CAL method," developed by Schüller (1969), is now widely used in central Europe. In the case of K^+ and $NH_4^+ - N$, C denotes the concentration of exchangeable, and to some extent nonexchangeable, fractions (Mengel, 1985). Because very low concentrations in the range of 2.0 μM may be attained on the root surface for both phosphorus and potassium (Claassen and Barber, 1976; Claassen et al., 1981; Hendriks et al., 1981), Nair and Mengel (1984) had to use electroultrafiltration to quantify C_l.

Thus, the nutrient depletion around the roots that is caused by the diffusive flux of the nutrient toward the root surface is related to both the quantity and the intensity parameters, and a quantifiable relationship between both represents the buffer power specific to the nutrient and the soil. A growing root will at first encounter a relatively high concentration of phosphorus, which is in the range of the concentration of the bulk soil solution (Nair and Mengel, 1984). As uptake continues, depletion will occur at the root surface. This depletion profile gets flatter with enhanced nutrient uptake (Claassen et al., 1981; Hendriks et al., 1981; Lewis and Quirk, 1967). But it is the

Table 1.5 Comparison of Phosphorus Buffer Power of Eight Widely Differing Central European Soils (Determined by Two Different Techniques)

Soil	Regression		r	
	(1)	(2)	(1)	(2)
Benzheimer Hof	$Y = 18.8x + 7.94$	$Y = 0.23x + 8.98$	0.912	0.995
Hungen	$Y = 38.2x - 1.03$	$Y = 0.25x + 4.32$	0.967	0.997
Oldenburg B6	$Y = 49.8x + 0.52$	$Y = 0.26x + 0.72$	0.994	0.999
Wolfersheim	$Y = 70.3x + 0.03$	$Y = 0.27x + 0.11$	0.998	0.983
Obertshausen	$Y = 70.5x + 2.66$	$Y = 0.30x + 2.89$	0.966	0.998
Oldenburg B3	$Y = 73.6x + 2.07$	$Y = 0.31x + 0.61$	0.994	0.997
Klein–Linden	$Y = 75.0x + 0.38$	$Y = 0.32x + 1.81$	0.999	0.991
Gruningen	$Y = 75.4x + 0.89$	$Y = 0.36x + 3.62$	1.000	0.996

Note: The b values in the regression functions represent the phosphorus buffer power of each soil. In regression function (1) (after Nair and Mengel, 1984), Y = CAL-P (Schüller's method), and in regression function (2) (after Nair, 1992), Y = the author's method, x in both refers to electroultrafiltrable phosphorus. Note the very high r values in all the cases. The soils are arranged in their sequential increase in phosphorus buffer power.

capacity of the soil to replenish this depletion which ensures a supply of nutrient ions to the plant root without greatly depressing its average concentration on the root surface. It is the nutrient's buffer power that decides these depletion and/or replenishment rates. A soil with a high phosphorus buffer power implies that the phosphorus absorbed from the soil solution is rapidly replenished. In such a case, phosphorus concentration at the root surface decreases only slowly, and mean phosphorus concentration at the root surface remains relatively high. In soils with a low phosphorus buffer power, the reverse is true, and mean phosphorus concentration at the root surface is rapidly diminished and remains relatively low. This has been proved experimentally for phosphorus (Nair, 1992; Nair and Mengel, 1984). This phenomenon holds true for Zn^{2+} (Nair, 1984b), potassium (Nair et al., 1997), and $NH_4^+ - N$ as well (Mengel, 1985).

1.4.4 *Background Information on the Importance of Measuring Zinc Buffer Power*

There is a great paucity of published material on the effect of buffer power on availability of heavy metals. Plants obtain most of their fertilizer zinc from reaction products and not applied sources as such, implying that any source of zinc added to soil has to necessarily conform to a chain reaction involving adsorptive, desorptive, and resorptive processes that govern the maintenance of an equilibrium between adequate zinc concentration in the soil solution nearest to the zone of zinc depletion on the one hand and plant uptake on the other. The zinc buffer power defines this. Because the zinc concentration in soil solution is normally very low, the supply to plant roots by mass flow can only account for a very small fraction of plant demand. For instance, with a transpiration coefficient of $300\,l\,kg^{-1}$ dry matter and a corresponding zinc concentration of 10^{-7}M in the soil solution, approximately 2 mg of zinc can be supplied by mass flow against a demand of $10–30\,mg\,kg^{-1}$ zinc dry weight of plant tissue.

In calcareous soils, as the zinc concentration is of a much lower order of approximately 10^{-8} M, the supply by mass flow could be very much lower (Marschner, 1994), indicating that mass flow can only contribute very negligibly to meet plant needs for zinc. Hence, zinc movement to the plant root surface is principally by diffusion and is essentially confined to a zone around the plant root that hardly extends beyond the root hair cylinder (Marschner, 1994). In a review on the mechanism of zinc uptake, Marschner (1994) indicated that flow culture experiments with various species showed adequate ranges of zinc concentration in the range of 6×10^{-8}–8×10^{-6} M, which are concentrations greater than those that would be expected in the solution of most soils. He further pointed out that although work using chelate-buffered solutions has indicated adequate zinc concentrations between 10^{-10} and 10^{-11} M, extremely low adequate zinc concentrations required a concomitant excess of about 100 μM zinc chelate as buffer at the plasma membrane of the root cells. This implies a need for an unlimited zinc pool for replenishment of Zn^{2+} at the plasma membrane. When plants grow in soil, it is impossible to expect a zinc buffer of this size to exist, and free Zn^{2+} and chelated-zinc concentrations will be at least threefold lower. Hence, critical deficiency or sufficiency concentrations obtained through research employing chelate-buffered solutions cannot be applied to soil-grown plants.

Most of the work on zinc availability to plants is based on chemical extractions, among which DTPA extraction is the most frequently used. The DTPA extraction quantifies a labile fraction of soil zinc comprising water soluble, exchangeable, adsorbed, chelated, and some occluded zinc. The critical soil level of DTPA-extractable zinc can vary from 0.3 to 1.4 mg kg^{-1} soil, which equates to about 900–4200 g ha^{-1} of zinc in heavy soils and about 600–2800 g ha^{-1} of zinc in light soils in the plough layer (0–20 cm). The crop requirements, on the other hand, are quite small, in the range of 100–300 g ha^{-1} for a total dry matter production of about 10 tons ha^{-1} (Marschner, 1994). The inadequacy of DTPA extraction to reflect plant zinc demand shows that other important factors, such as replenishment of soil-solution zinc (Nair, 1984b), mobility, and transport to the root surface (Wilkinson et al., 1968; Nair et al., 1984), and also the activity of the roots themselves (Wilkinson et al., 1968; Marschner, 1994) are involved. Because the zinc buffer power is intricately involved in all three factors, the focus of this review is mainly on that attribute.

As early as three decades ago, it was suggested that colloidal zinc was released by some specific process associated with root activity (Wilkinson et al., 1968). Conditions in the rhizosphere and particularly root-induced changes markedly affect zinc availability. A difference in rhizosphere pH of as much as 2 (higher or lower compared to bulk soil) can be expected to occur as a result of imbalance in ionic uptake. For instance, any acidifying fertilizer such as $(NH_4)_2SO_4$ can result in a net excretion of H^+ ions and others, such as NH_4NO_3, can result in a net excretion of HCO_3^- or OH^- ions. Additionally, secretion of organic acids and enhanced CO_2 production will affect rhizosphere pH, and all of the previously mentioned changes will markedly affect zinc availability. However, the scope of this review is confined to the kinetic/dynamic aspects of the changes occurring in the rooting zone mirrored in the zinc buffer power rather than changes in soil reaction in the rhizosphere per se on zinc availability.

The distribution of zinc between the solid and solution phases can be described by the buffer power. The availability of soil zinc to the plant depends on the initial zinc concentration, zinc buffer power, and effective diffusion coefficient (Barber, 1984). The Langmuir equation gives the relation between B and C_1 as

$$\frac{C_1}{(x/m)} = \frac{1}{aB} + \frac{C_1}{B}$$

where C_1 is the zinc concentration in the soil solution, x/m is the amount of zinc adsorbed per unit of soil, B is the adsorption maximum, and a is a constant related to the soil's bonding energy for zinc. A straight line is obtained when $C_1/(x/m)$ is plotted against C_1 with a slope of $1/B$ and intercept of $1/aB$. The inverse of $C_1/(x/m)$ is b, the zinc buffer power, where C_1 and x/m are both expressed in volume units (Barber, 1984). Using this approach, Shuman (1975) estimated the buffer power values varying from 5 to 100 for four soils representing different major physiographic regions of Georgia. Based on the diffusion model of Drew et al. (1969), Nair (1984) has argued that the C in the equation $U = 2\pi \alpha a C^{-} t$—where, U is the quantity of zinc absorbed per centimeter root length, a is the root radius in cm, α is the root-absorbing power, C^{-} is the average zinc concentration on the root surface, and t is the duration of the absorption period—in fact represents an indirect measure of the zinc buffer power. As we already know, the bulk of zinc uptake is by diffusion (Barber, 1984; Elgawhary et al., 1970; Wilkinson et al., 1968;). This diffusive process will maintain a concentration gradient in the root zone. This concentration gradient will directly affect zinc uptake because of its effect on the average zinc concentration on the root surface. The zinc buffer power will affect this concentration gradient, because the rate of zinc depletion and/or replenishment is mirrored by it. In a sense, the effective diffusion coefficient and the buffer power are analogous to each other for nutrients that are principally absorbed by the plant root through the diffusive process (Nair, 1989). Hence, the crucial question to examine would be the role of zinc buffer power in influencing zinc availability for plant uptake.

1.4.5 Quantifying the Zinc Buffer Power of Pepper-Growing Soils

Nair (1984b) has used electroultrafiltration to quantify zinc intensity in measuring the zinc buffer power of European soils. Because the procedure is very highly sophisticated and due to its nonavailability in most of the developing countries, a simple adsorption–desorption equlibrium technique was developed by the author (Nair, 2002) to quantify the zinc buffer power of pepper-growing soils. For this, 200 g of representative soil samples from three locations in the state of Kerala, where pepper is extensively grown, were incubated at 60% of the maximum water-holding capacity with graded rates of zinc over a fortnight, by maintaining the water regime to constancy. At the end of the period, the soil samples were extracted with 0.01 M $CaCl_2$ over a 24-h period and the extract tested for zinc using atomic absorption spectrometry. This represented zinc intensity. Separate extractions were made with DTPA, which represented zinc quantity. Data in Table 1.6 indicate that the zinc buffer power values of the experimental soils varied widely.

Table 1.6 Zinc Buffer Power of Pepper-Growing Soils

Soil	*r* value	*b* value
Peruvannamuzhi	0.8337[a]	0.7824
Thamarasseri	0.9304[a]	1.5786
Ambalavayal	0.9604[a]	3.0358

[a]Significant at a confidence level of 0.1%; *b* values represent the zinc buffer power.

Data in Table 1.6 bring to light two very important facts. First, the very highly significant correlation coefficients prove that the technique employed allows a precise determination of the zinc buffer power of the soils. Second, the soils varied widely in their zinc buffer power. It is this fact that has to be critically viewed against the existing zinc fertilizer recommendation, emanating from routine soil tests using the universally employed DTPA extraction, to understand the cruciality of the zinc buffer power for precise formulations of zinc fertilizer applications for pepper.

Data in Table 1.7 clearly indicate that integration of the zinc buffer power has improved the relationship between DTPA routine test versus both zinc concentration and zinc uptake. More remarkably, the overall relationship between DTPA test and dry matter production not only improved with zinc buffer power integration in the computations, but turned from negative to positive. This clearly shows that the commonly employed DTPA test could only be site specific, but on a larger scale, it is the zinc buffer power that determines plant performance. The experiment had only monitored dry matter production in the pepper plant without taking it to berry formation, because it takes about 3–5 years for berry formation. But, the data in Table 1.7 clearly substantiate the cruciality of the buffer power concept.

Data in Tables 1.6, 1.7, and 1.8 have to be viewed in conjunction to precisely understand the cruciality of the buffer power concept. Pepper is the economic mainstay of the state of Kerala. Of late, widespread zinc deficiency has been observed in the state, and the onset of the foot rot due to *Phytophthora* fungi is attributed, to a great extent, to this deficiency in the soils. The scientists at the Indian Institute of Spices Research have made a blanket recommendation of soil application of 23 kg $ZnSO_4$ ha^{-1} when the DTPA-extractable zinc is less than 1.6 ppm (parts per million). In terms of monetary input, this would translate to an equivalent of US \$25 ha^{-1} in terms of farmer investment. Because the soils are variously zinc buffered (Table 1.6) a single blanket dose, such as the one that is made, is totally unscientific because, where the farmer needs to apply only lesser quantities of zinc, he would end up applying more if he were to go by the routine recommendation based on the DTPA extraction. To elaborate, the soils from the Ambalavayal region would only require 25% of zinc needed in the soils of the Peruvannamuzhi region because the former has a zinc buffer power nearly fourfold more compared to the latter. In the case of the soils from the Thamarasseri region, the quantity required would only be 50%. In fact, the scientists at the Institute have made a gross underestimation of the zinc-supplying power of the soils in the state because their results are all based on the soils of the Peruvannamuzhi region alone, where the Institute has

Table 1.7 Correlation Coefficients (r) for the Interrelationship between Routine DTPA Soil Test versus Zinc Concentration, Zinc Uptake, and Dry Matter Production without (1) and with (2) Zinc Buffer Power Integration in Pepper

Details	1	2
Zinc concentration vs. DTPA test	0.88a	0.924[a]
Zinc uptake vs. DTPA test	0.782	0.862[a]
Dry matter production vs. DTPA test	−0.745	0.777[a]

[a]Significant at a confidence level of 0.1%.

Table 1.8 Pepper Yield from Farmers' Fields Weighted against Zinc Buffer Power

Region	Yield ($kg\,vine^{-1}$)		Deviation (%)
	Targeted	Actual	
Peruvannamuzhi	0.241	0.401	+66
Thamaraserri	0.490	0.487	+0.6

Note: Target weighting was done against the highest yield obtained from the Ambalavayal region. Note the remarkable closeness between targeted and actual yields in Thamarasseri region. Peruvannamuzhi soil is an atypical pepper soil.

its experimental farm. The soils here are obviously zinc impoverished, but that is not the case in the entire state of Kerala. This also shows that, more often than not, recommendations from experimental stations, when extrapolated on a large scale, can run into problems—especially when such recommendations are based on routine soil testing procedures. These results (Nair, 2003) substantiate the earlier ones (Nair, 2002) on the importance of measuring the zinc buffer, rather than the routine DTPA-extractable zinc alone, to make precise zinc fertilizer recommendations for wheat in central Asia (Turkey), where farmers have been advised to apply as much as $100\,kg\,ZnSO_4\,ha^{-1}$ as a blanket recommendation, without obtaining any tangible and consistent response in wheat yield. Hence, the economic importance of the buffer power concept, as against the routine soil testing procedures, needs hardly any underscoring. Nair et al. (1997) have conclusively demonstrated the importance of the buffer power concept in the case of potassium nutrition for another very important perennial crop and spice, cardamom (*Elettaria cardamomum* M.) in the state of Kerala. Both pepper and cardamom are the staple spices of the state.

1.5 Establishing a Pepper Plantation

1.5.1 The Indian Experience

From a commercial viewpoint, establishing a good pepper plantation is of great significance. Unlike in countries such as Thailand, Malaysia, and Brazil, where the life span of a pepper plantation is only 10–15 years, at the end of which replanting is done, in India a pepper plantation can last for more than a quarter century. The prime reason for the shorter duration of the pepper plantations in other countries, is because pepper is trailed on nonliving standards (support), while in India pepper is always trailed on live support, such as *E. indica*. The starting point to establish a good pepper plantation is the establishment of good planting material. With the exception of India, most other countries use the orthotropic (upward-climbing) shoot as the planting material. This is the best planting material because it results in vigorous plants and bear fruiting laterals right from the base itself and yields a lot earlier. In India, however, the use of orthotropic shoots is greatly restricted because of nonavailability.

In earlier times, pepper planters used either runner shoots or climbing orthotropic shoots to coincide with the onset of the southwest monsoon in India. These days, prerooted cuttings from runner shoots are used instead. When hanging shoots are used as planting material, they result in the formation of weak plants. Quite often, pruned material—pepper is pruned 5–6 months after planting, and once again, after a year—is also used as planting material. These are pruned stem cuttings.

Because pepper is only propagated through the type of planting material described earlier, there is a huge worldwide demand (running into millions) for ready-to-use planting material. Production of prerooted cuttings in light polyethylene bags, PVC (poly vinyl chloride) bags is the surest way of mass producing material once the demand for planting material escalated in India. Their production is as follows. Runner shoots from high-yielding and healthy plants are kept coiled at the base of wooden pegs fixed at the bottom of the plant so as to prevent the growing shoots from spreading on the soil and striking roots. These runner shoots can be separated from the mother plant in winter (January–February) by snapping them using scissors and are disinfected by dipping them in a fungicide solution, such as oxychloride or a Bordeaux mixture for 1 min; they are later surface dried and cut into three noded strips after clipping the leaves; they are then planted either in raised soil beds or PVC bags filled with a mixture of soil, sand, and some farmyard manure, which will provide enough nutrients in a growth substrate. Application of growth regulators, such as Indole Butyric Acid (IBA), by dipping the cuttings in the solution has been found to enhance root proliferation (Pillai et al., 1982; Suparman and Zaubin, 1988). Care must be taken to minimize fungal infection arising from the nursery bed Photo 1.1.

In Sri Lanka, growing planting material on split bamboo led to the production of good planting material on a large scale. Popularly known as the *bamboo technique* (Bavappa and Gurusinghae, 1978), the method consists of first digging a trench ($60 \times 40\,\text{cm}^2$) of convenient length, filling it with a rooting medium—preferably forest soil, sand, and farmyard manure in a 1:1:1 ratio mixture; split-bamboo poles (1.25–1.5 m long) or PVC pipes—both of about 10 cm in diameter—are then sunk into the trenches at 45° angles using a strong middle support. Rooted cuttings, one each per bamboo, are planted in the trench and allowed to trail onto the bamboo or PVC poles. The lower portion of the pole is filled with a rooting medium (of weathered coir dust and farmyard manure in a 1:1 ratio), and as the vine grows, it is tied to the support. Vines are irrigated daily, and to stimulate rapid growth, a

Photo 1.1 Production technology for *Trichoderma* (biological control of "quick wilt" caused by *Phytophthora* fungus).

nutrient solution consisting of nitrogen, phosphorus, potassium, and magnesium (1 kg urea, 0.75 kg superphosphate, 0.5 kg muriate of potash [MOP], and 0.25 kg magnesium sulfate all mixed in 250 l of water) can be applied at the rate of 0.25 l per vine every fortnight. It takes 3–4 months for the vine to reach the top of the pole, and when it does, the terminal bud is nipped off and the stem is crushed about three nodes above the base to activate the axillary buds; after about 10 days, each vine is cut at the crushed point and this is ready for transplanting into PVC bags, filled with the rooting mixture as detailed earlier. Care should be taken, while planting in PVC bags, not to injure the roots and not let the axil be below the rooting medium. The PVC bags are placed in a cool place and covered with a light PVC sheet (200 gage) to retain high internal humidity. Buds start to develop in about 3 weeks' time, and then the PVC bags can be shifted to a place with shade. Regenerated shoots from the stumps can be trained as before, which permits the production of a continuous stream of rooted cuttings as planting material. Four harvests are possible in 1 year, with a multiplication ratio of 1:40. A prolific root system develops, and field establishment is very good. In Sarawak (Malaysia), on average, 54 rooted cuttings could be obtained (Ghawas and Miswan, 1984), and the method has been successfully field tested in the experimental farm of the Indian Institute of Spices Research. Figure 1.5 shows the bamboo technique developed in Sri Lanka for vegetative pepper propagation.

Pepper can be planted on flat surfaces and slopy land. If it is the latter, slopes facing south should be avoided. On slopy land, it is advisable to plant along the contour lines to arrest topsoil loss due to runoff water during monsoon. An important feature of pepper plantations in India is that the vines are trailed on living standards, mainly *E. indica* or *E. lithosperma* (a thorny tree and the trailing vines have strong footholds

Figure 1.5 Split-bamboo technique developed in Sri Lanka for vegetative pepper propagation.

on the thorns), *Glyricidia*, *G. robusta* (Silky Oak), *G. pinnata*, and *Ailanthus* spp. In homestead plantations, coconut and arecanut trees act as the main crop in addition to being the standards. In the southern state of Karnataka, pepper is grown along with coffee, where coffee is the main crop and the forest trees act as the standards. *Glyricidia* is the most common standard in Sri Lanka and Malaysia. Pepper planting should precede the onset of monsoon. Two or three 3-noded rooted cuttings, prepared as explained earlier, are planted in shallow pits on the northern side of the standard. If the cuttings are unrooted, four to five could be planted. At least one node should be below the soil for the establishment of roots. To accelerate growth, pits can be filled with topsoil, farmyard manure, rock phosphate, and neem cake mixture (Pillai, 1992). Plant density is around 1970–2000 ha^{-1} (George, 1982) and can go up to 5000 ha^{-1} when nutrient and carbon stress can be observed (Reddy et al., 1992).

In order to support the climbing vine on the standard, it should be tied as frequently as necessary using jute thread. Pruning of the terminal part, though not extensively practiced in India, encourages growth and bearing (Kurien and Nair, 1988). When a pepper plantation is established, starting from the beginning of the year, specific operations are to be followed. In the first 2 months, starting from January, when harvesting is completed, pruning the hanging shoots, tying vines, and mulching the basin are done. In the following 2 months, diseased vines are removed and the pits could be filled with topsoil, 1 kg of lime, and 5 kg of farmyard manure mixture. Southwest monsoon begins in May and June, and with this a mixture of 5 kg of farmyard manure, 250 g of rock phosphate and 1 kg of neem cake should be applied. In June and July, a Bordeaux mixture (1%) is applied, and drenching the pits with copper oxychloride (0.2%) is done. The damaging pollu beetle (*Longitarsus nigripennis*), a deadly insect that leaves the berries hollow (in the regional language, *Malayalam*, of Kerala State, "pollu" means hollow), can be controlled by spraying endosulfan (1.5 $ml\,l^{-1}$). Application of 50 g of urea and 120 g of MOP should also be done. In September and October, 50 g of urea, 120 g of MOP and 50 g of $MgSO_4$ are applied. Additionally, a Bordeaux mixture spray (1%) is also applied. Tying of vines should be continued. In November and December, phytosanitary measures, such as removal of infected plants, drenching with oxychloride (0.2%), mulching, and shading the young vines must be done.

It is uncommon to irrigate a pepper crop, except in Thailand where it is grown as an irrigated crop. Nevertheless, pepper plants can seriously suffer due to moisture stress in the summer months, especially in India, when daytime temperatures in April and May can soar to 35 °C or higher. A field study conducted at the Pepper Research Station in Panniyur, Kerala State, over an 8-year period (1988–1996), indicated that irrigation at an IW/CPE ratio of 0.25 increased pepper yield as much as 90%, the positive effect being manifested most in Panniyur-1 and Karimunda (Cultivar), both released from this station (Satheesan et al., 1997).

An interesting aspect of pepper cultivation is the possibility of growing other crops, in addition to pepper, in the same field or plot of land. This practice is more in common when it is the case of homestead farming, where income from the accompanying crop adds to the profit from the main pepper crop. However, this is not a general practice on large pepper plantations because raising an additional crop makes demands of its own, in terms of nutrient input, pest management, and so forth. It is important to note that companion crops to pepper must best be raised in the prebearing stage of pepper, which at best can extend from 3 to 5 years (at the most). This is because, when fully grown, pepper canopy could completely shade the accompanying crop and the latter would hardly yield. An experiment at the Indian Institute of Spices Research (Sadanandan, 1994) reported that an accompanying crop of banana (cv. *Mysore poovan*) fetched an additional income of approximately US \$460 ha^{-1}. Pillai et al. (1987) have obtained comparable results.

The recent glut in the pepper market and slump in prices have prompted many farmers to look to companion crops to generate additional income. Simultaneously, the falling prices of coconut have prompted coconut farmers to use the coconut trees as a live standard to raise pepper. In this connection, an interesting experiment conducted at the Central Plantation Crops Research Institute (CPCRI) at Kasargode, Kerala State, which has a mandate for research—primarily with regard to coconut—merits mention.

Pepper vines (Panniyur-1) were trailed on 60-year-old coconut palms, and when the vines reached a height of 4–5 m, further growth was restricted to enable farmers from climbing the palms to pluck the nuts. On average, 2 kg of dry pepper $vine^{-1}$ were obtained. Companion cropping with arecanut has also been tried. An arecanut–pepper combination gave 3832 kg ha^{-1} of dry dehusked arecanut and 1418 kg ha^{-1} of dry pepper (Nair and Gopalasundaram, 1993) from 1000 vines ha^{-1}. Companion cropping is common in Brazil as well, where an array of companion crops—such as rubber, cocoa, orange, lemon, and clove—are extensively tried. Though growing mixed companion crops in pepper gardens is a practice in many situations, it needs to be clearly understood that when inter-specific crops are grown in association, the biological implications could be diverse. This diversity might originate not only in nutrient absorption patterns, moisture stress relationships, and the like, but can also

Figure 1.6 A field view of an Indian pepper plantation.

extend to other areas, such as susceptibility/resistance to specific pests and diseases. These are the areas that deserve thorough scientific scrutiny before value judgment can be made on the suitability or otherwise of growing companion crops with the main crop of pepper. Figure 1.6 shows a field view of an Indian pepper plantation.

1.5.2 The Indonesian Experience

Within Asia, next to India, Indonesia is the major pepper grower. Pepper was the first spice Indonesia traded with Europe through Persia and Arabia and prior to World War II, 80% of world production was controlled by Indonesia. It was the Japanese occupation during the war that left many pepper plantations uncared for, with the resulting decline in production. Compared to the pre-war period when the country could boast of more than 20 million vines, after the war it came down to just about a lakh (one hundred thousand) of vines. The most interesting characteristic of pepper cultivation in Indonesia is that most of the pepper gardens are owned by small farmers, unlike in India where large plantations can be found. Currently, the main producing areas are Lampung and Bangka, with Lampung in the lead. During the last quarter of the past century, there was a perceptible decline in cultivated area in Lampung, with production shifting to Bangka, east and west Kalimantan, and Sulawesi. In Bangka, pepper was first grown in Muntok; Muntok white pepper had a good world market.

In Indonesia, pepper is grown in various types of soil, such as andosol, grumsol, latosol, podsol, regusol. Well-drained alluvium rich in humus, pH above 5.8, is ideal for pepper (Zaubin and Robbert, 1979). In general, pepper soils in Indonesia are poor in fertility, texture, and structure and low in organic matter. The two main pepper-growing areas—Bangka and west Kalimantan—have poor, reddish brown, and sandy soil. Soils in Lampung are reddish brown latosols and contain more organic matter than the soils of Bangka. In Indonesia, the planting material is made up of cuttings from orthotropic shoots. Normally, cuttings with five to seven internodes are used,

Figure 1.7 A field view of an Indonesian pepper plantation.

with two to four nodes buried in the soil. Cuttings are planted at an angle of 35° to 45° with reference to the standard. For the first three to four weeks, the cuttings must be shaded from the sun. Cuttings are planted in pits 0.6 × 0.8 × 0.6 m in size. Rooted cuttings are also used. Precautions explained earlier must also be adhered to in order to protect the cuttings. The procedures used in the case of prerooted cuttings, except that they have a better ability to withstand heat exposure from sunshine, are the same as described earlier. Such planting materials are selected from the nursery, and care should be taken that only well-rooted and well-grown cuttings in PVC bags are transplanted in the main field.

Unlike in India, Sri Lanka, and the Philippines, in Indonesia the types of standards used are both living and nonliving. In the former category, we have *Glyricidia*, *Dadap*, and *Kapok*, while in the latter category, we have hard timber poles (made of iron or wood) or concrete poles. Experience shows that concrete poles result in poor growth and productivity. In Lampung, small farmers use living standards, while nonliving ones are used in Bangka and east and west Kalimantan. There is a gradual shift to living standards because nonliving ones, of late, have become more expensive.

Pepper plantations are usually owned by small farmers, with holdings ranging between 0.2 and 1.5 ha. On average, the size is 0.65 ha. In general, pepper cultivation is intensive in Bangka and south Sumatra, while in Lampung it is extensive, where pepper is planted under live standards and input rates, in terms of fertilizer, pest management, and the like, are meager.

The Indonesian pepper industry is mostly owned by small farmers, and planting pepper on a plantation scale did not last long in Java. The nature of pepper cultivation and the degree of involvement of labor and capital have played a very important role in determining the mode of the pepper farming system in Indonesia. However, it is a tribute to the pepper industry in Indonesia that the IPC office is headquartered in Jakarta. Figure 1.7 shows a field view of an Indonesian pepper plantation.

1.6 Pepper Pests and Their Control

The pepper plant is attacked by several diseases, caused by fungi, bacteria, virus and mycoplasma, insects, and nematodes. Nutritional disorders aggravate the impact of the pests—be they caused by any one of the preceding disorders or soil and/or environmentally triggered. However, it is important to make a distinction between, for example, a deadly disease that is caused by a fungus, such as *Phytophthora* spp. (from which specific symptoms in the plant will originate), and a nutrient deficiency in the soil that may lead to an imbalance of the specific nutrient in the plant and that will aggravate the symptoms of the disease subsequently. It is important to recognize that of all the diseases a pepper plant may suffer from, that caused by the *Phytophthora* fungus is the most devastating. Hence, this review will primarily focus on the foot rot disease caused by *Phytophthora*, which is the major disease in India, Malaysia, and Indonesia causing serious damage to pepper production (Holliday and Mowat, 1963; Kueh and Sim, 1992b; Manohara et al., 1992; Sarma et al., 1992). In India, although wilt disease was reported as the major cause of pepper crop loss, *Phytophthora*, as its causal agent, was first reported only in 1966 by Samraj and Jose (1966). There are, in fact, 17 diseases that cause loss in yield, but among these *Phytophthora* foot rot and slow decline—which were originally referred to as quick wilt and slow wilt, respectively—cause the most damage (Das and Cheeran, 1986; Nambiar, 1978; Nambiar and Sarma, 1977). *Phytophthora capsici* is the causative fungus. The infection occurs aerially as well as through soil, and the severity of infection depends on the site of infection (Anandaraj and Sarma, 1995; Mamootty, 1978). When it is a foliar infection, one to many dark spots appear (with characteristic fimbriate margins), which advance and later coalesce, leading to defoliation even before the lesions spread to the entire lamina. The base of the vines can also be infected. On tender shoots, the fungus profusely sporulates forming a white covering, and when the infection reaches the stem, abrupt wilting of the entire plant takes place. Spikes, when infected, lead to the formation of blackened berries. When the infection is below the soil on roots, rotting and degeneration start, leading to yellowing, defoliation, and drying up of the plants. Feeder root infection reaches the collar through main roots, which causes the characteristic foot rot, hence the name (Anandaraj et al., 1994) Photo 1.2.

Because foot rot results in sudden wilting followed by death of the plant, the disease was earlier referred to as wilt or quick wilt and the disease-causing fungus was identified as *Phytophthora palmivora* var *piperis* by Samraj and Jose (1966) and as *P. palmivora* by Holliday and Movat (1963). In India, though pepper is traditionally grown in the Western Ghats in the states of Kerala, Karnataka, and Tamil Nadu, it has been introduced to nontraditional pepper-growing areas, such as the Andhra state and also the northeastern states, and the incidence of foot rot has been reported from these places as well (Sarkar et al., 1985). Up to 30% crop loss has been reported (Nambiar and Sarma, 1977; Samraj and Jose, 1966). The onset and spread of foot rot has been found to be closely correlated to a wet and cloudy atmosphere prevalent during the southwest monsoon period in the state of Kerala. A combination of factors, such as daily rainfall of 1.6–2.3 cm, ambient temperature ranging from 22.7 to

Photo 1.2 Pepper harvesting by men labourers (harvested berries in back pack).

29.6 °C, relative humidity of 81–99% and daily sunshine duration of 2.8–3.5 h favor the spread of the disease (Ramachandran et al., 1988b, 1990).

Inasmuch as infection through soil substrate is concerned, the prime source of infection is contaminated soil where new planting is done. The inoculum can survive in the soil for up to 19 months in the absence of a host plant (Kueh and Khew, 1982). The pathogen is normally concentrated to a depth of 30 cm (Ramachandran et al., 1986) and the severity of concentration diminishes as one moves deeper and away from the infected vines. Management of foot rot is, perhaps, the most challenging aspect of pepper cultivation. There are various aspects to the control measures, such as cultural, phytosanitary, biological, and chemical. The most important among the cultural measures refer to maintenance of good drainage. Stagnant water leads to anaerobic conditions triggering germination of pathogen propagules. A decrease in the production of phenol oxidase, phytoalexin, and fixed nitrogen, combined with suppressed mycorrhiza, leads to host plant susceptibility in anaerobic conditions (Drew and Lynch, 1980). Good drainage blocks the buildup of *P. capsici*. Phytosanitation assumes an important role in controlling the disease. Because the infected plants serve as the foci of infection (Zadocks and Van den Bosch, 1994), their removal is most crucial in controlling the disease. The initial occurrence and spread of the disease is nonrandom and tends to cluster around the previously infected plants (Anandaraj, 1997); this is the reason such infected plants have to be removed forthwith. Further, as pepper grows on living standards, they develop a canopy of their own, generating a microclimate different from that of the ambient situation, with high humidity and low temperature, which is ideal for *P. capsici* to multiply fast and infect.

Branches of the living standards should be regularly lopped, especially during the rainy season, which facilitates the penetration of direct sunlight that leads to heating the soil and thereby reducing the humid surroundings, thus acting as a check for the multiplication of the pathogen. The loppings could be used as surface mulch that, while adding to the organic matter of the substrate soil, can also check weed growth. Because the prime source of the initial inoculum of *P. capsici* in pepper plantations is the contaminated soil, infection of the foliage can occur through soil splashed onto tender shoots trailing on the ground and from these shoots onto other parts of the vines through rain splashes (Ramachandran et al., 1990). In order to arrest soil splashes, legume or grass cover mulch is ideal (Ramachandran et al., 1991; Sarma et al., 1992). However, studies in Malaysia (Ahmed, 1993) indicate that pepper grows better in clean-weeded plots, unlike in those with a cover crop of *Desmodium trifolium*. The Indian experience shows (Anandaraj, 1997) that after the cessation of the southwest monsoon, it would be ideal to clean up the pepper plots of the existing weeds and rake up the soil, which was seen to check the spread of the pathogen.

Chemical control of *Phytophthora* has revolved around the use of systemic fungicides, such as metalaxyl formulations like Ridomil granules or Ridomilziram (Kueh and Sim 1992a; Kueh et al., 1993; Ramachandran, 1990; Ramachandran and Sarma, 1985; Ramachandran et al., 1988a, 1991; Sarma et al., 1992) and/or the classical Bordeaux 1% spray mixture. Pasting the collar with Bordeaux mixture and drenching the basin/trench around the vine with oxychloride have also been effective (Lokesh and Gangadarappa, 1995; Malebennur et al., 1991; Nair and Sasikumaran, 1991; Ramachandran et al., 1991). Owing to its high cost, however, chemical control is not very popular with small farmers. Further, Coffey (1991) has expressed doubts about the efficacy of chemical control of the soil-borne diseases caused by *Phytophthora*.

In addition to the preceding measures of control, there is also the possibility of the use of organic amendments and biological control. Inasmuch as the addition of organic amendments to the soil are used to control the *Phytophthora* spp., the objective is to raise the growth of saprophytes (Kueh and Sim, 1992a), and it has been observed that the enhanced activity of saprophytes checks the growth of *P. capsici* dramatically (Anandaraj, 1997). Oil cakes—such as those obtained after expelling oil from neem seed, groundnut, and coconut—can beneficially be used to control the growth of the disease-causing fungus (Nair et al., 1993; Sadanandan et al., 1992). In addition, even chicken manure has been found to be useful. Applications of organic amendments such as the ones cited earlier will have a twofold beneficial effect: (1) enhancing the nutrition of the vines and (2) encouraging the growth of saprophytes that check the population of *P. capsici*. Both complement each other in the overall control of the fungal growth.

A recent development in the control of *Phytophthora* is the employment of biological organisms, primarily from a fungal background. Mention must be made of vesicular arbuscular mycorrhizae (VAM), *Trichoderma*, and *Gliocladium* (Anandaraj and Peter, 1996; Anandaraj and Sarma, 1994b; Datta, 1984; Manjunath and

Bagyaraj, 1982; Ramesh, 1982). Soil-borne pathogens are amenable to biological control (Cook and Baker, 1983). Because *P. capsici* is a soil-borne fungus and the prime source of contamination is infected soil, growth of antagonistic fungi would check the buildup of the pathogenic fungus. Interesting work on isolation and mass multiplication of inexpensive carrier media for field application of microorganisms having biological control properties has been carried out (Anandaraj and Sarma, 1995; Sarma et al., 1996). An inexpensive carrier medium for the culture of these beneficial organisms is water contained in mature coconuts that is thrown away after shelling. Both *Trichoderma* and *Gliocladium* have been found to grow well in this medium (Anandaraj and Sarma, 1997).

Interest in VAM has been mainly focused on its ability to solubilize native phosphorus. Its use in the control of pepper diseases was started when it was first observed that VAM suppressed the foot rot caused by *Phytophthora* in citrus (Davis and Menge, 1981; Davis et al., 1978). Soil incorporation of VAM can also be done along with *Azotobacter* and *Azospirillum*, which, besides controlling the pathogen, have beneficial growth effects (Bopaiah and Khader, 1989; Govindan and Chandy, 1985) The positive effect of VAM is through the alteration of the nature of root exudates and the rhizosphere microflora—which is generally termed the "mycorrhizosphere effect"—and there is extensive research to substantiate this view (Dehne, 1982; Graham, 1982, 1988; Linderman, 1988). The positive effects of VAM have not only enhanced plant growth, but have also suppressed disease-causing pathogens, as reported by Ewald (1991) and Graham and Egel (1988). In view of the overall positive effect of VAM, it has been recommended that VAM be incorporated into the soil right from the nursery stage and onward so that the beneficial effects of enhancing growth and suppression of pathogenic effects is obtained from the early stages (Sarma et al., 1996). Field experience shows that strict phytosanitation in the nursery is a very important prerequisite in subsequently installing a disease-free pepper plantation.

Next in importance to diseases caused by pathogenic fungi are the diseases caused by nematodes. Crop productivity is seriously hampered by plant-parasitic nematodes, and pepper is no exception. Of the 15,000 nematode species described so far, 2200 are plant parasitic. Quite often, nematode damage goes unnoticed. Because nematode infestation starts at the root surface, pepper vines so infected will subsequently suffer from a number of secondary complications and quite often these symptoms can be mistaken for nutritional or other physiological disorders. When susceptible crops are continuously grown (in the following year), they become a good breeding ground for disease-causing nematodes. A compilation of plant-parasitic nematodes associated with pepper in the major pepper-growing countries listed 48 species that belong to 29 genera (Sundararaju et al., 1979), while 54 species that belong to 30 genera were listed by Ramana and Eapen (1998). In India, 17 genera of nematodes associated with pepper were listed from the two primary pepper-growing states of Kerala and Karnataka (Sundararaju et al., 1980), while in Indonesia, 14 genera were listed (Bridge, 1978; Mustika and Zainuddin, 1978). Among the nematode species, *Meloidogyne incognita* and *Radopholus similis* are the most important and cause the most severe damage. The former is commonly known as the "root knot nematode," while the latter as the "burrowing nematode." On a global scale, the former is the most devastating.

The root knot nematodes have a specialized and complex relationship with the host plant, and they are sedentary obligate endoparasites. Their infestation of the vine results in the formation of elongated swellings on the thick primary roots due to multiple infections and typical knots or galls on either secondary or fibrous roots because of hypertrophy and hyperplasia of the infected tissues. The name *root knot* is derived because of these typical knots on the roots. In thick primary roots, a number of adult females with egg masses localize deep below the epidermis, and the entire length of the root turns into a gall and appears smooth with infrequent swellings (Mohandas and Ramana, 1987). Because nematodes feed on vascular tissues, a disruption in the arrangement and continuity of the vascular bundles leads to impaired movement of both water and nutrients, and as a consequence, the plant is vitally affected. When infestation is severe, a large amount of root mass is lost due to eventual decay of the galled roots, which in turn very adversely affect the entire vine (Mohandas and Ramana, 1991; Siti Hajijah, 1993). When root knot infestation takes place, yellowing of the foliage occurs, resulting in stunted growth and eventual decline of the vine. Dense yellowing of interveinal areas—with the deep green veins prominently visible (Ramana, 1992)—can often be mistaken for nutritional deficiency. When vines are infested with *M. incognita*, certain impaired physiological reactions, such as lowered absorption and translocation of phosphorus, potassium, zinc, magnesium, copper, calcium, and manganese and their accumulation in leaves, have been observed (Ferraz et al., 1988) as well as a reduction in total chlorophyll content of leaves (Ferraz et al., 1989). These changes lead to stunted growth of the vines. When vines were inoculated with *M. incognita* inoculum, a high concentration of total phenols without expression of any resistance to the pest was observed (Ferraz et al., 1984). Additionally, several changes in amino acids, organic acids, and sugars were also observed in vines infested with the nematode (Freire and Bridge, 1985a).

The burrowing nematode (*R. similis*) is an obligate and migratory endoparasite, is extensively found in both tropical and subtropical regions of the world, is a serious pest of pepper (Holdeman, 1986), and has a wild range of hosts (about 370 plant species). The existence of *R. similis* was first reported on the banana host by Nair et al. (1966) in Kerala State. The nematode feeds on cortical tissues and produces elongated dark brown necrotic lesions on the roots at the site of infection. The nematode pushes through the cell wall of each cell after draining its contents, and this burrowing phenomenon results in the formation of tunnels in the root tissues. The nematode derives its name from this trait of burrowing. When infestation is severe, many lesions coalesce and encircle the root cortex; because of this damage to the cortical cells, the root portion distal to these lesions gradually disintegrates. The vines tend to produce new roots, which in turn get infected, leading to the formation of a bunch of decayed root mass (Mohandas and Ramana, 1987). Yellowing of leaves, defoliation, and generally stunted growth result from infestation.

Another important disease, popularly known as the "slow decline disease," caused by a combination of the two nematodes *M. incognita* and *R. similis* and the fungus

Fusarium spp., has been the cause for widespread destruction of pepper in several countries. The disease is also known as "yellows." It was the cause of major pepper devastation in the Bangka Islands of Indonesia in the 1950s (Christie, 1959) and also in Guayana (Biessar, 1969), India (Nambiar and Sarma, 1977), Malaysia (Kueh, 1979, 1990), Brazil (Ichinohe, 1975; Sharma and Loof, 1974), and Thailand (Bridge, 1978; Sher et al., 1969). The disease was first observed on the islands of Bangka (Van der Vecht, 1950) in Indonesia and later spread to the islands of Belantung. The disease caused losses of up to 30% annually (Sitepu and Kasim, 1991). The disease is widely prevalent in Sarawak (Malaysia). The total life span of the vines can be reduced to 8–10 years, compared with the normal 2530 years (Varughese and Anuar, 1992). In Cambodia, where pepper is predominantly found in the Kampot region, the disease was responsible for bringing down vine population from 2.5 million in 1942 down to 0.5 million in 1953 (Hubert, 1957). In the predominantly pepper-growing state of Kerala in India, there is no precise estimation of crop loss, but Menon (1949) reported a 10% loss of vines. It has been observed that plants infested by root knot nematodes are more susceptible to the fungus *Phytophthora* (Winoto, 1972).

Foolproof management of nematode infestation is an elusive target. It is important to recognize that the primary source of infestation is nurseries. Solarization and fumigation are effective in controlling nematode infestation in nurseries. Additionally, incorporation of biocontrol agents, such as VAM, into solarized soil has been found to be an encouraging method for the control of nematode infestation (Anandaraj and Sarma, 1994a, 1994b; Sarma et al., 1996). There are different angles to the control regime. For instance, in plantations it has been observed that chemicals released during decomposition of organic manure—for instance, Azadirachtin from neem cake and Ricin from castor cake—are toxic to nematodes (Stirling, 1991). Mulching has been found to have a positive effect on nematode control (Ichinohe, 1980, 1985; Wahid, 1976). Hubert (1957) obtained very encouraging results in nematode control using *Eupatorium* mulch. Of the various aspects of developing resistance/tolerance to nematodes, the most reliable is the development of plant resistance. But, to date, there are no varieties or cultivars that are totally resistant to nematode attack. Obviously, the focus of pepper breeding has not been on developing absolute nematode-resistant varieties. In fact, polygenic-horizontal resistance is more important than monogenic-vertical resistance (Fassuliotis and Bhatt, 1982). Existence of physiological races or pathotypes in nematodes, in particular *Meloidogyne* spp., is another factor that must be keenly focused on in any breeding program in pepper for nematode resistance. Currently, efforts focus on developing tolerance to nematode infestation rather than developing absolute resistance. Also, efforts are being made to screen the currently available cultivars for their reaction to nematodes. For instance, in Sarawak (Malaysia), Kueh (1986) found cv. *Uthirancotta* to be the most susceptible cultivar, while cultivars such as *Balancotta*, *Belantung*, *Cheriaka-niakkadan*, *Jambi*, and *Kalluvally* are less susceptible to root knot nematodes under field conditions. Mustika (1990, 1991) found cv. *Kuching* to be tolerant to *M. incognita* and *R. similis*, compared with *Kalluvallay*, *Jambi*, and *Cunuk*. In India, none of the cultivars tested were found to be resistant to *M. incognita* and *R. similis* compared with *Kalluvally*, *Jambi*, and *Cunuk*. In India, none of the cultivars tested were found to be resistant to *M. incognita* (Jacob and Kuriyan 1979; Koshy and Sunderaraju, 1979; Ramana and Mohandas, 1986) or *R. similis* (Ramana et al., 1987; Venkitesan and Setty, 1978). However, it has been found that among the Indian cultivars, *Pournami*, which is a selection from the germplasm collection at the Indian Institute of Spices Research, is tolerant to *M. incognita*. This has been supported by field evidence that shows that where this cultivar is cultivated, only a smaller population of nematodes could thrive.

Another possibility in breeding resistance to nematode infestation is the development of resistance from related wild species into the cultivated ones. The related wild species *P. colubrinum* and *P. aduncum* are highly resistant to *M. incognita* (Paulus et al., 1993; Ramana and Mohandas, 1986). Additionally, *P. hymenophyllum* and *P. attenuatum* showed remarkable resistance to *R. similis* (Venkitesan and Setty, 1978). Though *P. colubrinum* has been found to be immune to *R. similis* (Ramana et al., 1994), crosses between *P. nigrum* and *P. colubrinum* have been unsuccessful until now. The solution for cross transfer of resistance from resistant to cultivated species might be found through the biotechnological route, but, as of now, it still is very challenging. The gene transfer mechanism in pepper is still very poorly understood.

There is also the possibility of employing biological organisms, as in the case of controlling diseases, to control nematode infestation. Sewell (1965) defines biological control as "the induced or natural, direct or indirect limitations of a harmful organism or its effects by another organism or group of organisms." Because a variety of microorganisms inhabit the soil, some of which are either predatory or antagonistic to plant-parasitic nematodes, such a course of action could be taken. However, the question of how to biologically control nematodes is still very open. Nevertheless, this approach could be a part of overall integrated pest management in pepper. As of now, the research efforts in the area of biocontrol are, indeed, very sparse. A few examples, given next, illustrate the efficiency of biocontrol.

In the area of biocontrol, fungal agents are the most important. Among them, an opportunistic fungus, namely, *Paecilomyces lilacinus*, has great potential in the control of root knot and cyst nematodes. The fungus, on contact with the egg masses of the nematodes, colonizes and grows rapidly. The chitinolytic enzymes produced by the fungus help penetration of the eggs and cause the suppression of nematodes. The efficacy of the fungus in controlling root knot nematodes has been reported by Jatala (1986). Inoculation of the fungus into rooted cuttings or seedlings of pepper significantly reduced the growth of the root knot nematode, which in turn reduced damage to the roots (Freire and Bridge, 1985b; Ramana, 1994; Sosamma and Koshy, 1995). However, it was less efficient against the control of the borrowing nematode (Geetha, 1991; Ramana, 1994).

The next fungus that is effective against nematodes is *Trichoderma* spp. It has been found effective in controlling root knot nematode (Eapen and Ramana, 1996), in addition to suppressing the *P. capsici*, as detailed earlier in this review. *Verticillium chlamydosporium* is another fungus for control of root knot and cyst nematodes (Kerry, 1990). The fungus colonizes the rhizosphere, infects adult females and egg masses, and reduces nematode multiplication by inhibiting the hatching of eggs. Parasitization of root knot nematode eggs by the fungus in Brazil has been reported (Freire and Bridge, 1985b). The occurrence of the fungus in association with *Trophotylenchulus piperis* in pepper plantations was first reported in India by Sreeja et al. (1996). In an *in vitro* test, the fungus was found to colonize the egg masses of *M. incognita* and reduce hatching by almost 50% within 5 days of inoculation (Sreeja et al., 1996). In addition to these, the VAM also has a suppressive effect on nematode population through the indirect effect of enhancing plant vigor, which has been referred to earlier in this review. Occurrence of VAM fungi on pepper roots was first reported in India by Manjunath and Bagyaraj (1982). Bacteria come after fungus as nematicidal agents. Within the bacteria is *Pasteuria penetrans*, which is an obligate parasite of some nematodes. It infects the nematodes by direct penetration through the cuticle by germinating spores sticking to the body surface of the nematodes. It has been found to suppress both *M. incognita* and *R. similis*. In addition, *Bacillus pumilis*, *B. macerans*, and *B. circulans* also suppress *M. incognita* (Sheela et al., 1993). Eapen et al. (1997) reported the inhibitory effect of fluorescent pseudomonas, such as *Pseudomonas fluorescens*, on *M. incognita*. It is important to realize that research in the area of biocontrol of nematodes is in its infant stage; a number of studies are done under *in vitro* conditions, and careful evaluation has to be performed before large-scale field recommendations can be made that are applicable to pepper plantations. The experience, so far, has been promising, though in a limited sense, and much more work needs to be done to probe the area further.

Perhaps the most effective means of controlling nematode infestation is through the use of chemicals. Nematicides—chemicals used to kill nematodes—consist of two groups: systemic nematicides and soil fumigants. Of the two, the latter is more effective because nematodes generally inhabit the soil, and once the nematicide is applied to the soil, it is absorbed by the roots of the host plant and translocated to shoots where they act as systemic inhibitors entering into biochemical reactions within the host plant cell. Despite the effectiveness of nematicides, their widespread use is not in vogue because of high costs and the environmentally adverse fallout from their use. Various nematicides, such as Phenamiphos (Nambiar and Sarma, 1980), Aldicarbsulphone (Venkitesan and Setty, 1978), Phorate, and DBCP (Venkitesan and Charles, 1980), were found to be effective. Among the fumigants, DD, methyl bromide, and ethylene dibromide have been found to be effective. With the advent of organic farming and greater awareness among consumers of products of chemical-free agriculture, use of nematicides is being increasingly restricted. Hopefully, with more research to follow, better and safer nematicides will enter the market.

The next group of pepper pests (following plant pathogenic diseases and plant-parasitic nematodes) is infestation by insects. Though a number of insect pests attack the pepper plant, the most devastating is the pollu beetle, *L. nigripennis* Mots—though, to a lesser extent, others, such as top shoot borer (*Cydia hemidoxa* Meyr.), leaf gall thrips (*Liothrips karnyi*. Bagn.), and scale insects (*Lepidosaphes piperis* Green and *Aspidiotus destructor* Sign.), could also be considered as pests of major importance. However, this review will primarily focus on the pollu beetle. The insect pest derives the name *pollu* because of the hollow berry that results from its infestation (*pollu* means "hollow" in the regional Malayalam language of Kerala). The status and control of major insect pests in India have been reviewed (Devasahayam et al., 1988; Premkumar et al., 1994).

The extent of attack by the pollu beetle varies, and reports of damage vary from 6% to 21% (Thomas and Menon, 1939), while Rehiman and Nambiar (1967) report damage as much as 30–40% in the state of Kerala. A survey of one of the major pepper-growing districts of Kerala, the Kannur district (where incidentally the first research station in pepper in the world was established in the town of Panniyur), showed that the single greatest cause for loss in pepper yield was due to infestation by the pollu beetle, amounting to as much as 13% (Prabhakaran, 1994).

The beetle measures about 2.5×1.5 mm and feeds on tender shoots, leaves, and spikes. When infestation is severe, leaves and spikes rot and drop down. The grubs do the most damage. They bore into tender berries and empty their contents; the infested berries in turn become chlorotic first and then turn black and crumble when pressed. Among the pepper cultivars, wide variation has been observed inasmuch as susceptibility or resistance is concerned. Among the four cultivars in which field observations were recorded, *Kalluvally* and *Karimunda* were least susceptible to the pest in the northern and southern regions of the state, respectively (Premkumar and Nair, 1988).

Concerning control of the insect pest, both biocontrol and chemical control assume importance in addition to plantation management. Only few natural organisms exist that have a control trait on the pollu beetle. Among these, an unidentified entomophagous nematode (Mermithidae family), a predatory spider (Araneae family), and *Oecophylla smaragdina* Fabr. (Formicidae family) have been identified. The extent of parasitization or predation by these on the pollu beetle is rather limited, and their efficacy in controlling the insect pest in field conditions is open to question (Devasahayam and Koya, 1994). Within the ambit of plantation management, application of insecticides is the only viable proposition to effectively control the pollu beetle. A number of insecticides have been evaluated in the field for their efficacy in controlling the pest. On balance, endosulfan spray (0.05%) has been found to be most effective in Kerala (Nandakumar et al., 1987; Premkumar and Nair, 1987; Premkumar et al., 1986). It must, however, be noted that of late, there has emerged widespread opposition from environmental activists in the state of Kerala to the use of endosulfan because of suspected human health hazards, such as Mongolian births, central nervous system (CNS) disorders, and even blood cancer. These maladies have been noted in the northern part of Kerala, and the insecticide has been currently banned from use.

Application of insecticides, fungicides, or nematicides leave pesticide residues in the berries, and with increasing health awareness within the country and outside of it, consumers are opting for organically grown pepper. Though biocontrol is a promising avenue, results obtained so far are of a very preliminary nature, and no hard and fast recommendations have emerged. A number of by-products have been used, both *in vitro* and in situ in the field. Among them, mention must be made of neem oil, neem seed extract (*Azadirachta indica* A. Juss.), nuxvomica (*Strychnos nux-vomica* L.), and custard apple (*Anona squamosa* L.), and the success rate has not always been very consistent (Devasahayam and Leela, 1997), though they show potential for further research (Devasahayam and Anandaraj, 1997).

As far as Malaysia and Indonesia are concerned, the pepper weevil (*Lophobaris piperis*) is the most serious insect pest. The grubs bore into the nodal region of the climbing and flowering shoots, which result in infested plants. As the infestation advances, the entire aerial portion of the vine wilts and collapses. The infestation of the pest in the lower part of the vine results in the most damage, amounting to a yield loss of approximately 5–50% in the upper part (Deciyanto and Suprapto, 1992). A number of natural predators have been identified, of which *Spathius piperis* has been found to be the most effective in controlling the pest both in the laboratory and in the field. As much as 40% control has been reported (Deciyanto and Suprapto, 1992). A number of insecticides have been recommended for the control of the pepper weevil. These include the spraying of endosulfan (0.2%), parathion (0.2%), or permithrin (0.2%). Removal of affected branches and stems would also help in reducing the level of pest infestation in the field (Deciyanto and Suprapto, 1992; Kueh, 1979).

1.7 The Processing of Black Pepper on Farm

The processing of pepper starts from the harvesting operation, which is a long, drawn-out process. With the onset of the southwest monsoon in the months of May and June, the pepper plants start to flower. The harvest commences in December and spills over into January of the following year, when there is ample sunshine and the rains completely cease. Varietal differences and the prevalent climatic pattern decide the harvesting process. In the plains of the state of Kerala, the early maturing varieties reach the harvest stage by November and the late maturing ones by December–January. Harvesting of pepper is important in view of the end products, which have a very great industrial significance. Govindarajan (1979) has made a detailed and systematic analysis of the harvesting dates against the background of the end products as detailed in Tables 1.9 and 1.10. Data in Table 1.10 give a clear idea how the chemical composition varies in relation to maturity dates.

Harvesting commences based mainly on the visual appearance of the berries. When a couple of berries in a spike turn orange or red in color, harvesting commences. Only the mature spikes are harvested. When dry, the immature spikes tend to produce shriveled berries, and this in turn will reduce the quality of the produce. However, in Sri Lanka, premature harvesting is done, and the produce is generally known as "light pepper" (Purseglove et al., 1981), which garners a premium in the market because it contains a higher amount of oleoresin and volatile oils. Harvesting is done manually by climbing up the vine using a bamboo pole; the produce is collected in a jute bag tied to the back of the climber. Only in flat plantations can mechanical ladders attached to trolleys be used for harvesting. In India, pepper plantations are usually on slopy terrain, and such harvesting can only be done by manual labor.

Once harvesting is completed and the produce brought to the farmyard, post-harvest operations have to begin. Proper handling of the post-harvest produce is very crucial to maintain the quality of the produce.

The first step in post-harvest handling is "decorning" or removal of the berries from the stalk. It is synonymous to the threshing of wheat or rice grains. In most

Table 1.9 Harvest Schedule in Relation to the End Products of Pepper

End Products	Time to Harvest
Black pepper	Fully mature and nearly ripe
White pepper	Fully ripe
Canned pepper	4- to 5-month-old berries
Dehydrated green pepper	10–15 days before full maturity
Oleoresin	15–20 days before full maturity
Pepper oil	15–20 days before full maturity
Pepper powder	Fully mature berries when starch content peaks

Table 1.10 Changes in Chemical Composition of Two Important Cultivars in Relation to Maturity Dates

Cultivar	Maturity Date									
	Months after Fruit Setting									
	Panniyur-1					Karimunda				
End products	3.0	4.5	5.5	6.5	7.5	3.0	4.5	5.5	6.5	7.0
Volatile oil (%)	6.4	7.6	6.3	2.8	2.0	6.8	10.4	8.2	4.4	3.6
NVEE[a] (%)	8.7	8.8	8.7	8.1	7.8	10.3	9.7	8.6	7.5	7.4
Piperine content (%)	1.9	2.6	2.7	3.1	3.5	1.9	2.4	2.4	3.1	3.1
Starch content (%)	2.5	3.7	5.1	10.2	16.8	2.6	4.9	6.2	15.3	15.3

[a]Nonvolatile ether extract (NVEE).

situations, especially homestead farming and large-scale plantations, decorning is still done manually by spreading the produce on a clean floor in the farmyard, where laborers, mainly women, use their feet to trample the spikes to separate the berries from the stalk. It is not very hygienic, yet in most Indian situations, there seems to be no alternative. Mechanical threshers are a rarity, though they are in use in Indonesia and Malaysia. When mechanical threshers are used, the green spikes are slowly fed into the thresher, in which a rotating drum with aluminum blades removes the berries from the stalks. The speed of the rotation is adjusted to be neither too slow nor too fast. The spikes are fed into a moving drum with aluminum plates with suitable gaps; due to the mild stretching action, the berries are separated from the stalk. A thresher powered by a 3-hp motor can handle 1.5 tons/h. Before decorning, it should be ensured that the spikes are properly washed clean of all adhering dirt. Once the berries are threshed, they should be graded using a mesh; normally, three grades are in vogue, namely, >4.25 mm (large), 3.25–4.25 mm (medium), and <3.25 mm (small). The first released hybrid Panniyur-1 belongs to the large size.

After grading, the berries are blanched by dipping the berries carried in bamboo baskets in boiling water contained in troughs. This is done to clean the produce of all adhering dust particles and any other extraneous particles, such as bird excreta. The produce is then dried to a uniform moisture content (8–10%) in open sunlight for 3 consecutive days. In India, blanching is done in boiling water for 1 min; in Papua New Guinea, Indonesia, and Micronesia it is done at 82 °C for 2 min (Pruthi, 1992). The entire operation of washing, blanching, and grading can be mechanized by spraying boiling water onto pepper fruits as they move in a conveyor belt fitted with mechanical brushes; the blanching time is adjusted through the speed of the conveyor belt. At the end of the conveyor belt, the blanched and washed berries pass through sieves of various dimensions to separate "pinhead" (good and well-sized berries) and light pepper. The graded pepper is later flushed with dry air to remove surface moisture and then transported to the drying yard or to artificial driers.

The phenolase enzyme, which imparts the black color to the berries, is activated by blanching. It also ruptures the cells and thereby accelerates the escape of moisture from the inner core, simultaneously enhancing the black color with the resinoids inside the berry. Therefore, blanched pepper will shine more and dries at a faster rate. The black color the berries acquire on drying is due to the oxidation of colorless phenolic compounds present in the skin. Polyphenolase (0-diphenol oxidase) present in the fruit wall converts colorless phenolic substrates (3,4-dihydroxy phenyl ethanol glycoside) present in the cells to black polymeric compounds (Variyar et al., 1988).

1.7.1 Sun Drying of Pepper

Sun drying reduces the initial moisture content in the berries from around 65% to approximately 10%. Most of the pepper-growing countries practice sun drying. To check the development of mold, the berry heap, which is spread on a clean floor for drying, is periodically turned over. On average, sun drying leads to a recovery of 29–38% of the moisture content. Nonuniformity and contamination by microorganisms are the major disadvantages of sun drying. Pepper is dried on different surfaces, such as bamboo mats; cement floors; or black, low-density, polyethylene sheets. It has been observed that the black surface of the sheet, due to absorption and retention of heat, results in faster drying than when other surfaces are used (Krishnamoorthy and Zachariah, 1992). In the Wayanad district in Kerala, a special type of drying is practiced where after 2 days of sun drying on bamboo mats or a cement floor, the dried berries are collected in jute sacks and stacked upright for 2 days in rooms. During this process of storing, a fermentation process begins that imparts a uniform black color to the berries, leading to enhanced flavor. After 2 days of storage the berries are, once again, spread on the cement floor or bamboo mats and dried in the conventional manner. It is for this special type of drying that the pepper of this region is highly regarded for its unique quality of color and flavor.

1.7.2 Solar Drying of Pepper

Solar driers have been developed to dry agricultural products in India (Kachru and Gupta, 1993) at the Central Institute for Agricultural Engineering in India and in

Table 1.11 Chemical Composition of Dried Pepper (in Percentage)

Chemical Constituent	Composition
Moisture	8.7–14.0
Total nitrogen	1.5–2.6
Volatile ether extract	0.3–4.2
Nonvolatile ether extract	3.9–11.5
Alcohol extract	4.4–12.0
Starch (acid hydrolysis)	28.0–49.0
Crude fiber	8.7–18.0
Crude piperine	2.8–9.0
Piperine	1.7–7.4
Total ash	3.6–5.7
Acid insoluble ash	0.03–0.55

Source: After Pruthi (1993).

Germany at the Institute of Agricultural Engineering for the Tropics and Subtropics, University of Hohenheim (Esper and Muhlbauer, 1996). In the solar drier, the change occurs mainly in the final drying phase, whereas sun drying requires several days to reach the desired moisture level. Solar drying accelerates the drying process, leading to a considerable reduction in drying time. Additionally, the maintenance of a constant temperature throughout the drying process ensures a clear product free of microbial contamination or any other extraneous matter.

Drying of berries can also be done using mechanical driers. Different types are in use, such as the "copra drier," which is used to dry coconut shells and husks; a convection drier, which is a forced draft drier; and a cascade-type drier using indirect heating by kerosene or gas.

After drying, moisture recovery normally does not exceed 70%. Dry recovery varies from 29% to 38%, depending on the variety. Table 1.11 shows the composition of dried pepper.

1.7.3 Garbling, Cleaning, and Fractionation

Dried pepper is cleaned to get rid of dirt, grit, stone, leaves, and any other extraneous matter before packing. Pneumatic separators equipped with magnetic separators are used to remove metallic contaminations, such as iron filings, nails, and the like. For destoning of spices, a combination of air-classification and vibratory conveying using inclined docks is very efficient (Ramanathan and Rao, 1974). Some of the well-established processing houses clean and grade pepper with the help of multiple sieve-*cum*-air classifier machines, whereby dust, stalks, pinheads, hollows, immature pepper, red pepper, and extra-bold pepper are removed (Pruthi, 1992, 1993). Separated pepper is then washed and dried to make it free of adhering fungus and other extraneous matter. Hence, to obtain good-quality pepper of international standard, the following steps are necessary: (1) drying, (2) separation of various fractions, (3) size segregation or grading, (4) physical cleaning (washing and drying), and (5) packing. In India, farmers as well as intermediaries bring the produce to the market, usually in an ungraded form, and garbling and grading are generally performed at the exporters' premises. Most of the exporters clean and grade using garbling machines, which remove dust and chaff and grade the pepper according to densities. However, most of the machines do not remove iron filings and heavy metals. These are generally handpicked. Pepper collected by the large merchants (exporters) is first cleaned manually by picking out the contaminants, which is mostly done by women. After this, the pepper goes for garbling. A spiral separator has been developed by Madasamy and Gothandapani (1995) that ideally removes all extraneous materials. Traditionally, pepper is manually cleaned by winnowing and then packed into jute bags. On some plantations, cleaning is done through a funnel-like arrangement in which dried pepper is fed and cleaned using a blower at the outlet point when the pinheads and light berries are blown away.

The Directorate of Marketing and Inspection, Government of India, has classified pepper into the following grades:

1. Malabar garbled (MG grades 1 and 2)
2. Malabar ungarbled (MUG grades 1 and 2)
3. Tellicherry garbled black pepper special extra bold (TGSEB)
4. Tellicherry garbled extra bold (TGEB)
5. Tellicherry garbled (TG)
6. Garbled light pepper (GL special, GL grades 1 and 2)
7. Ungarbled light pepper (UGL special, UGL grades 1 and 2)
8. Pinheads (PH grade special and PH grade 1)
9. Black pepper (nonspecified) (NS grade X)

Graded pepper fetches a higher price than ungraded, and in the international market, TGSEB fetches a premium. Good garbled pepper should have a bulk density of 500–600 $g\,l^{-1}$. Light berries should be less than 10% and pinheads (unfertilized berries) less than 4%. When bulk density is low, it indicates the presence of more lightweight berries, with subsequently less starch content, leading to poor milling quality. Only when the end product has a good aroma and a biting taste will it be considered good. The end product should contain at least 1.5% volatile oil and 3% piperine (Lewis, 1984). Table 1.12 contains data pertaining to the physico-chemical composition of different grades of black pepper.

There is another grade of pepper known as "half pepper," which is placed between normal pepper and light pepper. The berries are slightly undermature and, therefore, contain more of the active principle piperine. Due to immaturity, the berries may appear slightly wrinkled, and this is ideal for the extraction of oleoresin. The concentration of piperine in oleoresin is more important than the aroma, which is determined by the oil content.

Using appropriate sieves, the pepper is graded into different sizes. Very small and underdeveloped berries are classified as pinheads. The bulk of the pepper belongs

Table 1.12 Physicochemical Composition of Different Grades of Black Pepper (in Percentage)

Grade	Moisture	Volatile Oil	Piperine	NVEE[a]	Starch	Crude Fiber
MG	13.0	3.7	5.0	12.3	39.7	11.8
MUG	12.0	2.8	5.0	11.4	41.8	12.5
TGSEB	10.0	3.2	4.9	10.3	40.9	9.2
TGEB	13.0	2.2	4.4	9.1	39.7	11.8
Light pepper	13.0	2.9	4.1	13.5	14.6	27.8
Tellicherry ungarbled	12.0	4.0	6.3	13.5	39.3	11.0
Pin heads	13.0	0.6	0.8	7.1	11.5	27.4
High-range ungarbled	12.0	2.6	4.0	11.1	41.8	10.5

[a]Nonvolatile ether extract (NVEE).

to the average size, known as Malabar garbled. The larger-sized ones are classified as Tellicherry bold, Tellicherry extra bold, Tellicherry special extra bold, and giant (Kachru et al., 1990; Mathew, 1992).

In normal pepper, starch accounts for 34%, while it is 56.5% in white pepper and 63.2% in decorticated white pepper. Of the 12% water-soluble nitrogen present in the berries, nonprotein nitrogen constitutes about 82%, and of this, more than 50% is made up of simple amino acids that act as a nutrient for humans when pepper is consumed as a culinary material.

One of the most worrisome aspects of pepper processing is the adulteration practiced by some of the more unscrupulous merchants involved in the pepper trade. When such adulterated pepper is exported and the adulteration subsequently detected at the port of entry, it takes not only a heavy toll on the credibility of the exporting country, but, with the recent provisions of the World Trade Organization (WTO), it can lead to international disputes. One of the most common practices of adulterating pepper is mixing it with papaya seeds, which look exactly like pepper (Hartman et al., 1973; Sen and Roy, 1974).

1.7.4 Packaging and Storing

Because it is hygroscopic in nature, pepper can easily absorb ambient moisture, and when such moisture-laden pepper is stored without care, it will lead to the formation of mold, followed by insect infestation because pepper contains a good amount of starch. Thus, great care must be taken when packaging pepper. Mold and insect infestation can lead to loss of aroma, and caking can result, along with hydrolytic rancidity. Whole pepper is generally packed and transported in jute bags (burlap bags) or polyethylene-lined, double burlap bags. Dried pepper having 10–11%

moisture can be stored without the mold infestation in jute bags with polyethylene lining or in laminated high-density polyethylene bags (Balasubramanyam et al., 1978). Storehouses should not be damp and should be free of rodent attack, with controlled ventilation and devices for both humidity and temperature control. It is important to fumigate the room before storage. The room walls should be white-washed with slaked lime and should be exclusively used for storing pepper.

Another aspect that affects the processing of pepper is microbiological contamination. Investigations conducted in Malaysia have indicated that pepper berries collected from various farms had a total viable count (TVC) between 10^5 and 10^7/g sample (Apun et al., 1993). The TVC dropped on drying. TVC was found to be less when the produce was mechanically threshed as compared to when threshing with feet. Washing the spikes before threshing and blanching in boiling water minimizes the microbial load.

White pepper is a major pepper product preferred by consumers all over the world. Soaking ripe berries in running water for about 7–10 days is done to make white pepper. The skin of the berries undergoes bacterial degradation, and the softened skin is removed by either decortication or trampling with feet as explained earlier (Pruthi, 1993). An investigation done in Sarawak indicated that pepper berries that have a specific gravity >1.12 are the best for conversion to white pepper and are not as good for making ordinary pepper (Anon., 1995).

1.8 An Account of Indonesian Pepper Processing

With Indonesia being a market leader in pepper, it will be informative to review, though briefly, the processing of pepper in that country, more so because the IPC office is located in Jakarta, the capital of the country. In Indonesia, most of the pepper is grown by subsistent farmers. To improve pepper-processing technology, the Research Institute for Spice and Medicinal Plants has carried out research on equipment design meant for pepper processing, which has relevance to on-farm processing. Immediate post-harvest treatment of pepper in Indonesia is quite similar to that practiced in India, but the final cleaning, grading, and packing of the dried pepper is carried out by exporters. This is mainly because the small-scale farmers do not have the infrastructure to carry out the various aspects of processing. Most of the exporters have a complete set of a winnowing and shaking machine (for removal of extraneous materials), a washing machine, a hot air drier, and a screw-type separation machine for size separation. The entire unit can process around 5 tons of black pepper per hour. Grading is normally done according to requirements of overseas buyer, and double-lined jute bags are used for packing. Packed bags are stored in "godowns" prior to shipment.

In the traditional method of making white pepper, hygienic and quality aspects are the major problems. The berries are usually soaked in rivers or ponds with poor-quality water. Microbial contamination is a major problem, including contamination with waste, and the product acquires a swampy odor that is difficult to eliminate later. In addition, soaking in water for a long time results in the loss of volatile oil and deterioration in aroma and flavor. The quality of white pepper in the whole and ground forms is imparted by appearance, aroma, and pungency. Blanching or steaming fresh green berries for 10–15 min for decortication and then removing the rind in a fruit-pulping machine has been tried (Purseglove et al., 1981). Decortication by soaking in boiling water for 15–25 min has been tried as well, and this led to the white pepper having a better aroma than when prepared by the traditional method (Risfaheri and Hidayat, 1996). Prolonged soaking in boiling water will lead to loss of volatile oil; it adversely affects the aroma of white pepper because prolonged soaking in boiling water is known to damage the flavor profile of pepper (Pruthi, 1992). Soaking in boiling water softens the pepper fruit wall and makes decortication easy.

Pepper oil extraction is an important aspect of pepper processing. It is produced by steam distillation of the pepper berries. Pepper oil does not have piperine, which imparts pungency, and so is valued for its aroma in the fragrance industry as well as the flavor industry. Pepper oil is used in high-grade perfumes and the toiletry industry. In the flavor industry, it is primarily used in foodstuffs requiring a high pepper aroma. Volatile oil derived from steam distillation ranges from almost colorless to slightly greenish. Mild in taste with no pungency at all, Indonesia has put much emphasis on pepper oil production and its export.

Oleoresin, another by-product of pepper having odor, flavor, and pungency, is obtained by extracting the berries using organic solvents. The organoleptic properties of oleoresin are determined by its volatile oil and piperine contents, and their abundance primarily depends on the raw material used for extraction. Ground pepper can be extracted with pure organic solvents, such as acetone, ethanol, and dichloroethane. When freshly made, pepper oleoresin is a dark green, viscous heavy liquid with a strong aroma, but, on standing, crystals of piperine appear and so the oleoresin requires mixing before use to ensure uniformity and consistency. The components and the quality of the extract depend on the raw material used, and research in Indonesia indicates that light black berries are better than pinheads for oleoresin, nonvolatile ether extract (NVEE) and piperine (Mapiliandri, 1989). Oleoresin offered for sale by some of the principal producers contains 15–20% volatile oil and 35–55% piperine. Purseglove et al. (1981) have reported that as much as 1 kg of oleoresin can be obtained from 8 kg of black pepper. From a review of research on extraction technology, it can be concluded that further research in this area is needed to minimize oil loss and fractionate oil for further blending to produce oil and oleoresin with specific quality characteristics.

1.9 Industrial Processing of Black Pepper

After on-farm processing, where drying and grading to the various standards takes place, pepper must undergo further industrial processing before the final product is ready for human consumption. The dried pepper is passed through mechanical sifters for removal of pinheads, vegetable seeds, fine dust, sand, and the like (all of which are extraneous materials very likely to be present) before winnowing and destoning for removal of stalks, dust, light foreign matter, and stones. Multiple sieve-*cum*-air classifier machines and gravity separators are used for this purpose. The pepper that is separated is then washed in mechanical washers fitted with brushes to remove mold and dust, and this gives a luster to the pepper berries. The product is centrifuged to remove adhering water and then dried in electric or diesel-fired indirect driers. The dried pepper is subsequently sent through spirals for final cleaning, followed by sterilization either by steam or gamma radiation before being packed. When a pneumatic conveyor-*cum*-drier-*cum*-grader is used, the entire operation of drying and grading can be done in sequential steps, saving a lot of time.

There are different grades of pepper. Grading is done on the basis of size and shape of the berries. As per the International Organization for Standardization (ISO), the following three grades of pepper exist:

1. Nonprocessed (NP).
2. Semiprocessed (SP).
3. Processed (P).

There are permissible limits, set by ISO, for contaminants or light berries, pinheads, broken berries, and the like, in the final product and its bulk density according to ISO specification (ISO/DIS 958-1, 1996). However, individual countries have their own grades, and these have been detailed earlier. Sterilization is important to ensure good quality and freedom from microbial contaminants. Several methods of sterilization are available, such as hot air/steam sterilization, extrusion, hydrostatic/pressure sterilization, ozone sterilization, compressed carbon dioxide treatment, irradiation, microwave heating, and alcohol treatment.

In hot air/steam sterilization, the product is preheated to 50–55 °C and subsequently sterilized by a combination of indirect heating and direct steam injection. The process substantially reduces microbial contaminants while not adversely affecting the flavor, aroma, or oil content (Schneider, 1993). Yet another process of steam sterilization involves exposure to high pressures and temperatures for a predetermined time period in a series of chambers. The material to be sterilized is placed in the first chamber; high pressure is exerted and later transferred to subsequent chambers, following a similar procedure for specific time periods; and finally, the sterilized material is depressurized in a chamber (Dudek, 1996). There is a modified steam sterilization process that consists of a steam-jacketed pressurized vessel with three temperature envelopes, namely, 50, 100, or 200 °C for different products. Because the system is pressurized, there will be no loss of oil or moisture and the appearance of the product remains unchanged. Steam, compressed air, and nitrogen gas may be used depending on the end product desired (Darrington, 1991). A continuous steam sterilization process developed involves subjecting the pepper to a rapid flow of superheated steam during a predetermined time period, followed by drying, rehumidification, and packaging. Both enzyme and microbial activities are significantly reduced to low levels, and no loss of flavor or aroma or oil was noted (Uijil, 1992). Leife (1992) describes a system of steam sterilization where pepper is subjected to rapid pulses of steam, the steam condenses on the surface of the berries, and contaminants are removed. Steam sterilization is the ideal method in situations where, due to national safety restrictions, both chemical and irradiation methods are prohibited.

Yet another method of sterilization is gamma irradiation. This method is particularly useful against insect infestation. Because the process is a cold treatment, loss of volatile oil, flavor, and aroma are practically nil (Nair, 1993). A low dose of 1 kGy (kiloGray) is sufficiently effective against insect infestation whereas a 10-fold increase is needed to control microbial infestation. The process does not lead to accumulation of any harmful residues, unlike fumigation. Depending on the source geometry and the conveyor system used, the size and shape of the containers will be decided. Codex General Standards for irradiated food and recommendations of the International Code of Practices for the operation of irradiation facilities give the guidelines for irradiation procedures. It is important that Codex Standards are followed to facilitate international trade. Pepper subjected to a radiation level of up to 10 kGy does not lose its aroma and flavor, and there are no significant changes in the oil or piperine contents. When the produce is heavily contaminated, irradiation up to 30 kGy is needed. Countries such as USA, France, Austria, and Germany have approved this irradiation procedure for sterilization, and a commercial application of the technology is followed. More than 95% of the gamma irradiation facilities operating in the world use cobalt-60. Gamma irradiation facilities are capital-intensive and the unit cost of processing is highly sensitive to the scale of operation. To minimize operational costs, facilities need to operate round the clock at the maximum rated throughput that the facility is designed for. A gamma irradiator is simple to operate and maintain. Chemiluminescence, thermoluminescence, and free radial dosimetry are employed as routine methods for the detection of irradiated pepper. Other methods in use are colored indicators on the containers that are to be irradiated, viscosity measurements of gelatinized pepper, and intensity measurements of reflected signals of the electron spin resonance (ESR) spectrum.

Two other methods of sterilization use chemicals and microwaves. Fumigation with ethylene dibromide to disinfect insect contamination followed by fumigation with ethylene oxide or propylene oxide for eliminating microbial contamination is the most commonly used chemical sterilization process. Moisture content in the berries, temperature, time of contact, and concentration of the fumigating gas decide the effectiveness of the sterilization process. Fumigation results in the reduction of both volatile and nonvolatile oil contents. The main problem with the use of these fumigants is their explosive, toxic, and irritant nature, and as such, they are unsafe when used without adequate safety measures for the workers who are involved. Additionally, harmful residues are left behind following fumigation. On account of these problems many countries have banned the use of these fumigants. The main advantage with microwave sterilization is that the microbial count is drastically reduced. Microwave heating is done at 2450 MHz and high frequency heating at 27.12 MHz. With an increase in moisture in the samples, microbial efficiency is increased, but there are losses of both volatile and nonvolatile oils. Both chemical and microwave fumigation are comparable in their efficiency.

A number of value-added pepper products of industrial origin exist. The first is dehydrated green pepper. This product is prepared by industrial processing from immature green pepper fruits of appropriate varieties. The fruit should be uniformly sized with the pungency, flavor, and color of green pepper. The fruits are blanched in boiling water for a few minutes; the water is drained; the fruits are cooled and soaked in SO_2 solution to fix the green color, followed by drying in a cabinet drier at 50 °C. Upon rehydration, the product will reconstitute to a good-quality product possessing the characteristic pungent spicy taste, color, and flavor of green pepper; reconstitution is achieved when 1 part by mass of dehydrated green pepper is cooked for 20 min in the presence of 10 parts by mass of a 1% concentration of NaCl solution. To conform to international standards (ISO/DIS 10621, 1996), the product should have a moisture content of less than 8%. The use of SO_2 is a health hazard, and efforts are under way at different research institutions to find an alternative (Sankarikutty et al., 1994).

The next in line among the value-added products is canned green pepper. Of late, there has been a surging demand for canned green pepper. After removing the spikes, the pepper fruits are washed in running water and kept soaked in water with a chlorine concentration of 20 ppm for 1 h. They are then covered with a 2% hot brine of 0.2% citric acid warmed to 80 °C, sealed airtight, and kept in boiling water for 20 min. After this, they are immediately cooled in a stream of running cold water. A better color is obtained using acetic acid instead of citric acid, but, as per international standards, only citric acid is permitted, not exceeding 0.6% by mass of the packing medium, and the covering brine must be 1–2% by mass of edible common salt. For making canned pepper, berries harvested a month prior to the actual date of maturity are the best.

Bottled green pepper, dry packed green pepper, and freeze-dried pepper are among the other directly edible pepper products. Bottled green pepper is made by first despiking fresh green pepper fruits of uniform size and maturity, immediately after harvest, followed by cleaning, washing, and steeping in 20% brine solution containing citric acid. This is then allowed to cure for 3–4 weeks. The liquid is then drained off, and fresh brine of 16% concentration together with 100 ppm SO_2 and 0.2% citric acid is added. The resulting product is stored in containers kept away from direct sunlight. As per international standards (ISO/DIS 11162, 1996), the product will have the characteristic odor and flavor of fresh green pepper, the color varying from pale green to green. Dry packed green pepper is produced in the same way as bottled green pepper, with the only exception being that the liquid at the final stage is drained off and packing is done in flexible pouches. Dry packed green pepper, which has similar qualities to that of canned or bottled green pepper, can replace the latter. Freeze-dried pepper retains the original color and shape of green pepper and fetches a premium price in the market. Its processing is a manufacturer's secret.

1.9.1 White Pepper

White pepper is produced from fully ripened fruit after removal of the outer pericarp either after or before drying. White pepper is preferred for use in food products, such as colored sauces, salad dressing, soups, and mayonnaise, where dark-colored (black) pepper is undesirable. In some European countries, it is white pepper that is in common use. It is made by any of the following methods: (1) steeping and retting in water, (2) steaming, (3) boiling, (4) chemical treatment, or (5) the simple decorticating process of ripened fresh or dry berries.

The water steeping and retting (or soaking) process is started when one or two fruits in the spike start yellowing when the crop is harvested and readied for treatment. After harvest, the produce is threshed and heaped in tanks through which water is allowed to run for 7–10 days. Lightweight pinheads and light berries accumulate on the surface and are removed, and the remaining mass is rolled over at least three times a day during the retting stage. On the 11th day, the outer skin of the berries is removed by gentle rubbing, and the deskinned fruits (the pepper seeds) are transferred to another tank containing a bleaching solution. The produce is left to stand in the bleaching solution for 2 days, after which the solution is drained and the seeds washed and sun dried. The process involves the steaming or boiling of the mature green fruits for 10–15 min when the outer skin of the fruits gets softened, which is then removed by passing through a pulping machine. The deskinned fruit (pepper seeds) are washed and treated with SO_2 or bleaching powder solution and subsequently washed in water and dried in the sun. The skin of the berries recovered in this process can later be used for the recovery of pepper oil by steam distillation, though it may not be economical. The process has been developed by the Central Food Technology Research Institute in Mysore, India.

There are a number of chemical processes that can be used to prepare white pepper by treating the berries with acids or alkalis. As of now, no commercially viable production unit exists. Manilal and Gopinathan (1995) have described a method by which skin of both dried and fresh pepper is removed by microbial decortication. By using decorticating machines, whole dried black pepper fruits can be processed to white pepper. Because loss due to breakage will be high, production of white pepper by this method will be expensive. The characteristic flavor and aroma also could be missing in the final product, and as such, this method of making white pepper is not recommended.

Among all the processes described herein, the traditional retting process is the most popular, and white pepper obtained by this method is preferred by consumers. According to international standards (ISO/DIS 959-2, 1996), the product should have a maximum of only 1% permissible extraneous matter, 4% of broken berries, and 15% of black berries. The minimum bulk density should be $600\,g^{-1}$. On a comparative basis, while 100 kg of mature green pepper will yield 33 kg dry pepper, in the case of white pepper, it will only be 25 kg. From this, it can be inferred that the cost of white pepper should be at least 35% above that of black pepper. The total world demand for white pepper is around 38,000 metric tons per annum, which is a quarter of the worldwide black pepper production.

There is yet another value-added pepper product known as ground pepper. It comprises both white and black pepper. Ground pepper is obtained by grinding the normal pepper without adding any external material. Pepper has to be ground to a specific size. For grinding, hammer mills with copper-tipped hammers are preferred to silica-tipped plate mills because ground pepper obtained from the latter can contain silica in excess of the permissible limit. The ground material, after sieving, is packed, and any material above the specified size is sent back to the mill for further grinding and size reduction.

When pepper is processed in modern spice-grinding units, it is first cleaned of any extraneous matter and then passed through a magnetic separator for the removal of metal particles. After this, it is passed through a vibrating screen for further removal of extraneous matter and then sent to the hammer mill. The ground product from the hammer mill is fed into a cyclone separator to recover pepper powder. It is further screened, using appropriate sieves, and the overflow is recycled. Bag filters are used after the cyclone separators to prevent outside escape of fine particles. The ground pepper is packed into airtight containers prior to shipping, and the packages are made to specifications provided by the customer on size, pungency, moisture content, ash content, crude fiber, volatile oil, piperine, starch content, and the like. White pepper is processed in the same manner as that described earlier, except, instead of black pepper, the starting material is white pepper. The ground pepper has a characteristic flavor—very aromatic and slightly sharp—and must be free of any extraneous odors and flavors, including moldy and rancid odors. The international standards for production of ground white pepper conform to ISO/DIS 959-2 (1996). White pepper can also be made from black pepper, using selective grinding followed by sieving. The skin of the berries is a by-product and contains essential oil and oleoresin, both of which can subsequently be extracted.

1.9.2 Cryoground Pepper

To cut down oxidative losses of essential oils and to assist in the fine grinding of pepper by making the raw product brittle at low temperatures, a new technique—known as *cryogrinding*—is used at temperatures very much lower than the normal 100 °C. The product disperses more uniformly in spice formulations. The process involves injecting liquid nitrogen into the grinding zone. A temperature controller maintains the desired product temperature by suitably adjusting the liquid nitrogen flow rate. The exhausted gas is recirculated for the precooling process.

1.9.3 Pepper Oil and Oleoresin

Pepper oil and oleoresin are the two most important ingredients of pepper and have great economic value. The first imparts the aroma to the pepper, while the second provides the pungency and flavor. The essential oil can be recovered by steam or water distillation. The essential oil contains mainly a mixture of terpenic hydrocarbons and their oxygenated compounds, having a boiling point in the range of 80–200 °C. Depending on the pepper cultivar, the agro-climatic situations in which the crop is grown, the management aspects, and the grades, the content of the oil varies. The pinheads contain the most sesquiterpenes. In order to meet different flavor requirements, it is possible to suitably blend oils from different grades of pepper.

In industrial processing to recover essential oils, the product is first flaked using roller mills and then steam distilled in stainless steel extractors. The product to be distilled is heaped into the distillation unit and then compacted near the walls to prevent any steam escape. Through the bottom of the still, dry steam is passed. The still also gets heated through the jacket provided for this purpose. When the steam comes into contact with the pepper, the temperature is raised and the oil present in the oil cells of the berries vaporize, rising along with the steam through the still. It is then condensed, and because the oil is lighter than water, it floats on the surface of the condensing water. Using an oil or water separator, the oil is then recovered. The oil recovered by steam distillation is allowed to remain on anhydrous Na_2SO_4 to exclude traces of moisture. When exposed to direct sunlight and air flow, the hydrocarbons and the sesquiterpenes present in the oil undergo oxidative changes, especially during long storage. Hence, it is always desirable to store the material in airtight containers, especially because the end use occurs after a long time as a result of export via shipping to different countries.

Oleoresin is usually obtained by extraction of ground pepper with acetone, ethanol, ethylene dichloride, and ethyl acetate. The main advantage of oleoresin is that it is uniform in composition and strength. Contaminants such as mold and fungus are absent in oleoresin and can be directly added to foodstuffs after adjusting the level of concentration. Oleoresin comes in oil-soluble, water-dispersible and dry forms. Currently, oleoresin is recovered either by a single-stage or by a two-stage process. The process involves size reduction of the ground pepper prior to solvent extraction done in stainless steel extractors.

In the single-stage process, the oil is recovered (along with the resins) by solvent extraction, whereas in the two-stage process, the ground pepper is first subjected to steam distillation for the recovery of the essential oil. The composition of the oil and the oleoresin content obtained in both the processes differ slightly from each other. Due to the moisture present in the wet cake obtained after steam distillation, there is a likelihood that the oleoresin yield obtained in the two-stage process will be less compared with that obtained in the single-stage process. Drying prior to solvent extraction can prevent such loss in oleoresin yield. In the case of the single-stage process, pepper is flaked to 1.0–1.5 mm thick and is packed in stainless steel extractors for extraction with organic solvent. Normally a solid-to-solvent ratio of 1:3 is maintained at an extraction temperature of 55–60 °C. The solvent is continuously recirculated to ensure efficient solid-to-solvent contact; after 3 h of extraction, the miscella is filtered and evaporated. The solids are further contacted with lean solvents from other extractors, and the entire extraction process is completed after six or seven stages of extraction. The miscella sent from the extractors are evaporated in a shell and tube evaporator at temperatures not exceeding 80 °C in a vacuum of 250 mmHg. As the temperature starts to rise and no further solvent is recovered from the concentrated miscella, it is pumped into a high-vacuum stripper. Final desolventization is done at a vacuum of less than 20 mmHg; at no stage of the operation is the temperature allowed to rise above 100 °C.

Initially, during vacuum distillation, the condensate recovered will mostly be the organic solvent; toward the end of the distillation, the essential oil will start coming out along with the organic solvent. At this stage, some entrainer, such as alcohol or the monoterpene fraction of the essential oil, is added to the desolventizer. This is done to remove the final traces of the organic solvent as well as to supplement the monoterpene fraction of the pepper oil that might have been lost during the final solventization. The product is pumped into storage tanks after confirming that the residual solvent levels are within limits. Suitable blending is used to meet the customer requirements in terms of both oil and piperine contents. Some customers specify the homogeneity of oleoresin through a homogenizer, which may be a colloid mill or sand mill. The oleoresin, which is dark in color, is bleached using activated carbon to obtain decolorized oleoresin. All over the world, oleoresin extraction is done mostly in batch extractors. Investigations conducted at the Regional Research Laboratory, Trivandrum, in Kerala State, India, showed that extraction can be done in ambient conditions and at low residence time (Sreekumar et al., 1993). Present-day extraction, which takes more than 18 h for the processing of one batch, can be dispensed with by employing continuous countercurrent extraction. Simultaneously, the quantity of solvent needed for extraction and its loss in the extraction process can be substantially reduced.

Research in the laboratory has shown that extraction by enzymatic breakdown of spice cell walls of the dried pepper can be accomplished. In this process, the product is mixed with water, adjusting the pH by the addition of citric acid, treated with enzymes, either individually or in combination, which is followed by centrifugation. The enzymes used are those that are commercially available, such as cellulase pectinase, hemicellulase, and liquefaction enzyme preparations. Extracts having good sensory and compositional properties were obtained with some of the enzyme combinations. Optimum results were obtained using a combination of cellulase and pectinase preparation. The addition of hemicellulase did not result in any improvement of flavor. This solvent-free extraction procedure is an alternate route for the recovery of flavor compounds that is not commercially possible at the moment. Another solvent-free extraction procedure that has shown much promise is supercritical carbon dioxide extraction. Low energy consumption, high purity of resultant extracts, environmental acceptability, and possible fractional separation of the components are some of the major advantages of this process.

The current demand for piperine is on the increase. Piperine can be produced from the oleoresin in concentrated form by centrifuging the oleoresin in a basket centrifuge. Part of the oil, along with some resin, can be collected after centrifugation, and the centrifuged cake, which contains as much as 60% piperine, is obtained. By washing the centrifuged cake with pepper oil and further centrifuging, the piperine concentration can be additionally enhanced.

A number of secondary products have been developed from oleoresin to improve solubility in food substances that are marked under various trade names by the manufacturers, with their flavor strength indicated on the respective label. Standardized seasonings, which are able to withstand almost all processing conditions, are some of these secondary products. Emulsions, solubilized spices, dry soluble spices, encapsulated spices, heat-resistant spices, and fat-based spices are some of the additional secondary products. Emulsions are liquid seasonings that are prepared by emulsifying blended pepper oil or oleoresins with gum acacia or other permitted emulsifying agents. A stabilizer is added to check creaming, and the products have only limited shelf life. Solubilized pepper is blended pepper oil and/or oleoresin that is mixed with one of the polysorbate esters in such a concentration to give a clear solution when mixed with fresh water. Dry soluble pepper is prepared by dispersing the standardized oleoresin onto an edible carrier-like salt, such as dextrose or rusk, in order to provide a product that has a flavor strength equal to that of good- or average-quality ground pepper.

In addition to the industrially produced, value-added by-products of pepper previously described, there are a few others, such as heat-resistant pepper, fat-based pepper, extruded spices, and microencapsulated flavors. Of these, microencapsulation is the most important. It is a process by which the flavor material is encapsulated in a solid matrix and is ready for release as and when required. Encapsulation can be achieved by a number of techniques, such as spray drying, coacervation, and polymerization. Of these, spray drying is the most popular. The process involves homogenization of the oil/water mixture in the presence of the wall material and later removal of water under controlled conditions in a spray drier. The advantage of the spray drying process over the others is that the product, though in contact with existing gas at a higher temperature, will never reach this high temperature within the short residence time. The oil/water emulsion is atomized in the spray drier. Commonly used wall materials for encapsulation are selected from among vegetable gums, starches, dextrins, proteins, sugars, and cellulose esters. The wall material is selected so as to meet, as closely as possible, the desired properties, such as low viscosity at a high solid state; ability to emulsify or disperse the active material; nonreactivity with the material to be encapsulated, both during processing and in prolonged storage; uniform film-forming; ready availability; and the like. Investigations indicate that the addition of surfactants during the process of encapsulation prior to spray drying will reduce the oil loss (Anon., 1989). Another process for microencapsulation is the CR-100 process (Anon., 1995; Findlay-Wilson, 1995; Mos, 1995), which overcomes the limitations of the spray drying process such as reduction in flavor quality and yield.

Heat-resistant pepper is a double-encapsulated product in which the capsules are rendered water insoluble by a suitable coating and the contained flavor is released only at high temperatures, such as in the case of baking. Fat-based pepper is a blend of pepper oil and/or oleoresin in a liquid edible oil or hydrogenate fat base formulated for use in products such as mayonnaise. Extruded spices are those sterilized, ground, and encapsulated in a single step. The process involves a combination of pressure changes, temperature shock, and shear. The process, accomplished in a twin-screw extruder, helps retain color and flavor in the original form because of the very short processing time. When fresh spice is fed into the twin-screw extruder, the starchy materials become gelatinized and form an encapsulated product. This process substantially reduces contamination by bacteria, mold, and yeast. The product emerges from the extruder as a "spice rope," which is then cut into pellets. Lucas Ingredient, Britain, markets the products under the brand name Master Spice (Scott, 1992).

1.10 The Future of the Global Pepper Economy

The global pepper economy has different facets: (1) supply side, (2) demand side, (3) price projections (both supply and demand included simultaneously), and (4) pricing and market equilibrium. Bade and Smit (1992) have devised precise models to study these facets. This review does not provide an elaborate discussion of these models, but the essential conclusions stemming from them will be discussed in this section. The various models are all mathematical in nature.

1.10.1 The Supply Side of the Pepper Economy

The supply side of pepper is approached on a national basis, examining the area, production, and exports from each country. Normal production value is obtained by multiplying the number of pepper vines by the average yield of a vine in a country. Although measurement by vines must be considered much better, a similar line of reasoning can be applied with area and average yield per hectare. The major drawback of this method in comparison with using vines is that it adds to a source of variation. Not only the per-vine yield fluctuates, but also the number of vines per hectare. In India, this is a very significant variation, but in other countries—for example, in Thailand—the number of vines per hectare is almost the same all over the country. If the intensity of cultivation is relatively constant and there is no significant change in cultivation patterns, it is possible to estimate a nationwide normal average yield per hectare. In case the average number of vines per hectare cannot be expected to be constant, an assumption is needed about the change. In case the intensity of cultivation increases or a superior variety is introduced, it can then be assumed that there will be an upward trend in the normal yield per hectare some 2 years after the start of the intensification when the planted vines become productive. If normal yield can be used as a basis for forecasting, the actual yield will deviate from it because of the influences of weather, amount of fertilizer applied, and level of maintenance. Unfortunately, these effects are not confined to only 1 year. When a year is very wet, the crop will be small, but, most importantly, there will be also the incidence of diseases such as foot rot. Invariably, such diseases will determine the yield in the succeeding year. Also, the effects of neglect or exceptionally good maintenance will spread over more than 1 year.

The preceding discussion illustrates that even when data on productive vines and average yield are available, and quite reliable, there would be sufficient scope for simulation and expert interpretation. However, global information is very expensive to procure, and gathering information on agriculture can be even more expensive and time consuming. This is not a priority of the developing countries. Planning depends largely on information, and a model cannot compensate for the lack of quality data. Inasmuch as quality of information is concerned, it can only interpret and detect inconsistencies. Thus, in conclusion, it can be said that the modeling of production and supply presented is only a preliminary step on the road to a more sophisticated model analysis based on superior data. In the following paragraphs, a brief review of the pepper economy of important pepper-producing countries, such as India, Brazil, Thailand, Malaysia, and Sri Lanka, is given.

1.10.1.1 India

It is a very important observation that the pepper-growing area (also known as the area "under pepper") in India would be enough to meet the entire global demand if the average yields were only a third of what is obtained in Sarawak (Malaysia). This points to very important lacunae in data collection, precision, and reporting. Except in large pepper plantations in the states of Karnataka and Northern Kerala, pepper is still grown in homestead gardens as an intercrop between coffee and cardamom and is quite often trailed using arecanut and coconut trees as standards. Up to 1986, a survey was conducted in Kerala—by field extension agents in randomly chosen parts of Kerala—in such a way that, within 5 years, every part was visited once. The total area under pepper from the population was then multiplied by the inverse of the sample area, divided by the state area ratio. The goal was to estimate area on the basis of 560 vines ha^{-1}. The method was applied, asking the same people, to get production estimates. From 1987 onward, the Department of Economics and Statistics of the state has used a more sophisticated system, especially to estimate production. Though the method used is rather elementary, other countries probably do not have any better system. A similar equation as that used in India is also used in Brazil.

On the market side, it is seen that an increase in price is an incentive for enhanced acreage as more farmers resort to the use of fertilizers and adopt better maintenance practices followed by more picking rounds. The previous year's price is the trigger for these added incentives. Farmers also keep pepper in stock, hoping to sell it for a better price in the following year if a price drop is encountered in the current season. Alternatively, when prices escalate and farmers release their stocks, it will seem as though production has gone up, which is not the case. Indian exports depend largely on the current season's crop. The other factor is the influence of price changes on stocks of traders. When there is a sudden increase in price, the total amount of exports may even exceed production. The year 1985 was exceptional. Again, 1991 and 1992 were also exceptional on account of the restructuring of the Indian economy when globalization and liberalization processes set in.

1.10.1.2 Brazil

In Brazil, there are no time-series data available on area by cultivation system, although there are different cultivation systems in the country. To be sure, it is still totally unclear how the Brazilian Pepper Exporters Association gathers the data presented at the IPC meetings. Apparently, data on area and production do indicate that pepper yields have gone up over the years in Brazil. But this is not the case. If production is divided by area, it would appear that there is a decreasing trend over time. In fact, between 1978 and 1979, yields plummeted—despite area expansion—suggesting that in those years data on area did not include area with immature vines. After 1979, the correlation between area and production was strong, although 1984 was an extraordinary year that had a much higher yield than could be explained. This may have been due to positive price correlation.

Pepper consumption in Brazil is negligible compared with production and presumably kept-out-of-production statistics, and because there are some other small pepper-producing regions outside of Para state, it can be expected that almost all pepper produced will be exported. However, there is a rather constant amount of about 1500 tons that is apparently not exported. Price is the most important variable in the current productive area under pepper.

1.10.1.3 Indonesia

Indonesian data on aggregate area under pepper are rather scanty. The official records claim that the total area did not change from 1983 until 1987. However, records for Lampung and Bangka show substantial changes over these years. The time-series data available on the areas of Lampung and Bangka are still too minimal. Further, there is hardly any information on area in Kalimantan. For future modeling research, regional disaggregation of supply of Indonesia is important, especially because of the special status of Bangka, which produces only white pepper. Prices have played only a minor role in area expansion. Clearing of new land for pepper production will play a predominant role in pepper production, especially in Kalimantan and Sulawesi. For Indonesia, the influences of price on stocks is far less important than on maintenance, a similar conclusion to that for India. The year 1993 was a bad year because of a serious setback in production and thus on exports, reflecting a serious situation for white pepper in Bangka.

1.10.1.4 Thailand

Thailand is a relative newcomer to pepper production. Until 1986, area under pepper was quite steady; thereafter it expanded, but once again dropped between 1991 and 1993 when low prices ruled the market. At this time, the government encouraged farmers to diversify from pepper. Changes in prices were influenced by prices prevailing in the previous year, while area was held steady. Productivity per hectare is very high, and the production is influenced by the previous year's prices. Over recent years, production has been on the decline. The country has a high domestic consumption, which is around 5000 tons per year. Exports take a fixed share of production, which is influenced by the previous year's shortage or surplus.

1.10.1.5 Malaysia

Malaysian data on area and production of pepper, of which nearly 98% refers to that of Sarawak, are rather unreliable. Foot rot strongly influences yield. Data on production are based on exports; pepper farmers from Sarawak speculate on the pepper market, so export is not a reliable index to show production trends because the farmers who speculate in the pepper trade are rich. Hence, large differences between production and exports are bound to arise. From 1989 onward, there is a decline in area arising from shortage of labor and its high cost, which did not allow for profitable production to the extent as was the case in the past. The price of the previous 2 years greatly influenced production per hectare, though this influence is through the planting of new vines.

1.10.1.6 Sri Lanka

A traditional pepper producer, there is only a slight increase in area over the years in Sri Lanka. Productivity is low, and despite an increase in area, production did not correspondingly increase, which is rather surprising. The production pattern in 1993—a year that showed an incredible increase in production and exports—is just the opposite that of Indonesia. Export in 1990 was exceptionally high, exceeding production. Export can be explained on the basis of production, the change in production, and domestic prices.

1.10.1.7 Other Countries

In addition to the countries specifically discussed, countries such as Madagascar and those on the African continent (Benin, Cameroon, Congo, Gabon, Ethiopia, Nigeria, Zambia, Zimbabwe) also produce pepper. Madagascar, with a production of about 2500 tons, leads the pack. Because long-term time-series data and information on area and production are needed, as well as information on internal markets, no tangible conclusions can be drawn from the existing information.

1.10.2 The Demand Side of the Pepper Economy

To obtain a clear analysis of the demand side of the pepper economy, precise data on the end use of pepper are needed. To date, precise data on differentiation of pepper use in food industries, institutional catering, and household consumption are unavailable, and as such, only rough estimates can be made. Further, a distinction between black and white pepper is needed. The following countrywise groupings will provide an insight into the demand side of the pepper economy.

1.10.2.1 European Union

As a shift in the food consumption pattern takes place—for example, consumption of more prefabricated food, fast food, and the like), the demand scene will fluctuate. The countries of the European Union are price conscious as far as stock formation is concerned. Also, the level of stocks at the end of the previous year has a reasonably significant influence.

1.10.2.2 Rest of Western Europe

Starting from 1998, Switzerland has import data on pure pepper; these figures previously included pimento and capsicum. Price changes influence stock formation.

1.10.2.3 Eastern Europe and CIS

The economic problems that overtook eastern Europe and the Commonwealth of Independent States (CIS) have had a very negative effect on import since 1990. Price changes influence stock formation.

1.10.2.4 North America (USA and Canada)

There is a negative relationship between price levels and starting stocks in the USA and Canada. When prices are low, traders will keep more stock, expecting prices to increase again sometime in the near future; if stocks are high, traders tend to sell.

1.10.2.5 *Japan*

Demand for pepper in Japan is pegged to the gross domestic product (GDP) in more than one way. A shift toward outdoor, ready-to-eat food, in particular Western food, goes along with a rising GDP. Changes in stocks paralleled trends in North America.

1.10.2.6 *Australia and New Zealand*

Stock levels are related to consumption in Australia and New Zealand, and no price influence is found on the demand side.

1.10.2.7 *Latin America*

Per capita import of pepper did not show any increase in Latin America. Between 1976 and 1980 and 1985 and 1988, significantly lower imports took place, and stock formation depended on prices and lagged stocks.

1.10.2.8 *Asia and Pacific Countries, Singapore, Australia, and New Zealand (Excluding China)*

Stock formation depended on prices and lagged stocks in the Asia-Pacific area.

1.10.2.9 *China*

China was a small importer of pepper, and the import spurted in the latter part of the 1970s. Of late, imports have dwindled.

1.10.2.10 *Middle East and North Africa*

The Middle East and North Africa have a rather irregular import pattern. Stock formation was influenced by price changes.

1.10.2.11 *Rest of Africa*

Per capita import of pepper did not show any increase in the rest of Africa, as was the case in Latin America. Since 1990, imports surged. Stock formation was influenced by price changes.

1.10.3 Prices and World Market

As far as the world market is concerned, Singapore holds a special place for pepper import and its commercial consequences. Of the total pepper exported from Malaysia, 69% moves *via* Singapore. The function of Singapore as an entrepot is accounted for by the variable total exports of producing countries less estimated world consumption, which is merely an estimate of the change of stocks outside producing countries. Singapore is expected to import some of these stocks while keeping the major part of these stocks as carryover. Some part of pepper imports is consumed, but no statistics are available on pepper consumption in Singapore. It is estimated that consumption is approximately 1% of imports. The price pattern indicates that pepper traders are presumably more interested in trade if prices are high, which does not seem unrealistic because margins will definitely be correlated with price peaks.

1.10.4 The Pepper Price Outlook by 2020

Perhaps the most important aspect of the price structure is the impact of the size of the Indian supply responses to price fluctuation. This aspect has not been sufficiently captured. For all countries, the investment side is not yet represented adequately for long-term analysis. Stock formation at various levels needs further work as well. Price cycles of around 7 years will continue in the future, showing that the 10-year cycles in income have very little impact. There will not be an increase in average real prices. Obviously, with inflation around 4% per year, nominal prices do increase on average.

1.10.5 The Pepper Supply Outlook by 2020

Projections indicate that production and exports are expected to decline in Brazil, Malaysia, and Thailand because of the increase in labor costs. (India and Indonesia, however, show strong growth.) Brazil shows a modest decline on average, and the same conclusion can be drawn for Malaysia and Thailand. These countries are becoming too rich to grow labor-intensive agricultural crops such as pepper. Sri Lanka is expected to level off on account of limited area. Exports from Vietnam and China, especially the former, are expected to grow rapidly, reaching an aggregate of more than 60,000 tons. A large quantity of pepper is imported into India from Vietnam, often clandestinely, in the garb of value addition. Because Indian pepper fetches a premium in the world market, unscrupulous traders mix the low-quality and cheap Vietnamese pepper with Malabar pepper and re-export it as Malabar pepper to the USA and Europe. On detection by the consumers, this adversely affects Indian pepper in the world market, pricewise, and also affects future imports.

The gap left behind as relatively rich countries such as Brazil, Malaysia, and Thailand decrease their production and exports is filled by Vietnam and China—but not by traditional major producers such as India and Indonesia. Exports from Madagascar and other African countries are expected to be steady. This inevitably leads to the important conclusion that almost all pepper exports will originate in Asia.

1.10.6 The Pepper Demand Outlook by 2020

On the consumption side, North America, Japan, and the European Union will show a steady growth. Developments in eastern Europe will affect the consumption pattern, so will the increased consumption in the Pacific countries and other upcoming Asian regions, such as the Philippines, Korea, and Cambodia. Another major consumption area is the Middle East and North Africa.

1.10.7 Countrywise Economic Growth Affecting Production and Consumption

It is interesting to analyze the GDP/growth rates that affect pepper production and consumption. The details of the estimates are presented by Burger and Smit (1997). Historical developments have affected consumption. For instance, the late 1980s were years of political turmoil for eastern Europe and the CIS, and this adversely affected growth rates and consumption. On the other hand, countries in eastern, far eastern, and southeastern Asia have enjoyed very buoyant growth rates, which positively affected consumption. However, such very high growth rates cannot be maintained for decades, as has been clearly shown by other countries with high growth rates in the past.

In the long run, the USA will average a growth rate of just below 2% per annum, while Japan, despite its recovery, may not reach the US growth rate after its period of high growth up to 1990. In western Europe, growth rate will hover around 1–3%. Eastern Europe, CIS, and former Yugoslavia are expected to recover, but, on average, at rather moderate levels of economic growth. Growth rates in Latin America will be somewhat higher than in Europe. In south Asia, India is maintaining a growth rate around 5%, though expectations were higher. Pakistan has a somewhat higher rate. In southeast Asia, the Philippines will show a moderate growth, while in the other countries, though showing higher rates, a declining trend has been observed. The same applies to China. Vietnam will grow around 4%. And, with the exception of Nigeria, Africa on the whole is showing only very moderate growth rates.

The preceding growth scenario indicates that the potential for pepper export from producing countries to the USA and Europe remains bright in the long run. This is primarily because the processed and fast-food cultures will unfold further in the years to come. Southeast Asian countries are also potential markets. There is not much scope for export into Africa. Latin America, which will have a higher growth rate, is a potential market, but countries such as Brazil are already important pepper-producing countries, and as such, export from other pepper-producing countries has only limited scope for export to Latin America.

1.11 Pepper Economy in India

Because India is the most important producer of black pepper in the world, a discussion on pepper economy and marketing patterns in the country is relevant. Currently, India accounts for 49.5% of the total acreage in the world, but in 1951 it was 70%. Though the country holds the largest pepper acreage in the world and contributes up to 29.8% of world production, its productivity is the lowest in the world—an abysmally low figure of 294 kg ha^{-1}. In fact, the total contribution of India to world production dropped from 66% in 1951 to the current 29.8%. This has been the case for exports as well. From 1951 to 1991, the India's share of the global pepper market fell from 56% to 23%. It is interesting to note that while India's position of preeminence in pepper production dwindled, other countries' performance was on the rise. The last decade of the past century witnessed a global downtrend in pepper production.

In India, pepper is primarily cultivated for export. From around 16,000 tons in 1950–1951, export has scaled up to 32,000 tons, a 100-fold increase, by the turn of the century. The 1990s witnessed a fluctuating market because of price fluctuations. The latter half of the 1990s witnessed a sharp increase in unit price, peaking at about US \$50 kg^{-1} in 1997, and though export per se witnessed a drop, it was more than compensated for by the escalating unit price. In the latter half of the 1990s, global pepper production was around 189,000 tons, of which more than 118,000 tons were exported as per the data provided by the IPC headquartered in Jakarta. In these estimates, India tops the list of producing countries with 65,000 tons, followed by Indonesia. India exports spices to more than 120 countries around the world; of these countries, export to 23 countries account for about 82% of the hard currency earned (in US \$) by the country. In the case of pepper, India is the major supplier to 83 countries; among the importers, USA tops the list with an α growth rate of 5% in quantity and 3% in value. Of the total spices imported into the USA, pepper accounts for about 15%, of which 50% orginates in India. USA is also the major importer of spice oils, primarily oleoresin, and India supplies close to 86%. However, the recent trend shows a decline in import into the USA. In the beginning of the last decade, pepper import from India to the USA was 56%, which declined to 30% by the middle of the decade.

Though USA tops the list of pepper importers from India, it is interesting to note that per-unit value realization is lowest from Russia, Canada, Italy, and Poland, with Russia topping the list. It is also interesting to note that there is a visible shift in the food habits of Americans, where the preference for hot spices like pepper in the last two decades accounts for nearly 75% of consumption. The US trade association projected spice consumption to peak to 1000 million pounds by the turn of the century (Cheriankunju, 1996a). Maintenance of quality assumes the greatest importance in this regard. The United States Food and Drug Administration (USFDA) detained 1164 consignments of Indian spices due to contamination with seeds of noxious weeds in 1995. In fact, there has been a steep escalation of detection of contaminated consignments from 140 in 1991 to 757 by 1994 (Sivadasan, 1996), which, indeed, is a very poor reflection of Indian trade. With the WTO tightening the noose on substandard-quality imported food articles from the developing countries, India would do well to clean its own facilities, lest a very adverse fallout in the pepper market follows. It is, however, encouraging to note that the Spices Board of India, the apex body attached to the Ministry of Commerce, Government of India, has initiated a number of measures to ensure that quality of pepper is guaranteed both at the producer as well as the exporter levels.

The former USSR was the largest pepper importer from India, with a total tonnage of 19,400 in 1989–1990; the tonnage dropped dramatically to 6060 tons by 1994–1995 primarily because of political turmoil in the region (Cheriankunju, 1996a). In spite of the economic instability, Russia continues to be a major consumer of pepper from India. Russia, with an import of about 3200 tons in 1994–1995, accounted for US \$6.05 million, followed by Italy at US \$3.62 million and Canada at US \$2.54 million. Traditional export of pepper from India to such countries as Hungary, North Korea, Maldives, Trinidad, Yemen, United Arab Emirates, and Yugoslavia has considerably decreased; in their place, new markets, such as Chile, Kazakhstan, Tajikistan, and Venezuela, have emerged. Of these, Kazakhstan imports the largest amount. Apart from these, Japan is fast emerging as a potential market for Indian pepper. This is mainly due to the fast-changing food habits of the country, where ready-to-eat fast foods are emerging along the lines of Western food habits, where pepper forms an important culinary component. In 1994–1995, Japan imported 41,911 metric tons of spices, of which 17.8% was pepper. Most of the pepper goes into industrial uses followed by domestic consumption and other services sectors. The annual rate of increase in spice import by Japan is estimated to be 5% (Cheriankunju, 1996b). The market share of Japan's import from India is stagnant at 5.5%, while that from Malaysia, the bulk exporter to Japan, stays at 68%. Indonesia contributes 31%. Indian pepper has a premium in the world market and if strategic advantages are exploited following the establishment of WTO, India can scale up its export to Japan. Additionally, Israel is a potential market for Indian pepper. In 1994–1995, Israel imported 6902 tons valued at US $7.42 million (Anon., 1996). Among the potential markets for Indian pepper, USA, Canada, Russia, UK, and Japan are on the top rungs because these countries have a lower level of instability in pepper imports (Jeromi and Ramanathan, 1983). According to the International Trade Center in Geneva, world trade in spices during the last decade averaged 450,000 metric tons valued at US $1500 million; pepper is the major component, which accounts for as much as 30–35% in quantity and 20–25% in value (Peter, 1996b).

1.11.1 Pepper Production Scenario in India

In 1950–1951, nearly 80,000 ha of cultivated area in India were under pepper, 98% in Kerala State. An analysis of the trends in growth with regard to area, production, and productivity is given in Table 1.13. It can be surmised from the table that during the last five decades, overall acreage, production, and productivity have declined all over India and, in particular, in Kerala State. The steepest fall was in productivity. On an all-India basis, while acreage slipped by only 22% from the 1950s to the 1990s, production and productivity declined by 54% and 89%, respectively. For Kerala, the respective figures for decline in area and productivity are 25% and 96%,

Table 1.13 Area, Production, and Productivity Growth Rates for Black Pepper in India and Kerala State

Details		1950s	1960s	1970s	1980s	1990s	Overall
Area	India	0.87	1.12	1.05	1.81	1.11	0.68
	Kerala	0.94	0.19	−0.19	1.85	1.00	0.70
Production	India	1.36	−0.50	0.34	2.43	−0.32	0.63
	Kerala	0.40	−0.88	−0.40	2.26	0.70	0.68
Productivity	India	0.44	−1.40	1.27	0.63	−1.40	−0.65
	Kerala	0.46	−1.07	−0.22	0.41	−0.01	−0.02

Source: Eaconomic Preview, Government of the State of Kerala.
Note: Data given as log $Y = a + bt$ in percent per annum.

while production increased by 70%. In fact, both on an all-India basis and for Kerala State, productivity showed very dramatic negative trends.

The reason for the steep fall in productivity, especially in Kerala, is due to the prevalence of diseases, of which the *Phytophthora* foot rot is the most serious; there is, as yet, no permanent remedy for this disease. As the question of area expansion has only very limited possibilities, if there is to be a breakthrough in pepper production in Kerala, it will have to be through breeding superior, disease-resistant varieties with good agronomy to match. Soil management is still based on book knowledge, and there is a very clear dearth of fresh ideas.

Pepper cultivation in Kerala is largely concentrated in certain districts. Kannur, Kozhikode, Kollam, and Thiruvananthapuram districts, in sequence, contribute more than 80% of the total pepper area in the state and 75% production of the state. By the early 1990s, the share of these four districts declined to 33% and 29% in acreage and production, respectively, and others took over. Currently, the districts of Idukki, Wayanad, and Kannur have emerged as the major pepper-producing districts of the state, contributing 60% of arca and 67% of production. Among the three, Idukki district leads, with 21% area and 27% production. It is followed by the Wayanad district, which has contributed 25% of pepper produced in the state during the past decade. A very interesting development of late is the emergence of the Idukki district as the frontrunner in the production of "organic" pepper—much valued in the USA and Europe. Based on the cool climate of the region and relatively lower impact by the chemical agriculture, the district holds out great promise to emerge as a global leader in the production of organic pepper.

Pepper as a monocrop is confined to the district of Idukki and Wayanad, which constitute only 3% of the total area (George et al., 1989). There is ample scope to cultivate pepper as an intercrop on coffee, tea, and cardamom plantations. The high-range districts of Kerala, such as Idukki and Wayanad, where currently most of these plantations are located, are potential areas where pepper can be grown as an intercrop. Silver oak, used as a shade tree in these plantations can be used as standards (support for the pepper vine to trail). There is a strong case to initiate systematic field trials to develop a clear-cut package of practices in growing pepper as an intercrop on these plantations. Isolated studies have been carried out. However, no efforts on a large scale have been initiated. This is a very promising area for enhancing pepper production in India, but until now, it has not been tackled in a very systematic and scientific manner.

Pepper is an important component of "homestead farming" in Kerala, and the vines are trailed on coconut trees, arecanut trees, mango trees, jackfruit trees, and tamarind trees, which are generally grown on small farms owned by small-scale and marginal farmers. The level of inputs—namely, fertilizers, pesticides, and the like—in these situations is low. Transformation of these homestead gardens due to various socioeconomic reasons largely reduced this practice. Coupled with this, the incidence of foot rot swept away a substantial number of pepper plants on homestead farms. There is a strong case for developing a package of practices at low input levels in these homestead farms. There is a need to reintroduce pepper in homestead farming, both in urban and rural areas. For this purpose, bush pepper offers much promise.

Photo 1.3 Bush pepper in pot.

1.11.2 Bush Pepper

Bush pepper holds out much promise in homestead farming. The plagiotropic, lateral fruiting branches of pepper exhibit sympodial growth, and these branches, when rooted and planted, grow into a bush. More fruiting branches are produced by these bushes, and flowering commences in the first year of planting itself. A common method developed at the Indian Institute of Spices Research (IISR) involves the following steps: (1) selecting healthy lateral shoots of the previous year's growth with two or three leaves; (2) dipping the cut end in a commercially available rooting hormone; and (3) planting them in moist rooting medium, which consists of weathered coir dust, in a 200-gauge PVC bag about 40 × 25 cm in size (later, the bags are kept under shade). Prior to closing the bag, air is blown into it. In about a month, the cuttings start producing roots. After about 2 months, the bags are opened and allowed to remain in that state for a few days before transplanting into clay pots of convenient size (Photo 1.3). Almost 80% success can be ensured provided the three steps are meticulously followed. Bush pepper can be planted at a minimum density of 2500 plants ha^{-1} spaced 2 × 2 m apart, and the population can be doubled if spacing is reduced. Record yields of 1960 kg ha^{-1} have been obtained using Panniyur-1 variety (Annual Report of IISR, 1997) and yields in the range of 1600–1700 kg ha^{-1} are quite common. Bush pepper is commonly grown in Indonesia, Malaysia, and the Philippines. It is still very far from catching on in India, and the home state of pepper, Kerala. Figure 1.8 shows bush pepper growth in India.

Figure 1.8 Bush pepper in India.

1.11.3 Economics of Pepper Production in the State of Kerala

In India, Kerala is the most important pepper-producing state and merits further discussion. Vinod (1984) and Santosh (1985) have attempted to obtain a dependable estimate of the cost structure of pepper production in the Idukki and Kannur districts, respectively, which are at the forefront of pepper production in Kerala State. Pepper production in Kerala is highly labor intensive, and labor costs account for more than 50% of the total cost of production. An important reason for the decline in pepper production in the state could be due to the nonavailability of timely labor and its very high wage structure on an all-India basis. Among all the states of India, Kerala's labor wages are the highest. Though Idukki district leads in cultivation, profitability is higher in Kannur district. The cost/benefit ratio was 1.09 for Idukki district and 1.16 for Kannur district as per the findings of Vinod (1984) and Santosh (1985). The payback period of pepper cultivation was estimated as 10 years in Idukki district and 11 years in Kannur district. The corresponding net present worth is approximately US $93 for Idukki district and US $148 for Kannur district, and the internal rate of return is 13.48 and 17.22%, respectively.

1.11.4 Marketing

Village traders dominate the local market inasmuch as small pepper farmers are concerned in Kerala. The trade is based, in many ways, on faith and long-term personal bonds, which is absent in the case of wholesale merchants who deal in large quantities. Additionally, transportation to the city or town market is quite often out of the reach of small farmers, and this also leads to the small farmer going to the village trader. By tradition, village traders collect the product directly from the farmers' yard or, if not possible, from neighboring village markets. Wholesale merchants, though, offer higher prices compared to the village trader. Village and internal wholesale merchants are important market intermediaries (Vinod, 1984). Small and marginal farmers generally affect village sales in the nature of a pre-harvest contract, whereby the farmer is obliged to sell his product at a predetermined price prior to harvest; if there arises any subsequent fluctuation in price, both parties are bound to go by the earlier agreed-upon contract of sale.

The upcountry wholesale merchants transport the produce to Kochi (which is the port city of Kerala and the prime business center of the state), in the central part of the state. Once the produce reaches Kochi, the commission agents negotiate with the exporters or wholesale merchants or brokers. The brokers act on behalf of the internal wholesale merchants or exporters and negotiate with the commission agents. At the exporters' premises, the produce is processed and graded. Prior to the official clearance of the produce for export, it is subjected to check sampling and all grading tests . The consignment thus cleared for export is then handed over to the clearing or forwarding agency after completion of the paperwork. Subsequent work is carried out by the forwarding agency. Four major marketing channels were identified in Idukki District, while there were five in Kannur District. The share of the producer on the free on board (FOB) price of pepper was estimated to be 86.06%. Compared to this, the producer is the beneficiary of a higher share, which is 88.80%, when the produce is internally consumed and moved through the domestic consumption channel. In either case, it is clear that intermediaries corner a good percentage of the value chain.

1.11.5 Pepper Futures Market

The origin of the futures trade is almost a century old in India. Futures trade is an insurance against price fluctuations of commodities handled by traders. It was the Indian Pepper and Spice Trade Association (IPSTA) that heralded the futures trade in pepper in India in Kochi (Sethuram, 1995). The move had a very positive effect on the transfer of risk and recovery of price. From hedgers, the risk is transferred to the speculators, who in turn provide the liquidity to the market.

Futures markets are standardized with regard to trading specifications and terms of delivery. The international pepper price shows a trend that is unfavorable to most of the pepper-producing countries. This is largely attributable to supply situations rather than production scenarios. The IPC, in an effort to ensure an efficient price management system, conducted a study on the viability of pepper futures contract, sponsored by the United Nations Commission for Trade and Development (UNCTAD). Based on this, a working group of representatives from IPC member-countries was set up to examine the requirements of a global pepper futures contract. Consequently, the IPSTA at Kochi initiated action for the establishment of the International Pepper Exchange (IPE) at Kochi. The expectations are that it will provide an incentive in pepper trading, both in the domestic and international markets; ensure better prices to producers; and extend a hedging facility to exporters of all member countries. Further, it is also proposed to link the Kochi pepper exchange with the Kuala Lumpur Commodity Exchange at a future time, when a second trading floor would be started. The globalization process, in general, and the increasing volatility of pepper prices, in particular, necessitated the establishment of a global pepper market to trade pepper futures contract. This also facilitates a free trade regime in pepper, because the product can be freely moved among the countries without tariff barriers and quota restrictions.

Due to the additional premium for Indian pepper, its price in the international market is relatively higher than that of pepper produced by other countries. Low yield combined with high unit cost of production combined with better prices in the domestic market leads to a higher international price. Also, the improved stock-holding capacity of the farmers influence price, because, in times of price decline, farmers can hold on to their produce without unloading it to the market. Both corporate procurement and speculation play a significant role in pushing up the price. Futures trade ensures a constant supply and steady price for the produce.

One of the main criticisms against the exchange is how it translates the benefits to the actual producers. Pepper production in Kerala is mainly scattered among small farms, and the market awareness of these small and marginal farmers is indeed scanty. At least 72% of the farmers belong to this ignorant category (George et al., 1989). These small and individual farmers neither have sufficient quantity to trade nor the technical expertise to benefit from the pepper exchange. This has resulted in the futures trading in Kochi being handled mainly by representatives of exporters. The history of futures trading in pepper has clearly shown that producers have almost no say in its operations, including the large producers. The Primary Agricultural Credit Societies (PACS) have not attempted to enter the futures market either. The state marketing agencies, such as the State Trading Corporation and the State Marketing Federation, are wary of their association with the exchange setup because they have met with large financial losses. This is only natural—governmental agencies in Kerala rarely have true accountability, which is a legacy of the earlier socialistic pattern of governance in the state. If fair trading is to be ensured, the possibilities of manipulative tactics—which create an artificial market to the detriment of the interests of the producers—by the powerful members of the exchange have to be fully scotched. Invariably, however, the speculator calls the shots rather than the toiling producer, no matter how big or small. Despite these shortcomings, the IPSTA is hopeful of bridging the disparity between farm gate price and market price, often quite wide in certain instances, by eliminating needless intermediaries involved in transactions, which would ensure better prices for the producers. Liquidity of traders is expected to increase because banks are now willing to advance as much as 80–90% of the value of the produce when it is hedged against adverse price fluctuations through the medium of futures contract. Reckless speculations will be prevented by the imposition of limits on open position, daily price band fluctuations bound both up and down, and collection of special margin deposits to curb speculative activity through financial restraints.

An important component of the economics of pepper growing in Kerala is the availability of reliable market information to the farmers. As of now, most pepper farmers either have no dependable access to reliable market information or are illiterate and cannot make use of it. Hence, the need to educate farmers on the importance of acquiring reliable market information and marketing processes in pepper is of paramount importance. One of the prime reasons for pepper farmers' inability to realize a good price, especially in the case of small and marginal farmers, is the lack of adequate storage facilities. Holding capacity of the farmers can be enhanced through the provision of credit support. There is good scope to attempt a cooperative marketing system, as is the case of other plantation crops in the states (such as rubber, coconut, and arecanut). In the case of these crops, there are cooperative marketing societies spread across the state—a similar attempt for pepper could be initiated.

1.12 Pepper Pharmacopoeia

The use of black pepper as a drug in both the Indian and Chinese systems of medicine, apart from its wide spectrum use as a food additive, is well documented (Atal et al., 1975; Kurup et al., 1979; Nadkarni, 1976). Pepper is elaborately described for its medicinal values in the ancient Indian system of medicine *Ayurveda*. It is described as *katu* (pungent), *tikta* (bitter), and *ushnaveerya* (potent). Further mention of pepper is also made with regard to the control of the principal sources of diseases (as described in *Ayurveda*—namely, *vata* and *kapha*, in the ancient Indian language Sanskrit) as the biological processes in the human system controlled by the CNS and the later formation and regulation of all fluids that have a preservative quality (such as mucus, synovia, and the like). Pepper is described as a drug that enhances digestive power in the body; improves appetite; and cures colds, coughs, dyspnea, diseases of the throat, intermittent fever, colic, dysentery, worm infestation, and hemorrhoids. Pepper is also prescribed to relieve toothache, muscular pain, inflammation, leucoderma, and even epileptic fits (Ayier and Kolammal, 1966; Kirtikar and Basu, 1975). Black pepper is called *maricha* or *marica* in Sanskrit, which implies its ability to dispel poison. In the Chinese system of medicine, pepper is used as an antidote against snake and scorpion bites.

The preceding descriptions explain the diverse actions of pepper being used in the Indian system of medicine, either as such, or through its active ingredients in many formulations. In sharp contrast to the use of pepper in the Indian system of medicine, pepper finds no use in allopathy or homeopathy. The beneficial effect of pepper in medical formulations can be attributed to piperine and other phenolic amides and essential oils. But in *Ayurveda*, the active ingredient–based specific activities are not taken into consideration. Hence, the pharmacological and toxicological aspects of pepper and its constituent secondary metabolites were not studied. It must be, without doubt, mentioned that many of the clinical uses of pepper are time-tested, over several generations, and there is implicit faith among the Indian masses on the unquestionable medicinal value of pepper. However, to scientifically establish the value of pepper as a unique medicinal product, it is crucial to investigate the pharmacological aspects of pepper and its active ingredients based on well-defined experimental protocols currently used in modern drug research and development. Some of the preliminary investigations carried out over the past two decades in India on these lines provide valuable information. Antipyretic, analgesic, anti-inflammatory, antimicrobial, and antineoplastic activities were reported after *in vitro* and *in vivo* investigations. Of late, there is a surge of interest in environmentally friendly use of products to control crop pests, and pepper has been found to possess insecticidal as well as insect-repellent properties. Developments of new pepper-based products as insecticides could be a boon to human use because they would be free from the toxic nature of most of the insecticides currently in use. Piperine is the major alkaloidal constituent of pepper. Systematic pharmacological studies on piperine have revealed its analgesic (pain relieving), antipyretic (fever relieving), and anti-inflammatory properties, in addition to rejuvenation of the CNS. The antimicrobial activities are provided mainly by the essential oils.

In addition to the ancient *Ayurveda* system of medicine, there are other native systems of medicines in India, such as *Siddha* and *Yunani*. Pepper is used in these as well. In the Chinese system of medicine, pepper and chenghan (*Dichora febrifuga* L.) are used in the treatment of malaria (Das et al., 1992). The active ingredients of chenghan are febrifugin and iso-febrifugin, both of which show a 50% higher anti-malarial property than the conventional quinine. As an antipyretic, pepper is also used in the treatment of malaria (Nadkarni, 1976). The curative effects claimed in the preceding cases with regard to pepper can be attributed to the antipyretic and analgesic actions of the active ingredients. A strong antipyretic effect on typhoid vaccinated rabbits at a dose of $30\,mg\,kg^{-1}$ body weight after oral administration has been reported by Lee et al. (1984). In these studies, acetaminophen was used as the reference drug. The antipyretic action of piperine was found to be stronger than that of acetaminophen. To determine the analgesic action, acetaminophen and aminopyrine were used as reference compounds. Piperine showed a strong activity, with an LD_{50} value of $3.7\,mg\,kg^{-1}$ dose using the writhing method and $104.7\,mg\,kg^{-1}$ using the tail-clip method.

1.12.1 Anti-inflammatory and Central Nervous System Depressant Activity of Pepper

Several researchers have reported the anti-inflammatory property of pepper. Piperene showed a significant inhibition of the increase in edema volume in a carrageenan-induced test with an oral intake of $50\,mg\,kg^{-1}$ body weight (Lee et al., 1984). The effect of different acute and chronic experimental models were studied by Mujumdar et al. (1990). These researchers evaluated the mechanism of anti-inflammatory activity by biochemical processes and concluded that piperine acted positively in the case of early acute changes in inflammatory processes. These studies have been further corroborated by those of Kapoor et al. (1993).

In the case of epileptic patients, pepper is used in *Ayurveda* to induce sleep (Kirtikar and Basu, 1975). In pharmacological investigations, pepper and piperine were found to have a CNS depressant effect. Both pentetrazole seizure and maximal electroshock seizure were inhibited by pepper and piperine (Won et al., 1979). The LD_{50} value of $287.1\,mg\,kg^{-1}$ i.p. was found for piperine and a corresponding oral value was $1638.8\,mg\,kg^{-1}$. In another investigation, Shin et al. (1980) showed that piperine at 1/10 its LD_{50} value had a strong potentiating effect on hexobarbitol-induced hypnosis in mice. Decreased passivity, ptotic symptoms, and decrease in body temperature were also observed. Protection against electroshock seizure and a muscle relaxant effect were observed at a relatively low intraperitoneal dosage range of an LD_{50} of 15.1 mg kg^{-1} (Lee et al., 1984). This dosage appeared almost equipotent to the reference compound phenytoin. Petrol extract of pepper leaves was found to potentiate phenobarbitone-induced hypnosis in mice (Sridharan et al., 1978). Majumdar et al. (1990) also reported that a high dosage of piperine potentiates the phenobarbitone sleeping time by inhibiting the liver microsomal enzymes. It also acted partially through stimulation of the pituitary adrenal axis.

The preceding investigations clearly show that piperine has the capacity to be an effective pepper ingredient in the treatment of petit mal (minor ailments). Antiepilepsirine (AE) isolated from white pepper is a compound of very great pharmacological interest. AE is 1-(3-benzodioxol-5yl)-1 (oxo-2-propenyl)-piperidide. Perhaps this is the only compound derived from pepper that is clinically used in the treatment of human epilepsy. This is an alternative to dilantin therapy, which is used in Chinese hospitals for the treatment of epilepsy (Ebenhoech and Spadaro, 1992). It is probable that AE may release 5-hydroxy tryptamine (5-HT) from nerve endings, which can intensify the anticonvulsive state (Liu et al., 1984). Studies using rats as test animals show that AE increases 5-HT concentration in their brains, intensifying the anticonvulsive state. AE also raises the tryptophan level in the brain, which causes elevation of serontin and monoamine levels, and leads to control of seizures.

1.12.2 Effect on Hepatic Enzymes

While pepper or piperine positively affects liver function, as evidenced by the research of many, they in no way cause hepatic toxicity. Piperine functions mainly as a chemopreventive substance through modulating enzyme function. Hepatic toxicity of piperine in rats was studied by estimating the hepatic mixed function oxidases and serum enzymes as specific markers of hepatotoxicity (Dalvi and Dalvi, 1991). An intragastric dose of 100 mg kg^{-1} body weight resulted in an increase in hepatic microsomonal enzymes 24 h after treatment. In the experiment, cytochrome p-450, cytochrome b5, NADPH-cytochrome C reductase, benzphetamine N-demethylase, aminopyrine N-demethylase, and aniline hydroxylase were estimated. An intraperitonial dose of 100 mg kg^{-1} body weight did not produce any effect on the activity of the drug-metabolizing enzymes. However, when the dose was enhanced an eightfold to 800 mg kg^{-1} body weight as intragastric and 100 mg kg^{-1} body weight as intraperitoneal, a significant decrease in the levels of the enzymes was noted. These treatments did not, however, affect those serum enzymes that are specific markers of toxic liver conditions. Piperine has been found to provide significant protection against chemically induced hepatotoxicity. Both *in vitro* and *in vivo* lipid peroxidation and prevention of depletion of GSH and total thiols were observed (Kaul and Kapil, 1993). Liquid peroxidation causes the production of free radicals, which in turn causes tissue damage. GSH conjugates xenobiotics, which are excreted by subsequent glucuronidation. In the previously cited study, the hepato-protective action of piperine was compared with a reference compound, silymarin, a known hepato-protective drug, and it was found that piperine has slightly lower activity. In a feeding experiment using Swiss albino mice fed with a diet containing 1, 2, and 3% black pepper, w/w, for 10 and 20 days, a dose-dependent increase in the levels of the hepatic biotransformation enzymes—namely, glutathione-s-transferase, cytochrome p-450, cytochrome b5, and acid soluble sulfhydril-S—was observed (Singh and Rao, 1993). Reen et al. (1993) found lowered levels of glucuronidation due to the inhibition of the enzyme UDP-glucose dehydrogenase in an *in vitro* study. Tripathi et al. (1979), while investigating the hypoglycemic action of several plants, found that pepper fruits are devoid of any significant hypoglycemic action in rabbits. The aqueous extract of pepper leaves at a dose of 10–20 $mg\,kg^{-1}$ body weight led to a moderate increase in the blood pressure of dogs (Sridharan et al., 1978). Both piperine and AE have detoxifying qualities that may lead to an increase in the bioavailability of other drugs, as a result of which they alter the pharmokinetic parameter of the epileptic (Bano et al., 1969).

1.12.3 Carcinogenic and Mutagenic Effects of Black Pepper

Positive effects of pepper as an anticarcinogenic and nonmutagenic source have been found by investigators. The Ames test shows its nonmutagenic quality. Chemical carcinogenesis has been found to be prevented by pepper because of its stimulatory effect on xenobiotic biotransformation enzymes. The antioxidant properties of piperine and associated unsaturated amides play a preventive role in carcinogenesis. Dietary intake of natural antioxidants can be a crucial aspect of building up the human body's natural defense mechanisms against the onslaught of diseases, which can be caused by degradative changes brought about by mutagens. Additionally, the essential oil constituents in pepper inhibit DNA adduct formation by xenobiotics. This observation shows that pepper has anticarcinogenic potential. However, pepper extracts showed enhanced incidence of tumor formation in mice and an elevated level of DNA damage caused by piperine in cell culture investigations. Hexane, water, and alcohol extracts of pepper were tested for mutagenicity on *Salmonella typhimurium* strains TA 98 and TA 100 by Ames assay, and the results indicated that the extracts were nonmutagenic. This investigation provides evidence that water extract exerts an antimutagenic action on carcinogen-induced mutagenesis (Higashimoto et al., 1993). The chemoprotective role of pepper, as well as its constituents, has been indicated because of its positive effect on the activity of biotransformation enzymes in the liver in a dose-dependent manner (Singh and Rao, 1993).

Nakatani et al. (1986) report a very significant antioxidant activity due to the phenolic amides present in pepper. It is clearly established that antioxidants exert a preventive role in carcinogenesis. The modulating effects of the essential oil constituents of pepper were investigated by Hashim et al. (1994), who showed that they have an inhibitory effect on carcinogenesis. The volatile oil and its constituents suppress the formation of DNA adducts with aflatoxin B1. The microsomal enzymes modulate this action. Feeding powdered pepper to mice at 1.66 w/w dose showed no positive impact on carcinogenesis, while feeding and painting of the mice body with solvent extract of pepper, 2 mg for 3 days per week for three months, showed an increased incidence of tumor formation (Shwaireb et al., 1990). The activity of methylcholanthrene, a potent carcinogenic compound, was found to be reduced by a pepper terpenoid D-limonene (Wuba et al., 1992). Two minor constituents of pepper—safrole and tannic acid—have minor carcinogenic activity.

In a tissue culture study using V-79 lung fibroblast cell lines, it was found that piperine-treated cell lines showed increased DNA damage compared with untreated ones (Chu et al., 1994). Piperine treatment lowered the activity of the enzymes glutathione-s-transferase and uridine diphosphate glucuronyl transferase, indicating cytotoxic potential. The *in vivo* formation of N-nitroso compounds from naturally occurring amines and amides contribute to the carcinogenic potential of certain foods and food additives. Piperine and other phenolic amides present in pepper are also known for their conversion to N-nitroso compounds in acidic conditions and, hence, are treated as carcinogenic (Lin, 1986). However, it can be inferred that the presence of a conjugated unsaturated system in the phenolic amide prevents the oxidation of the amide nitrogen to N-nitroso compounds to a great extent. Additionally, the essential oil constituents in pepper would also ensure DNA stability because of their anticarcinogenic potential.

All of the cited research results point to the ambivalent nature of pepper—both as an anticarcinogenic agent and as a procarcinogenic agent. But overwhelming evidence indicates that it is more of the former than the latter. It is because of this belief that pepper has been a crucial ingredient of the Indian systems of medicine, primarily *Ayurveda* and to a lesser extent *Siddha* and *Yunani*, for centuries, and continues to be so even today.

1.12.4 Pepper as an Antioxidant

Antioxidants are one of the most crucial biochemical compounds in the human system ensuring good health. They scavenge free radicals, which trigger many untoward biochemical reactions in the human system, and control lipid peroxidation in mammals. Of late, there is considerable interest in antioxidants. Lipid peroxidation is a chain reaction that is a constant source of free radicals, which initiate further peroxidation, which trigger deterioration of food, but also damage tissue (both of animal and plant origin), which in human beings causes many inflammatory diseases, aging, atherosclerosis, and cancer. Many investigations reveal that pepper and its phenolic constituents, such as amides, possess good antioxidant properties. Tocopherol and vitamin C are two important natural antioxidants.

Chiapault et al. (1955) have investigated the antioxidant property of spices in a two-phase aqueous fat system. The investigations indicated that pepper has an antioxidant index of 6.1, while turmeric and clove show values of 29.6 and 103.0, respectively. The antioxidant property of pepper has been attributed to its tocopherol content (Saito and Asari, 1976). Revankar and Sen (1974), however, attribute the antioxidant property of pepper to its polyphenolic content. These authors investigated the effect of pepper oleoresin on fish oil and arrived at this conclusion. Abdel-Fattah and El-Zeany (1979) observed that the autoxidation of unsaturated fatty acids and proteins was delayed by the addition of pepper, and substantial protection against oxidative degradation was obtained. Nakatani et al. (1986), while investigating the family Piperaceae, established the fact that the five phenolic amides present in *P. nigrum* possess very good antioxidant properties, which were found to be superior to synthetic antioxidants, such as butylated hydroxy toluene and butylated hydroxy anisole.

1.12.5 Pepper as an Antimicrobial Agent

Volatile oils, which are an active constituent of most spices (and of which pepper is the most important), impart antiseptic, antibacterial, and antifungal properties. The positive effect of volatile oils can be attributed to their terpenoid constituents. Pepper possesses both bactericidal and bacteriostatic properties. These properties aid in enhancing the shelf life of foods to which pepper has been added. Even the leaf extract of pepper possesses the antibacterial activity (Subramanyam et al., 1957). However, the extract was not inhibitory to the growth of *E. coli*, *Aerobacter aerugenes*, *Lactobacillus casei*, *Staphylococcus faecalis*, *S. aureus*, and *S. sonnei* (Subramanyam et al., 1957). The essential oil obtained from pepper was found to be inhibitory to the growth of a penicillin-C–resistant strain of *S. aureus*. Jain and Kar (1971) documented the inhibitory action of pepper oil on *Vibrio cholerae*, *S. albus*, *Clostridium dipthereae*, *Shigella dysenteriae*, *Streptomyces faecalis*, *S. pyogenes*, *B. pumilis*, *B. subtlis*, *Micrococcus* sp., *Pseudomonas pyogenes*, *P. solanacearum*, and *S. typhimurium*. The antibacterial action was determined by the agar well-diffusion technique using cephazolin as standard (Perez and Anesini (1994)). The mycelial growth and aflatoxin synthesis by *Aspergillus parasiticus* were inhibited by pepper oil at a concentration of 0.2–1% (Tantaoui and Beraoud, 1994). Antifungal activity against *Candida albicans* (Jain and Jain, 1972) was exhibited by pepper leaf oil. Rao and Nigam (1976) reported a similar effect of pepper leaf oil on *A. flavus*. The antibacterial effects of pepper extract, essential oil, and isolated piperine *in vitro* against sausage micro flora, *L. plantarum*, *Micrococcus specialis*, and *Streptococcus faecalis* have been reported by Salzer et al. (1977), in which the authors noted that only isolated piperine displayed microbial growth inhibiting effects at a normal dose. Pepper powder and extract were active only at high concentrations. An alcoholic extract of pepper was found active against the deadly food-borne bacterium *C. botulinum* (Huhtanen, 1988). Another potential activity was revealed by a study carried out using tincture of pepper by Houghton et al. (1994). These authors investigated the antibacterial activity against nine strains of *Mycobacterium tuberculosis* and found that growth of all nine strains was inhibited.

The mode in which pepper acts positively is through its effect on enhancing bioavailability of administered medicaments, when they contain pepper or its active ingredients, and uptake of proteins and amino acids from ingested food. In an investigation using *Trikatu*, which is an *Ayurvedic* preparation containing *P. nigrum*, *P. longum*, and turmeric (*Zinjiberus officinale*, another spice with great therapeutic value), it was observed that the pepper-containing preparation enhances the bioavailability of other medicaments (Johri and Zutshi, 1992). At a dose of 25–250 μmol^{-1}, piperine enhanced the uptake of L-leucine, L-iso leucine, and L-valine and increased lipid peroxidation in an *in vitro* study using intestinal epithelial cells of rats (Johri et al., 1992). Piperine probably interacts internally to enhance the permeability of intestinal cells. Protein uptake from pulses was enhanced by up to 1.5% with the addition of pepper (Pradeep and Geervani, 1994). The study mainly involved two pulses in the Indian vegetarian diet—winged bean and horsegram. The results clearly suggest that the Indian vegetarian diet, where spice is a common ingredient, has a decided advantage in human health.

1.12.6 The Pharmacological Effect of Pepper on Human Health

The precise role of pepper on human health can only be understood through well-structured studies, which are only very limited in number to date. However, from the preceding sections, it can be concluded that where extrapolations and comparisons can be made, pepper (as a whole product) or its active ingredients have been found to have a very positive role on human health. A brief description of the limited number of studies is given here.

When used in high doses, gastric mucosal injury caused by pepper is comparable with that of aspirin. This observation was made in a double-blind study of intragastric administration of pepper to human volunteers (Mayore et al., 1987). In this investigation, healthy human volunteers were given meals containing 1.5 g of pepper; 655 g of aspirin and distilled water were used as positive and negative controls. This was a short-term study, and it must be pointed out that long-term effects of daily pepper intake, through food or medicaments, are unknown. Vezyuez et al. (1992) investigated the effect of intestinal peristalsis by measuring the orocaecal transit time (OCTT) utilizing a lactulose hydrogen breath test on healthy human subjects. They were given 1.5 g of pepper in gelatine capsules, and the OCTT was measured after several days; it was found that OCTT increased significantly after administration of pepper. This finding has great clinical importance in the management of various gastrointestinal tract disorders. An equally important investigation reveals the effectiveness of pepper extract volatiles in the treatment of cessation of smoking. Results of the investigation on human subjects reveal that cigarette substitutes, which deliver pepper volatile compounds, alleviated smoking withdrawal symptoms. The results of both the investigations detailed have very great potential for further experimentation and, perhaps, pharmacological exploitation for further therapeutic benefits for human health.

1.12.7 Clinical Applications of Pepper

It is only in the case of the Indian system of medicine—primarily *Ayurveda* and, to a lesser extent, *Siddha* and *Yunani*—that pepper has been used in many clinical applications. No such use is made of pepper in allopathy. When clinically used, none of the active pepper ingredients are used as such. Dried black pepper with other medicinal plants, also in the dried form, is used in the preparation of specific formulations. The most widely used formulation in *Ayurveda* is *Dasamulakatutrayadi Kashayam*, a formulation that is used in the treatment of asthma. Then comes *Ashtacurnam*, which is used to treat dyspepsia, flatulence, and the like. The formulation also has stomachic and carminative effects. *Amritharishtam* is used as an antipyretic and also in the case of women with excessive menstrual bleeding. *Muricadi Thailam* is used as an anti-inflammatory agent and also to relieve rheumatic pain. To cure coughs and bronchitis, *Dasamularasayanam* is used, and *Gulgulutiktaka-Ghrtam* has both analgesic and anti-inflammatory properties. Though pepper as such is widely used in the preparation of *Ayurvedic* medicines, its antipyretic, analgesic, and anti-inflammatory properties merit further research for use in allopathy, because, to date, pepper is not used in the preparation of any of the allopathic medicines. Its antioxidant and antimicrobial activities are also worth investigating further for possible use in the manufacture of allopathic medicines. Additionally, pepper has a good dietary value. It has a high fiber content (15–33%), high iron content (54–62 $mg\,g^{-1}$), high calcium content (1–1.5%), and also appreciable amounts of essential amino acids (Uma Pradeep et al., 1993). As a good digestive, pepper enhances the secretion and flow of the salivary enzyme amylase.

1.12.8 Toxicological Effects

Published reports do not detail any toxic effects of pepper. This may be due to the relatively small amounts used in many medicinal formulations. Because, in most cases, the total contents of piperine and associated phenolic amides used would add up to only 7–9%, w/w, and that of the volatile oils 2–4%, which are negligible amounts, no untoward health hazard has been encountered so far. At this level, the actual doses of the different constituents available from the quantity of pepper powder, oleoresin, or extractive used will be very unlikely to trigger any toxic side effects in human body. In fact, the very pungency of piperine and the strong flavor of the volatile oils act as deterrents against excessive human consumption.

The FAO (Food and Agriculture Organization) and WHO Experts Committee on Food Additives do not prescribe any limit to acceptable intake of piperine and the volatile oils. The major untoward consequence of pepper use is gastric mucosal injury at a dose of 1.5 $g\,kg^{-1}$ of food. It enhances the DNA adduct formation. The extract of pepper produces enhanced cancer incidence in mice. When mice were fed with extract of black pepper, it resulted in the formation of heptocellular carcinoma, lymphosarcoma, and fibrosarcoma (El-Mofty et al., 1991). However, a large number of investigations show that pepper has anticarcinogenic attributes.

Before conclusive evidence can be arrived at—whether pepper and its active ingredients trigger anticarcinogenic or procarcinogenic reactions in human beings or not—much more detailed scientific scrutiny needs to be carried out. As of now, such studies are conspicuously lacking. The observation that pepper enhances the liver microsomal enzyme activity must also be investigated further. This will help decide whether or not constant use of pepper in the human diet leads to production of nonspecific enzyme induction. Pepper is a universal dietary component on the Indian subcontinent, and the quantity used is much more than used anywhere else in the world. Such information has very vital implications of health for a population that currently stands at more than a billion.

1.12.9 The Insecticidal Activity of Pepper

The primary advantage of using plant-derived insecticides in agriculture is that while the product is toxic to the pest in question, it is nontoxic to human beings (including those who handle it in the field), unlike most other chemically produced commercial insecticides, which have come under attack from the environmentalists. Plant-derived products have been used since time immemorial both in agriculture and for domestic purposes. The extract of pepper, volatile oil components, and the different phenolic amides present in pepper have shown insecticidal, insect repelling, and ovicidal activities to various plant insects and pests harmful to human beings. The major alkaloid of pepper—piperine—is more toxic than pyrethrine, a standard insecticide for house flies (Harvill et al., 1943). In a number of instances, the insecticidal or insect-repellent activity was obtained at low concentrations. Hence, this aspect needs to be further probed for positive field application.

The volatile oil of pepper was found to cause mortality of the cigarette beetle (*Lasioderma serricorne*) (Samuel et al., 1984). Several groups of researchers investigated the effectiveness of the alcohol extract of pepper and found it effective against cotton boll weevil, rice weevil, and *Drosophila* (Barakat et al., 1985; Scott and McKibben, 1978; Su, 1977). The LD_{50} value for the tropical application of the crude and purified extracts, obtained after 24 h, showed a mortality rate of 3.4 and 4.8 μg, respectively, per adult insect of *Sitophilus oryzae* and 4.5 and 7.2 μg, respectively, per adult insect of *Callosobrunchus maculatus*. The oleoresin, piperine, and other amides in pepper were found to be lethal to the culex mosquito (*Culex pipiens*) and also to the pulse weevil (Su, 1977). Pepper amides, such as piperonal, piperine, piperoline, pellitorine, pipercide, and dihydropipercide, also were investigated by Miyakado et al. (1979), Masakazu et al. (1980), and Deshmukh et al. (1982) for their effectiveness against insects and were found to be effective against the pulse beetle, rice weevil, and *Drosophila*, where larval growth inhibition was found. Desai et al. (1997) showed that 100% mortality within 24 h at a dose of 80 ppm was obtained using acetone extract of pepper in the case of *Anopheles subpietus* larvae.

1.13 Consumer Products Out of Black Pepper

Of all the spices used in the world, black pepper is, perhaps, the most widely used in the kitchen, in perfumery, in medicine, and in industry. Most consumer products that have their origin in black pepper are "value added." For instance, when pepper is used in the food-processing industry, it is not pepper as such, but the oleoresin extracted through solvent extraction, and its pungency and flavor are the elements that add to the value of the end product made by the food-processing industry.

1.14 Value Addition in Pepper

There has been a dramatic advancement in the field of value addition of black pepper and diversification of processed pepper products. The value-added pepper products are classified as follows:

1. Green pepper–based products.
2. Black and white pepper–based products.
3. Pepper by-products.

These products may be further classified, as discussed in the following sections (Pruthi, 1997).

1.14.1 Green Pepper–Based Products

1. Bottled green pepper in brine.
2. Canned green pepper in brine.
3. Bulk-packed green pepper in brine.
4. Cured green pepper (without any covering tissue).
5. Freeze-dried green pepper.
6. Frozen green pepper.
7. Green pepper pickles in oil, vinegar, or brine.
8. Green pepper mixed pickle in oil, vinegar, or brine.
9. Green pepper–flavored products.
10. Green pepper paste.
11. Sun-dried or dehydrated green pepper.

Pepper fruits that are not fully mature are used in the manufacture of green pepper–based products, such as pepper in brine, pepper in oil, pepper in vinegar, desiccated green pepper, freeze-dried green pepper, and pepper paste. Almost all green pepper products are used by the catering sector to be served with meat dishes, such as steak and pork. The food industry also uses green pepper in the production of certain specialized cheeses. The other green pepper–based products are pickled green pepper and pepper spike. Tender pepper spikes and fruits alone or in combination with tender cardamom, pickled in vinegar and salt or sugar, make very delicious dishes that can be served with dinner. Green pepper paste is another product and is now a common sight in supermarkets, in polypacks or bottles. It can be used as a substitute to pepper powder, and it gives a refreshing taste with flavor and a unique "bite," which enhances the palatability of fish and meat dishes. It is a common choice of well-trained chefs.

1.14.2 Black and White Pepper–Based Products

1. Black pepper powder.
2. White pepper powder.
3. White pepper whole.
4. Pepper oleoresin.
5. Pepper oil.
6. Other pepper products.
7. By-products from pepper waste.
8. Miscellaneous forms used in medicine, culinary, and industrial applications.

The Pepper Marketing Board of Malaysia (PMBM), as well as many privately owned industrial houses, have brought to the market a number of innovative end products of pepper and have facilitated planning and product development of black pepper. The following structure has been proposed by PMBM (Abdulla, 1997). The pepper products are whole black pepper in retail packs for table use and ground pepper in retail packs and dispensers. Pepper-based products are flavored ground pepper, such as lemon pepper, and garlic pepper; sauces; marinades, which have pepper as the primary component; and pepper paste. Spice mixtures and blends are curry powders and spice blends for specific cuisines, such as five-spice powder and soup blends. In addition, there are pepper-flavored products, which are pepper mayonnaise, pepper tofu, pepper cookies, and pepper keropok (which are prawn or fish crackers). Products that use pepper extracts are pepper candy, pepper perfume, and the like. Pepper is also put to other auxiliary uses, such as in the case of pepper stalk, which is used as a substrate for growing mushrooms.

There are a number of common black pepper products, of which the most important is ground pepper. This is the exclusive form in which black pepper is used on the table. It is also the only spice served along with food (in flight) on board and used in restaurants specializing in fast foods. Pepper is very commonly used for seasoning food at different stages of preparation. The following factors have to be kept in mind in the case of ground pepper use, which is the easiest to manufacture and market: (1) Freedom from bacteria and mold is extremely important; otherwise, there is a potential health hazard. (2) During the grinding process, volatile oil content can be adversely affected. Care must be taken to ensure that this does not happen. (3) Because high moisture content in the berries will adversely affect shelf life, moisture during storage should be kept to a minimum. (4) Optimum particle size should be ensured for free flow during the period of storage and active use. (5) Packaging is also very important. Without good packaging, the product will not have a good market value and could deteriorate during storage. (6) Above all, marketable pepper should be free from all extraneous matter.

Flavored pepper is an important product with a good market for culinary purposes. Lemon and garlic pepper are examples. The former contains lime powder, pepper powder, and salt, and the latter is a blend of dehydrated garlic powder and dried pepper powder. Both are table condiments, especially in nonvegetarian dishes, containing chicken and fish. The PMBM has developed these products. Pepper sauce is another product of PMBM that can be used as a marinade, especially in meat-based dishes such as steak and lamb; it is also used in stir-frying. The preparation of this product involves the use of black pepper, soybean extract, garlic, and other ingredients to obtain a relatively mild dark sauce. Pepper is also used extensively in the preparation of a variety of sauces including Worcestershire sauce.

Spice mixtures and blends are extensively used in continental kitchens as well as in regional kitchens. Pepper-based spice mixes, available in departmental stores and supermarkets, are widely used in different meat-based preparations. There are also soup mixes that contain pepper. Private industrial houses and PMBM have developed a variety of pepper-flavored products, such as pepper-flavored mayonnaise, egg tofu, savory pepper cookies, traditional biscuits with pepper flavor, and pepper-flavored prawn and fish crackers. When dressed with pepper, many traditional dishes and snacks will result in enhanced taste and flavor. Pepper-flavored vinegar and pepper salts are excellent taste enhancers.

Pepper is widely used in industrially manufactured food items. Almost all fish- and meat-based products use pepper for seasoning, which enhances both flavor and pungency. Industrially manufactured foods, such as soups and pickles, use pepper extracts or powdered pepper. Additionally, there are pepper-flavored beverages such as pepper tea, pepper coffee, pepper-flavored milk, and the like. Some brands of brandy contain piperine and pepper oil, which impart a pungent taste and an "exotic" flavor. In the perfumery industry, volatile oil from pepper is used to impart an exclusive spicy "oriental" touch. Some of the more popular brands of perfume that contain pepper volatile oil are Revlon's "Charlie" and Christian Dior's "Poison" and Malaysian perfumes, such as "Sensai" and "Amila" (Ng, 1993).

1.14.3 Other Ancillary Pepper-Based By-products

An industrial area worth greater research and development is that of ancillary pepper-based by-products, which are of uncommon use. For instance, Malaysia has succeeded in producing paper and board from pepper stalks. These materials—especially paper made of pepper stalk—make good invitation cards, having a unique texture and color pattern (Ng, 1993). Pepper stalk makes a good substrate for oyster cultivation when grown in a 1:1 mixture of pepper stalk and shredded paper (Siti Hajijah and Bong, 1993). In Malaysia, the remnants of the pepper-processing industry, which consists mainly of stalk and pericarp, are powdered and used as an organic manure called pepper dust. It can enrich soil fertility when used in combination with other organic agricultural wastes (Ng, 1993).

In addition, there are a number of other pepper-based commercial formulations. Though a number of commercial formulations, such as canned green pepper, green pepper in brine, pepper oil, and oleoresin, are readily available in the market, they have yet to find widespread acceptability by the consumer. The most recent use of spice-processing technology is encapsulation of the flavoring components—namely, spice oils and oleoresin. These can be fully exploited commercially on a large scale, provided capital and entrepreneurship are readily available. This is yet to happen.

There is now a changing consumer preference for white pepper from black pepper in some Western countries. White pepper is used both as table pepper and also for enriching the flavor and taste of crab soups, seafood salads, casseroles, chicken, fish, and egg preparations. Sauces like mayonnaise also contain white pepper as an ingredient. One of the primary reasons for this shift in preference is because the black specks from the skin of black pepper are unseemly to the eye on a dining table.

The pepper product that is becoming increasingly popular in European countries is the green pepper. Because the piperine and pepper oil levels are high in slightly immature corns, they are preferred in the preparation of green pepper. This gives a characteristic flavor and is soft with a "bite"—a unique feel on the palate. These qualities make it ideal for garnishing meat dishes. The major products are green pepper in brine, canned green pepper, dehydrated green pepper, freeze-dried green pepper, and green pepper paste. While an increasing number of consumers prefer green pepper in brine and green pepper paste, and while demand for such is on the rise, demand for canned pepper is not high because its cost of production is quite prohibitive. India, Brazil, and Malaysia are the major producers of these value-added pepper products. The catering industry is the main source of demand for these products, followed by the food manufacturing industry.

Among the flavoring agents, the most important are oleoresin, pepper oil, and encapsulated pepper. The most extensive use of oleoresin is in the flavoring of meat. The other end uses of oleoresin are for preparing pickles, sauces, gravies, dressing "chutneys" (unique Indian spicy eatables), soups, and snacks. In most meat preparations, pepper is used to impart flavor. The same task is also accomplished by oleoresin. One of the most recent advances made by the Regional Research Laboratory in Trivandrum, Kerala State, India, is in developing the spray-drying encapsulation technique for oleoresin. The encapsulated flavor powder is used in a variety of products, such as cake mixes, dry beverage mixes, desserts, soup mixes, dusting on potato chips, and nuts. These have an emerging market all over the world.

1.15 Conclusions and a Peep Into Pepper's Future

Pepper has certainly had a checkered history. Within the past, present, and future, pepper can indeed be termed the "King of Spices." Though beset by many problems, both economic and agronomic, it is a safe bet that pepper will emerge as the world's most sought-after spice. It has been estimated that by the year 2020, global demand for pepper will be about 280,000 metric tons, which is projected to escalate to 360,000 metric tons by the year 2050. This would entail almost doubling the current production in the first half-century of this millennium. Where will the additional output come from? Increase of the pepper-growing area could only be marginal on a global basis. India, the major producer, will, perhaps, experience only marginal area increases. On the Asian subcontinent, it is only in China and Vietnam (the latter already emerging as a key producer) that marginal area increases can be expected. Like India, only a marginal increase in area can be expected in Indonesia. Malaysia and Thailand, the other two pepper producers of Asia, are already experiencing a decrease in area. This is also true for Brazil, a key producer on the Latin American continent. Labor is becoming increasingly difficult to obtain and very expensive when found; hence, countries such as Malaysia and Thailand on the Asian continent and Brazil on the Latin American continent are moving away from the labor-intensive pepper production. The impact of globalization is clearly seen in these countries, where market forces are shying away from labor-intensive crops such as pepper in preference to more mechanically managed crops. By the very nature of the pepper plant and the pepper canopy, mechanization has only very limited possibilities.

From the foregoing review, it can clearly be seen that one of the most pressing needs of pepper production is to have a pepper ideotype that combines many positive attributes in terms of boosting production potential and, at the same time, resisting many ravages of nature—especially those of diseases, among which the most predominant is foot rot, caused by the *Phytophthora* fungi. There is yet no single pepper variety that is universally resistant to this dreaded disease, though some lines have been identified that have shown regional promise. Conventional breeding, which will take years to accomplish because pepper is a long-duration perennial, will not be the answer to this major problem. Biotechnology is a promising area, but there are vast gaps in knowledge. Behavior of the pepper plant at the molecular level is far from clearly understood. Though tissue culture is a promising area in terms of cutting down the long gap in producing a desirable ideotype, this will not give final answers because there are far too many gray-area variables in the technique. Gene sequencing, currently a very highly sophisticated technology with very high levels of investment, opens up a door in the case of pepper as well. At the current level of global scientific knowledge and available technical expertise in the case of pepper, it is a safe bet to say that among those who exhibit a great fascination with crops, especially pepper, it is a very long road to traverse. Sophisticated instrumentation is only part of the answer; the availability of superlative technical skills is of great importance, but the pepper-producing countries, India included, clearly lack such skill. What one might expect on the pepper front, at least for the next few decades, is the following.

As science advances and global populations become more and more health conscious, the demand for organically grown food will become greater. One of the most promising areas for future pepper development will be in the area of producing "organically grown" or what one might term "clean" pepper. The environmental scare generated by the so-called green revolution—the hallmark of which has been extensive and indiscriminate use of chemical fertilizers and pesticides—is driving more and more people to organically grown food, and the demand for organically grown pepper is no exception. The potential market for organically grown food products, including pepper, is most promising in the USA, followed by Germany. The rigors of growing organic foods are encompassed in the Codex Alimentaris stipulations developed by France, and the USA has its own stipulations. In a country like India, where pepper production on the farm and its further processing is not subject to the rigors of quality control, export of organically grown pepper can only be put on sound footing if production and processing methods are well streamlined and rigorously controlled for quality, sanitation being most important. The same would hold for the situations in other pepper-producing countries like Indonesia, Vietnam, Malaysia, and Thailand.

Another area that has much promise is the development of pepper-derived food additives. The potential of this is gradually unfolding on an industrial scale. Heavy investments are necessary to make the various products price competitive. Some regional success stories do exist, though. For instance, there are pepper-based products that have a ready regional market. Yet, these products do not have a global reach. The potential market for these products would be Europe and the USA, where a gradual change in food habits is taking place. Both post-harvest technology and product development would undergo remarkable changes in the future. Imaginative product diversification would lead to the development of an array of novel products for the consumer, currently unheard of. Pepper with high flavor but low pungency will begin to be increasingly used in the manufacture of liquors, fruit juices, bakery products, and even choice perfumes. A possible challenge to the development of pepper varieties with high flavor will be the creation of the *in vitro* pepper flavor from bioprocessors. If this comes about, the only way natural pepper can be salvaged is a preference of the consumer for the clean, natural spice, rather than the laboratory-created spice.

Another important area that needs to be thoroughly researched is management of the fertility of pepper soils. Almost the entire approach to fertility management in pepper soils is based on classical textbook knowledge, where empirical recommendations generated from micro plots are extrapolated to large-scale field conditions. These experimental micro plots are nothing but artifacts, and many of the recommendations that emanate from such studies, when applied to large-scale plantations, turn out to be quite off the mark in reproducing accountable results. This often shatters farmers' confidence. A very significant departure is being made after the concerted efforts of the author, who has developed an entirely new approach to soil testing and fertilizer management, based on nutrient buffering. The concept is now universally known as the nutrient buffer power concept. A detailed discussion on the concept and its relevance in pepper nutrition is given in Section 1.4. In India, in the state of Kerala, where pepper is grown to the largest extent, it has been observed that fertilizer input can be significantly reduced by taking into consideration the buffer power of the nutrient concerned. Experimental and field results (Nair, 2002) show that when compared with routine soil testing and fertilizer recommendations, addition of zinc fertilizer, a very crucial input in pepper production, can be reduced by almost 75% in some soils.

There is an urgent need to extend the concept to other crucial nutrients, such as phosphorus and potassium and possibly nitrogen as well. Of the first two, phosphorus is more important in view of the fact that soils of Kerala State, the home of pepper, are lateritic, and much of the applied phosphate fertilizer, based on routine soil testing is not available to the crops. It is in this context that this new concept holds much hope for pepper farming. However, it must be emphasized that the success of a new approach, to a great extent, rests with the ingenuity of those applying it to suit the demands of a new situation. This principle is no exception to making the nutrient buffer power concept succeed in the case of pepper production, as has been the case with other crops, such as maize, rye, white clover, and even a perennial crop like cardamom, tested by the author (Nair et al., 1997). The fact that pepper is a perennial crop makes it all the more important because, unlike an annual one, where a mid-course correction can be effected in the following season, the fertilizer regime has to be correct from the very beginning because pepper grows for upward of 25 years. Unlike routine soil testing, the new approach calls for an accurate determination of the buffer power of the nutrient in question at the very start. Once this is accomplished, the buffer power factor can be integrated into the computations with the routine soil test data, and accurate fertilizer recommendations can be made on the basis of this new information. This implies that in addition to obtaining routine soil test data, one also needs to know the buffer power. The author has obtained very encouraging results with the new concept in pepper production in Kerala with regard to zinc fertilization, which signals a very promising change in the fertilizer practices. Hopefully, the new concept could very successfully be extended to other important plant nutrients.

Agriculture is the engine for development, nationally and globally, as is being increasingly proved against the unfolding scenario of globalization. A nation with a strong agricultural base can be expected to be in the forefront of economic development. Pepper is the crop of the tropics and, in that sense, is very much a part of the development of the third world economy. A lot has been achieved in pepper production, but before it moves to center stage as a commanding crop of third world economy, much more needs to be done. This calls for the concerted efforts of not just agronomists, soil scientists, or plant breeders dealing with pepper, but of a whole range of visionary thinkers and planners, who can really make the crop, not just the King of spices, but the Monarch of the third world agricultural economy.

Acknowledgment

With love I dedicate this chapter to my wife, Pankajam, my all, who sustains me in this very difficult journey that is life.

References

Abdel-Fattah, I.E., El-Zeany, B.A., 1979. Effect of spices on the autooxidation of fatty foods. Rev. Ital. Soztanze Grasse 56, 441–443.

Abdulla, A. (1997). Product development and promotion of pepper products. In: Proceedings of the 22nd Pepper Tech. Meeting. Spices Board, Cochin, Kerala State, India, pp. 1–6.

Adams, F., 1971. Ionic concentrations and activities in soil solutions. Soil Sci. Soc. Am. Proc. 35, 420–426.

Adzemi, M.A., Hamdan, J., Sjahril, J.S., 1993. Micronutrient removal studies on black pepper (*Piper nigrum* L.) in Sarawak. In: Ibrahim, M.Y., Bong, C.F.J., Ipor, I.P. (Eds.), The Pepper Industry: Problems and Prospects (pp. 104–109). Universiti Pertanian Malaysia, Sarawak.

Agarwal, S., 1988. Shoot tip culture of pepper and its micropropagation. Current Sci. 57, 1347–1348.

Ahmed, K., 1993. The effect of Nitrogen on the growth of *Piper nigrum* L. planted with *Desmodium trifolium* cover. In: Ibrahim, M.Y., Bong, C.F.J., Ipor, I.P. (Eds.), The Pepper Industry: Problems and Prospects (pp. 97–105). Universiti Pertanian Malaysia, Sarawak.

Anandaraj, M., 1997. Ecology of *Phytophthora capsici* (Leonian 1922, emend A. Alizadeh and P.H. Tsao). In: Causal Organism of Foot Rot Disease of Black Pepper (*Piper nigrum* L.), Ph. D. Thesis, University of Calicut, Kerala State, India, p. 154.

Anandaraj, M., Peter, K.V., 1996. Biological Control in Spices. Indian Institute of Spices Research, Calicut, Kerala State, India. p. 52.

Anandaraj, M., Sarma, Y.R., 1994a. Effect of Vesicular Arbuscular Mycorrhiza on rooting of black pepper (*Piper nigrum* L.). J. Spices Aromatic Crops 3, 39–42.

Anandaraj, M., Sarma, Y.R., 1994b. Biological control of black pepper diseases. Indian Cocoa Arecanut Spices J. 18, 22–23.

Anandaraj, M., Sarma, Y.R., 1995. Diseases of black pepper (*Piper nigrum* L.) and their management. J. Spices Aromatic Crops 4, 17–23.

Anandaraj, M., Sarma, Y.R., 1997. Mature coconut water for mass culture of biocontrol agents. J. Plantation Crops 25, 112–114.

Anandaraj, M., Sarma, Y.R., Ramachandran, N., 1994. *Phytophthora* root rot of black pepper in relation to age of the host and its culmination in foot rot. Indian Phytopath. 47, 203–206.

Annual Report, 1997. Indian Institute of Spices Research, Vision 2020. IISR Perspective Plan, Calicut, Kerala State, India.

Anon., 1989. Technology transfer document of microencapsulated spice flavors Regional Research Laboratory, Council of Scientific and Industrial Research, New Delhi, India, Trivandrum, India, pp. 9–10.

Anon., 1995. Annual Report of the Research Branch, Department of Agriculture, Sarawak, Malaysia.

Anon., 1996. Israel—A promising market for Indian spices. Spice India 9, 2–3.

Anon., 1997. Spice export grows to an all time record. Spice India 10, 2–4.

Apun, K., Siti Hajijah, A.S., Yusof, I.M., 1993. Microbiological properties of pepper during processing. In: Ibrahim, M.Y., Bong, C.F.J., Ipor, I.P. (Eds.), The Pepper Industry: Problems and Prospects (pp. 337–341). Universiti Pertanian Malaysia, Sarawak.

Atal, C.K., Dhar, K.L., Singh, J., 1975. The Chemistry of Indian Piper. Lloydia 38, 250–264.

Ayier, K.N., Kolammal, M., 1966. Pharmacognosy of Ayurvedic Plants, 9. Kerala University, Trivandrum, Kerala State, India. p. 63.

Bade, J., and Smit, H. P., 1992. An annual model of the world pepper economy, Seminar on Commodity Analysis, Jakarta, Indonesia, ESI-VU, Amsterdam, p. 63.

Balasubramanyam, N., Mahadevan, B., Anandaswamy, B., 1978. Packaging and storage studies on ground black pepper (*Piper nigrum* L.) in flexible consumer packages. Indian Spices 15 (4), 6–11.

Bano, G., Alma, V., Raina, U., Zutshi, V., 1969. The effect of piperine on the pharmokinetics of phenytoin in healthy volunteers. Planta Medica 12, 566–569.

Barakat, A.A., Family, H.S.M., Kandil, M.A., Ebrahim, N.M.M., 1985. Toxicity of the extract of black pepper, cumin, fennel, chamomile and Lupin against *Drosophila ceratetis* and *Spodoptera*. Indian J. Agric. Sci. 55, 116–120.

Barber, S.A., 1984. Soil Nutrient Bioavailability: A Mechanistic Approach. Wiley, New York.

Barber, S.A., 1995. Soil Nutrient Bioavailability: A Mechanistic Approach. Wiley, New York.

Bavappa, K.V.A., Gurusinghae, Rapid multiplication of black pepper for commercial planting. J. Plantation Crops 6, 92–95.

Bergman, A., 1940. De pepper culture en-handel op Bangka. Landbouw 16, 139–256.

Biessar, S., 1969. Plant parasitic nematodes of crops in Guyana. PANS 15, 74–75.

Bopaiah, B.M., Khader, K.B.A., 1989. Effect of bio-fertilizers on growth of black pepper (*Piper nigrum* L.). Indian J. Agric. Sci. 59, 682–683.

Bridge, J., 1978. Plant nematodes associated with cloves and black pepper in Sumatra and Bangka, Indonesia. In: O.D.M. Technical Report, U.K. Ministry of Overseas Development, U.K., p. 19.

Broome, O.C., Zimmerman, R.N., 1978. *In vitro* propagation of black pepper. *Hortic. Sci.* 43, 151–153.

Burger, W.C., 1972. The evolutionary trends in the Central American species of *Piper*. Brittonia 24, 356–362.

Burger, K., Smit, H.P., 1997. The Natural Rubber Market. Woodhead Publishing Limited, Cambridge. p. 354.

Chaudhary, R., Chandel, K.P.S., 1994. Germination studies and cryo preservation of seeds of black pepper (*Piper nigrum* L.) a recalcitrant species. Cryo-Lett 15, 145–150.

Cheriankunju, N.E., 1996a. Know your market for spices—Japan. Spice India 9, 5–7.

Cheriankunju, N.E., 1996b. Know your market for spices—USA. Spice India 9, 2–5.

Chiapault, J.R., Mizuna, G.R., Hawkins, J.M., Lundberg, W.O., 1955. The antioxidant properties of spices in oil-in-water emulsions. Food Res. 20, 443–451.

Christie, J.R., 1959. Plant Nematodes—Their Bionomics and Control. University of Florida, Gainesville, Florida, USA. p. 256.

Chu, C.V., Chang, J.P., Wang, C.J., 1994. Modulatory effect of piperine on benzopyrene cytotoxicity and DNA adduct formation to V-79 lung fibroblast cells. Food Chem. Toxicol. 39, 373–377.

Chua, B.K., 1981. Studies on *in vitro* propagation of black pepper (*Piper nigrum* L.). MARDI Res. Bull. 8, 155–162.

Claassen, N., Barber, S.A., 1976. Simulation model for nutrient uptake from soil by growing plant root system. *Agron. J.* 68, 961–964.

Claassen, N., Hendriks, K., Jungk, A., 1981. Rubidium—Verarmung des wurzelnahen Bodens durch Maispflanzen. Z. Pflanzenern. Bodenk. 144, 533–545.

Coffey, M.D., 1991. Strategies for the integrated control of soil-borne *Phytophthora* species. In: Lucas, J.A., Shattock, R.C., Shaw, D.S., Cooke, L.R. (Eds.), Phytophthora (pp. 411–432). Cambridge University Press, Cambridge.

Collins, L., Lapieree, D., 1976. Freedom at Midnight. Vikas Publication House, New Delhi, India.

Cook, R.J., Baker, R., 1983. The Nature and Practice of Biological Control of Plant Pathogens. American Phytopathological Society, St. Paul, Minnesota, USA. p. 53.

Dalvi, P.R., Dalvi, P.S., 1991. Comparison of the effect of piperine administered intragastrically and intraperitonially on the liver and liver mixed function oxidases in rats. Drug Metabol. Drug Interact. 9, 23–30.

Darrington, H., 1991. Herbs and spices. Food Manuf. 66 (11), 21–23.

Das, T.P.M., Cheeran, A., 1986. Infectivity of *Phytophthora* species on cash crops in Kerala. Agric. Res. J. Kerala 24, 7–13.

Das, P. C., Das, A., Mandal, S., and Chatterjee, A., 1992. On the validity of the ethnic use of *P. nigrum* and *M. indica* L. In: Proceedings of the International Seminar on Traditional Medicine. Calcutta, India, p. 84.

Datta, P. K., 1984. Studies on two *Phytophthora* diseases (*Koleroga* of arecanut and black pepper wilt) in Shimoga District, Karnataka State, India, Ph.D. Thesis. University of Agricultural Sciences, College of Agriculture, Dharwad, Karnataka State, India, p. 121.

Davis, R.M., Menge, J.A., 1981. *Phytophthora parasitica* inoculation and intensity of vesicular–arbuscular mycorrhizae in citrus. *New Phytol.* 87, 705–715.

Davis, R.M., Menge, J.A., Zentmyer, G.A., 1978. Influence of vesicular arbuscular mycorrhiza on *Phytophthora* root rot of three crop plants. Phytopathology 68, 1614–1617.

Debrauwere, J., Verzele, M., 1975a. Constituents of pepper VI. Oxygenated fraction of pepper essential oil. Bull. Soc. Chem. Belg. 84, 167–177.

Debrauwere, J., Verzele, M., 1975b. New constituents of the oxygenated fraction of pepper essential oil. J. Sci. Food Agric. 26, 1887–1894.

Debrauwere, J., Verzele, M., 1976. The hydrocarbons of pepper essential oil. J. Chromatogr. Sci. 14, 296–298.

Deciyanto, S., Suprapto., 1992. Research progress on important insect pests of black pepper in Indonesia. In: Wahid, P., Sitepu, D., Deciyanto, D., Suparman, U. (Eds.), Proceedings of the International Workshop on Black Pepper Diseases, 3–5 December, 1991, Bandar Lampung, Indonesia. Research Institute for Spice and Medicinal Crops, Bogor, Indonesia, pp. 237–244.

Dehne, H.W., 1982. Interaction between vesicular–arbuscular mycorrhizal fungi and plant pathogens. Phytopathology 72, 1115–1119.

Desai, A.E., Ladhe, R.U., Deorag, B.M., 1997. Larvicidal activity of acetone extract of *Piper nigrum* seeds to a major vector, *Anopheles subpietus* from Nashik region. Geobios 24, 13–16.

Deshmukh, P.B., Chavan, S.R., Renapurkar, D.M., 1982. A study of insecticidal activity of twenty indigenous plants. Pesticides 16, 7.

Devasahayam, S., Anandaraj, M., 1997. Black pepper (*Piper nigrum* L.) research. In: Narwal, S.S., Tauro, P., Bisla, S.S. (Eds.), Neem in Sustainable Agriculture (pp. 117–122). Scientific Publishers, Jodhpur.

Devasahayam, S., Koya, K.M.A., 1994. Natural enemies of major insect pests of black pepper (*Piper nigrum* L.) in India. J. Spices Aromatic Crops 3, 50–55.

Devasahayam, S., Leela, N.K., 1997. Evaluation of plant products for antifeedant activity against pollu beetle (*Longitarsus nigripennis* Motschulsky), a major pest of black pepper Abst. First Natn. Sym. Pest Management in Horticultural Crops. Environmental Implications and Thrusts, Bangalore, p. 85.

Devasahayam, S., Premkumar, T., Koya, K.M.A., 1988. Insect pests of black pepper *Piper nigrum* L. in India—a review. J. Plantation Crops 16, 1–11.

De Waard, P.W.F., 1964. Pepper cultivation in Sarawak. World Crops 16, 24–30.

De Waard, P.W.F., 1969. Foliar Diagnosis, Nutrition and Yield Stability of Black Pepper (*Piper nigrum* L) in Sarawak. Royal Tropical Institute, Amsterdam, The Netherlands. p. 150. Comm. No. 58.

De Waard, P.W.F., 1979. Yellow disease complex in black pepper on the island of Bangka, Indonesia. J. Plantation Crops 7, 42–49.

De Waard, P.W.F., 1984. *Piper nigrum* L. In: West Phal, E., Jansen, P.C.M (Eds.), Plant Resources of South East Asia (pp. 225–230). Rijksherbarium, The Netherlands.

Drew, M.C., Lynch, J.M., 1980. Soil anaerobiosis, microorganisms and root function. Annu. Rev. Phytopath. 18, 37–67.

Drew, M.C., Nye, P.H., Vaidyanathan, L.V., 1969. The supply of nutrient ions by diffusion to plant roots in soil. I. Absorption of potassium by cylindrical roots of onion and leek. Plant Soil 30, 252–270.

Dudek, D. H., 1996. Sterilization method and apparatus for spices and herbs, US Patent US 5523053 [US 260068 (940615)].

Eapen, S.J., Ramana, K.V., 1996. Biological control of plant parasitic nematodes of spices. In: Anandaraj, M., Peter, K.V. (Eds.), Biological Control in Spices (pp. 20–32). Indian Institute of Spices Research, Calicut, Kerala State, India.

Eapen, S.J., Ramana, K.V., Sarma, Y.R., 1997. Evaluation of *Psuedomonas fluorescens* isolates for control of *Meloidogyne incognita* in black pepper (*Piper nigrum* L.). In: Edison, S., Ramana, K.V., Sasikumar, B., Nirmal Babu, K., Eapen, S.J. (Eds.), Biotechnology of Spices, Medicinal and Aromatic Plants. Indian Society for Spices, Calicut, Kerala State, India.

Ebenhoech, A., Spadaro, O., 1992. Antiepilepsirine: A new Chinese anticonvulsant herb drug. J. Eco. Tax. Bot. 16, 99–102.

Elgawhary, S.M., Lindsay, W.L., Kemper, W.D., 1970. Effects of EDTA on the self-diffusion of zinc in aqueous solution and in soil. Soil Sci. Soc. Am. Proc. 34, 66–70.

El-Mofty, M.M., Khudoley, V.V., Shwaireb, M.H., 1991. Carcinogenic effect of force-feeding an extract of black pepper in Egyptian toads. Oncology 48, 347–350.

Esper, A., Muhlbauer, W., 1996. Solar tunnel dryer for fruits. Plant Res. Develop. 44, 61–79.

Ewald, S., 1991. Vesicular–Arbuscular Mycorrhiza Management in Tropical Agrosystems. Technical Cooperation, Federal Republic of Germany, p. 371.

Fassuliotis, G., Bhatt, D.P., 1982. Potential of tissue culture for breeding root knot nematode resistance in vegetables. *J. Nematol.* 14, 10–14.

Ferraz, E. C. A., Lordello, L. G. E., and de Santana, C. J. L., 1988. Nutrient absorption of black pepper vine (*Piper nigrum* L.) infested with *Meloidogyne incognita* Kofoid and

White., 1919, Chitwood (1949). In: Boletin Tecnico Centro de Pesquisas do Cacau, Brazil, No. 160, p. 34.

Ferraz, E.C.A., Lordello, L.G.E., Gonzaga, E., 1989. Influence of *Meloidogyne incognita* (Kofoid and White 1919) Chitwood 1949 on chlorophyll content of black pepper (*Piper nigrum* L). Agrotropica 1, 57–62.

Ferraz, E.C.A., Orchard, J.E., Lopez, A.S., 1984. Reactions of black pepper to *Meloidogyne incognita* in relation to total phenol. Revista Theobroma 14, 217–227.

Findlay-Wilson, I.Z.F.L., 1995. Microencapsulated flavours for a great taste. Int. Zeitsch. Lebensmittel—Techn. Market. Verpack. Anal. 46 (10), 76–78.

Fitchet, M., 1988a. Establishment of black pepper in tissue culture. In: Information Bulletin No. 189. Citrus and Subtropical Fruits Research Institute, South Africa.

Fitchet, M., 1988b. Progress with *in vitro* experiments of black pepper. In: Information Bulletin No. 196. Citrus and Subtropical Fruits Research Institute, South Africa.

Freire, F.C.O., Bridge, J., 1985a. Biochemical changes induced in roots and xylem sap of black pepper by *Meloidogyne incognita*. Fitopatologia Brasileira 10, 483–497.

Freire, F.C.O., Bridge, J., 1985b. Parasitism of eggs, females of juveniles of *Meloidogyne incognita* by *Paecilomyces lilacinus* and *Verticillium chlamydosporium*. Fitopatologia Brasileira 10, 577–596.

Fresco, L.O., Kroonenberg, S.B., 1992. Time and spatial scales in ecological sustainability. Land Use Policy, 155–167.

Gamble, J. S., 1925. Flora of the Presidency of Madras. In: Bot. Sur. India, Vol. II. Calcutta (Repr.).

Geetha, S. M., 1991. Studies on the biology, pathogenicity and biocontrol of different populations of *Radopholus similis*, Ph.D. Thesis, Kerala University, Trivandrum, India, p. 196.

Geetha, S. P., Manjula, C., and Sajina, A., 1995. *In vitro* conservation of genetic resources of spices. In: Proceedings of the Seventh Kerala Science Congress, January 27–29. Science Technology and Environment Committee, Kerala Government, pp. 12–16.

Geetha, C.K., Nair, P.C.S., 1990. Effect of plant growth regulators and zinc on spike shedding and quality of pepper. Agric. Res. J. Kerala 14, 10–12.

George, C.K., 1982. Pepper cultivation in Malaysia. Indian Cocoa Arecanut Spices J. 5, 75–76.

George, P.S., Nair, K.N., Pushpangadan, K., 1989. The Pepper Economy of India. Oxford and IBH Publishing Co. Pvt. Ltd., New Delhi, India.

Ghawas, M., Miswan, J., 1984. Satu Kaedah memprecepatkan Pembiakn lada hitam (*Piper nigrum* L.). Teknol. Pertanian. MARDI Res. J. 5, 50–54.

Gopalakrishnan, M., Menon, N., Padmakumari, K.P., Jayalakshmi, A., Narayanan, C.S., 1993. GC analysis and odor profiles of four new Indian genotypes of *Piper nigrum* L. J. Essential Oil Res. 5, 247–253.

Govindan, M., Chandy, K.C., 1985. Utilization of the diazotroph, *Azospirillum* for inducing rooting in pepper cuttings (*Piper nigrum*). Curr. *Sci*. 54, 1186–1188.

Govindarajan, V.S., 1977. Pepper—chemistry, technology and quality evaluation. Crit. Rev. Food. Sci. Nutr. 9, 115–225.

Govindarajan, V.S., 1979. Pepper—chemistry, technology and quality evaluation. CRC Crit. Rev. Food. Sci. Nutr. 9, 1–115.

Graham, J.H., 1982. Effect of citrus root exudates on germination of chlamydospores of the vesicular–arbuscular mycorrhizal fungus *Glomus epigaeum*. Mycologia 74, 831–835.

Graham, J.H., 1988. Interactions of mycorrhizal fungi with soil borne plant pathogens and other organisms: an introduction. Phytopathology 78, 365–366.

Graham, J.H., Egel, D.S., 1988. *Phytophthora* root rot development on mycorrhizal and phosphorus-fertilized non-mycorrhizal sweet orange seedlings. Plant Dis. 72, 614–615.

Guenther, E., 1950. Essential oils of the plant family Piperaceae The Essential Oils, vol. 5. Robert Krieger Pub. Co., New York. pp. 135–161.

Harden, H. J., and White, J. T. H. (1934). Quoted by De Waard (1969).

Hartman, C.P., Divakar, N.G., Nagaraja Rao, U.N., 1973. A study of identification of papaya seeds in pepper. J. Food. Sci. Technol. 10, 43–45.

Harvill, E.K., Hartzell, A., Arthur, J.M., 1943. Toxicity of piperine solution to house flies control. Boyce Thompson Inst. 13, 87–92. Merck Index, 1976, p. 1266.

Hashim, S., Aboobaker, V.S., Madhubala, R., Bhattacharya, R.K., Rao, A.R., 1994. Modulatory effects of essential oils from spices on the formation of DNA adducts by aflatoxin B *in vitro*. Nutr. Cancer 21, 169–175.

Hasselstrom, T.F., Hewitt, E.J., Konigsbacher, K.S., Ritter, J.J., 1957. Composition of oil of black pepper. *J. Agric. Food Chem.* 5, 53–55.

Hendriks, L., Classen, N., Jungk, A., 1981. Phosphatverarmung des wurzelnahen Bodens und Phosphataufnahme von Mais und Raps. Z. Pflanzenern. Bodenk 144, 486–499.

Higashimoto, M., Purinatrapiban, J., Katoka, K., Kinouchi, T., Vinitkekumnuen, U., Akimoto, M., Dinishi, M., 1993. Mutagenicity and antimutagenicity of extracts of three medicinal plants. *Mutat. Res.* 303, 135–142.

Holdeman, Q. L., 1986. The Burrowing Nematode, *Radopholus similis* sensu lato. Division of Plant Industry, Department of Food and Agriculture and the Agriculture Commissioners of California, California, U.S.A., p. 52.

Holliday, P., and Mowat, W. P., 1963. Foot rot of *Piper nigrum* L. (*Phytophthora palmivora*). In: Phytopathological Paper no. 5, Commonwealth Mycological Institute, Kew, Surrey, p. 62.

Hooker, J.D., 1886. The Flora of British India,, vol. 5. L. Reeva and Co., London. pp. 79–95.

Houghton, P.J., Astaniou, A., Grange, J.M., Yates, M., 1994. Antibacterial effects of extracts and constituents of *Piper nigrum*. L. Pharm. Pharmacol. 46 (Suppl. 2), 1042.

Hubert, F.P., 1957. Diseases of some export crops in Indonesia. Plant Dis. Reptr. 41, 55–63.

Huhtanen, C.N., 1988. Inhibition of *Clostridium botulinum* by spice extracts and aliphatic alcohols. J. Food Prot. 43, 195–196.

Huitema, W. K., 1941. Quoted from De Waard (1969).

Humboldt, A. von, Bonpland, A., Kunth, C., 1815. Cited by Burger (1972).

Ichinohe, M., 1975. Infestation of black pepper vines by the root knot nematode, *Meloidogyne incognita* at Tome-Acu, Para, Brazil. Jap. J. Nematol. 5, 36–40.

Ichinohe, M., 1980. Studies on the root knot nematode of black pepper plantation in Amazon. In: Annual Report of the Society of Plant Protection of North Japan, No. 31, pp. 1–8.

Ichinohe, M., 1985. Integrated control of root knot nematode, *Meloidogyne incognita* on black pepper plantations in the Amazonian region. Agric. Ecosyst. Environ. 12, 271–283.

IISR. 1997. Annual Report of Indian Institute of Spices Research, 1996–97, Calicut, Kerala State, India, pp. 50–51.

Ikeda, R.N., Stanley, W.L., Vannier, S.H., Spitler, F.M., 1962. The monoterpene composition of some essential oils. *J. Food Sci.* 27, 455–458.

Jacob, J.A., Kuriyan, K.J., 1979. Screening of pepper varieties for resistance against root knot nematode (*Meloidogyne incognita*). Agric. Res J. Kerala 17, 90.

Jain, S.R., Jain, M.R., 1972. Antifungal studies on some indigenous volatile oils and their combinations. Planta Medica 22, 136.

Jain, S.R., Kar, A., 1971. Antibacterial activity of some essential oils and their combinations. Planta Medica 22, 118.

Jatala, P., 1986. Biological control of plant parasitic nematodes. Annu. Rev. Phytopath. 24, 453–489.

Jennings, W.G., Wrolstad, R.E., 1961. Volatile constituents of black pepper. *J. Food Sci.* 26, 499.

Jeromi, P.D., Ramanathan, A., 1983. World pepper market and India: an analysis of growth and instability. *Ind. J. Agric. Econ.* 48, 89–97.

Johri, R.K., Thusu, N., Khajuria, A., Zutshi, U., 1992. Piperine mediated changes in permeability of rat intestinal epithelial cells. *Biochem. Pharmacol.* 43, 1401–1407.

Johri, R.K., Zutshi, U., 1992. An Ayurvedic formulation of "Trikatu" and its constituents. *J. Ethnopharmacol.* 37, 85–91.

Jose, J., Sharma, A.K., 1984. Chromosome studies in the genus Piper L. J. Indian Bot. Soc. 63, 313–319.

Joseph, B., Joseph, D., Philip, V.J., 1996. Plant regeneration from somatic embryos in black pepper. Plant Cell Tiss. Org. Cult. 47, 87–90.

Kachru, K.P., and Gupta, R.K., 1993. Drying of spices-status and challenges. In: Proceedings of the National Seminar of Post Harvest Technology of Spices, Trivandrum, India, pp. 15–27.

Kachru, K.P., Patil, R.T., Srivastava, P.K., 1990. Post harvest technology of pepper. Spice India 3 (8), 12–16.

Kapoor, V.K., Chawla, A.S., Manoj, K., Pradeep, K., 1993. Search for antiinflammatory agents. India Drugs 30, 481–493.

Kaul, I.B., Kapil, A., 1993. Evaluation of liver protective potential of piperine: an active principle of black pepper. Planta Medica 59, 413–417.

Keating, M., 1993. The Earth Summit's agenda for change. In: A Plain Language Version of Agenda 21 and the Other Rio Agreements. Centre for our Common Future, Geneva, Switzerland.

Keerthisinghe, G., Mengel, K., 1979. Phosphatpufferung verschiedener Böden und ihre Veränderung infolge Phosphatalterung. Mitt. Dtsch. Bödenk. Ges. 29, 217–230.

Kelker, S.M., Krishnamoorthy, K. V., 1996. Control of bacterial contamination in *in vitro* cultures of pepper (*Piper* sp.). In: Proceedings of the National Seminar on Biotechnology of Spices and Aromatic Crops, April 24–25, 1996, Calicut, Kerala State, India.

Kerry, B.R., 1990. An assessment of progress towards microbial control of plant parasitic nematodes. *J. Nematol.* 22 (Suppl), 621–631.

Kirtikar, K. R., Basu, B. D., 1975. Indian Medicinal Plants III, 2133–2134 Bishan Singh Mahendrapal Singh. Dehra Dun, India.

Koshy, P.K., Sunderaraju, P., 1979. Response of seven black pepper cultivars to *Meloidogyne incognita*. Nematol. Medit. 7, 123–125.

Kovar, J.L., Barber, S.A., 1988. Phosphorus supply characteristics of 33 soils as influenced by seven rates of phosphorus addition. Soil Sci. Soc. Am. J. 52, 160–165.

Krishnamoorthy, B., John Zachariah, T., 1992. Drying black pepper on polyethylene materials. Indian Cocoa Arecanut Spices J. 15, 75.

Kueh, T.K., 1979. Pests, Diseases and Disorders of Black Pepper in Sarawak Semongok Agricultural Research Centre. Department of Agriculture, Sarawak, Malaysia. p. 68.

Kueh, T.K., 1986. Pests and diseases of black pepper-a review. In: Bong, C.F.J., Saad, M.S. (Eds.), Pepper in Malaysia (pp. 115–133). Universiti Pertanian Malaysia, Sarawak, Malaysia.

Kueh, T.K., 1990. Major diseases of black pepper and their management. The Planter 66, 59–69.

Kueh, T.K., Khew, K.L., 1982. Survival of *Phytophthora palmvivora* in soil and after passing through alimentary canals of snails. Plant Dis. 66, 897–899.

Kueh, T.K., Fatimah, O., Lim, J.L., 1993. Management of black berry disease of black pepper. In: Ibrahim, M.Y., Bong, C.F.J., Ipor, I.B. (Eds.), The Pepper Industry: Problems and Prospects (pp. 162–168). Universiti Pertanian Malaysia, Sarawak, Malaysia.

Kueh, T.K., Sim, S.L., 1992a. Etiology and control of *Phytophthora* foot rot of black pepper in Sarawak, Malaysia. In: Wahid, P., Sitepu, D., Deciyanto, S., Suparman, U. (Eds.), Proceedings of the International Workshop on Black Pepper Diseases. Bander Lampung, Indonesia (pp. 155–162). Institute for Spice and Medicinal Crops, Bogor, Indonesia.

Kueh, T.K., Sim, S. L., 1992b. Important pepper diseases in Sarawak, Malaysia. In: Wahid, P., Sitepu, D., Deciyanto, S., Suparman, U. (Eds.), Proceedings of the International Workshop on Black Pepper Diseases Bander Lampung, Bogor, Indonesia, pp. 101–117.

Kurien, S., Nair, P.C.S., 1988. Effect of pruning on yield in pepper (*Piper nigrum* L.). Agric. Res. J. Kerala 26, 137–139.

Kurup, P.N.V., Ramdas, V.N.K., Joshi, P., 1979. Handbook of Medicinal Plants, Indian Council of Agricultural research, New Delhi, India p. 143.

Lawrence, B.M., 1981. Major Tropical Spices—Pepper (*Piper nigrum* L.). Essential oils 1979–80. Allured Publications Corporation, USA. pp. 141–228.

Lee, E.B., Shin, K.H., Woo, W.S., 1984. Pharmacological Study of Piperine. *Arch. Arch. Pharm. Res.* 7, 127–132.

Leife, A., 1992. Sterilization of spices with pulsed steam. Livsmedelsteknik 34 (1/2), 24–27.

Lewis, Y.S., 1984. Spices and Herbs for Food Industry. Food Trade Press, England. pp. 69–77.

Lewis, Y.S., Krishnamurthy, N., Nambudiri, E.S., Sankarikutty Amma, B., Shivshankar, S., Mathew, A.G., 1976. The need for growing pepper cultivars to suit pepper products Proceedings of the International Seminar on Pepper. Spices Export Promotion Council, Cochin, Kerala State, India. pp. 4–9.

Lewis, Y.S., Nambudiri, E.S., Krishnamurthy, N., 1969. Composition of pepper oil. Perfum. Essent. Oil Rec. 60, 259–262.

Lewis, D.G., Quirk, J.P., 1967. Phosphate diffusion in soil and uptake by plants III ^{31}P-movement and uptake by plants as indicated by ^{32}P-autoradiography. Plant Soil 26, 445–453.

Lin, J.K., 1986. Food-borne amines and amides as potential precursors of endogenous carcinogens. Proc. Natl. Sci. Council Rep. China. (B) 10, 20–34.

Linderman, R.G., 1988. Mycorrhizal interactions with the rhizosphere microflora: The micorrhizosphere effect. Phytopathology 78, 366–370.

Linnaeus, C., 1753. Species Plantarum. London, UK.

Lissamma, J., Nazeem, P.A., Mini, S.T., Shaji, P., Mini, B., 1996. *In vitro* techniques for mass multiplication of black pepper (*Piper Nigrum* L.) and *ex vitro* performance of the plantlets. J. Plantation Crops 24 (Suppl), 511–516.

Liu, G., et al., 1984. Quoted from Ebenhoech and Spadaro (1992).

Lokesh, M.S., Gangadarappa, P.M., 1995. Management of *Phytophthora* foot rot and nematode diseases in black pepper (*Piper nigrum* L.). J. Spices Aromatic Crops 4, 61–63.

Lu, S., Miller, M.H., 1994. Prediction of phosphorus uptake by field-grown maize with Barber-Cushman model. Soil Sci. Soc. Am. J. 58, 852–857.

Madasamy, M., Gothandapani, L., 1995. Spiral separator for cleaning. Spice India 8 (11), 7–8.

Madhusudhanan, K., Rahiman, B.A., 1996. *In vitro* response of Piper species on activated charcoal supplemented media. In: Edison, S., Ramana, K.V., Sasikumar, B., Nirmal Babu, K., Eapen, S.J. (Eds.), Biotechnology of Spices, Medicinal and Aromatic Plants (pp. 16–19). Indian Society for Spices, Calicut, Kerala State, India.

Mahindru, S.N., 1982. Spices in Indian Life. Sultan Chand & Sons, New Delhi, India.
Majumdar, A.M., Dhuley, J.N., Deshmukh, V.K., Raman, P.H., Thorat, S.L., Naik, S.R., 1990. Effect of piperine on pentobarbitone induced hypnosis in rats. *Indian J. Exp. Biol.* 28, 486–487.
Malebennur, N.S., Ganadarappa, P.M., Hegde, H.G., 1991. Chemical control of foot rot of black pepper caused by *Phytophthora capsici* (*Phytophthora palmivora* MF 4). Indian Cocoa Arecanut Spices J. 14, 148–149.
Mamootty, K. P., 1978. Quick wilt disease of pepper (*Piper nigrum* Linn.)-1: symtomatological studies on the quick wilt disease of pepper, M.Sc. (Ag.) Thesis, Kerala Agricultural University, Thrissur, Kerala State, India, p. 87.
Manilal, V.B., Gopinathan, K.M., 1995. Annual Report of Regional Research Laboratory, Council of Scientific and Industrial Research, Trivandrum, Kerala State, India.
Manjunath, A., Bagyaraj, D.J., 1982. Vesicular arbuscular mycorrhizas in three plantation crops and cultivars of field bean. *Curr. Sci.* 51, 707–709.
Manohara, D., Kasim, R., Sitepu, D., 1992. Current research status of foot rot disease in Indonesia. In: Wahid, P., Sitepu, D., Deciyanto, S., Suparman, U. (Eds.), Proceedings of the International Workshop on Black Pepper Diseases Bander, Lampung, Indonesia (pp. 144–154). Institute for Spice and Medicinal Crops, Bogor, Indonesia.
Mapiliandri, I., 1989. The extraction of black pepper oleoresin. In: Thesis from Diplome. IV. Ministry of Industry, Indonesia, Ministry of Agriculture (1996). Statistik Perkebunan, Jakarta.
Marschner, H., 1994. *Micronutr. News Information.* 14, 3–5.
Masakazu, M., Isamu, N., Hirosuki, Y., 1980. The Piperaceae amides Part III. Insecticidal joint action of pipercide and co-occurring compounds isolated from *Piper nigrum*. Agric. Biol. Chem. 44, 1071–1073.
Mathew, P.M., 1958. Studies on Piperaceae. J. Indian Bot. Soc. 37, 155–171.
Mathew, P. M., 1972. Karyomorphological studies on *Piper nigrum*. In: Proceedings of the National Symposium on Plantation Crops. J. Plantation Crops 1 (Suppl.), 15–18.
Mathew, A.G., 1992. Chemical constituents of pepper. Int. Pepper News Bull. 16 (2), 18–21.
Mathews, V.H., Rao, P.S., 1984. *In vitro* responses in black pepper (*Piper nigrum*). *Curr. Sci.* 53, 183–186.
Mayore, M., Smith, J.L., Garhaur, D.Y., 1987. Effect of red pepper and black pepper on the stomach. *Am. J. Gastroenterol.* 82, 211–214.
McCown, H.B., Amos, R., 1979. Initial trials of micropropagation with birch. Proc. Int. Plant Prop. Soc. 29, 387–393.
Mengel, K., 1985. Dynamics and availability of major nutrients in soils. *Adv. Soil Sci.* 2, 65–131.
Menon, K.K., 1949. The survey of pollu and root diseases of pepper. Indian J. Agric. Sci. 19, 89–136.
Miquel, F.A.W., 1843. Systema Piperacearum. Rotterdam, The Netherlands.
Miyakado, M., Nakayama, I., Yoshoka, H., Nakatani, N.N., 1979. The Piperaceae amides: structure of pipercide. *Agric. Biol. Chem.* 43, 1609–1611.
Mohanakumaran, B., Cheeran, A., 1981. Nutritional requirement of pepper vine trailed on living and dead standards. Agric. Res. J. Kerala 19, 3–5.
Mohandas, C., Ramana, K.V., 1987. Slow wilt disease of black pepper and its control. Indian Cocoa Arecanut Spices J. 11, 10–11.
Mohandas, C., Ramana, K.V., 1991. Pathogenicity of Meloidogyne incognita and Radopholus similis on black pepper (*Piper nigrum* L.). J. Plantation Crops 19, 41–43.
Montaya, L.A., Sylvain, P.G., Umana, R., 1961. Effect of light intensity and nitrogen fertilization upon growth differentiation balance in *Coffea arabica* L. Coffee (Turrialba) 3, 97–115.

Mos, A.M., 1995. Microencapsulation enhances flavourings and ingredients. Voedingsmiddelen-technologie 28 (23), 64–65.

Mujumdar, A.M., Dhuley, J.N., Deshmukh, V.K., Raman, P.H., Naik, S.R., 1990. Antiinflammatory activity of piperine. *Jpn. J. Med. Sci. Biol.* 43, 95–100.

Muller, C.J., Creveling, R.K., Jennings, W.G., 1968. Some minor sesquiterpene hydrocarbons of black pepper. *J. Agric. Food Chem.* 16, 113–117.

Muller, C.J., Jennings, W.G., 1967. Constituents of black pepper. Some sesquiterpene hydrocarbons. *J. Agric. Food Chem.* 15, 762–766.

Murashige, T., Skoog, F., 1962. A revised medium for rapid growth and bioassays with tobacco tissue culture. Physiol. Plant 15, 473–497.

Murni, A.M., Faodji, R., 1990. The combination effect of KCl with two nitrogen sources on the growth of black pepper. Bull. Penelitian Tanaman Rempah dan Obat. 2 (2), 79–84.

Mustika, I., 1990. Studies on the interactions of *Meloidogyne incognita, Radopholus similis* and *Fusarium solani* on black pepper (*Piper nigrum* L.). Ph.D. Thesis, Wageningen Agricultural University, Wageningen, The Netherlands, p. 127.

Mustika, I., 1991. Response of four black pepper cultivars to infection by *Radopholus similis, Meloidogyne incognita* and *Fusarium solani*. Industrial Crop Res. J. 4 (1), 17–22.

Mustika, I., Zainuddin, N., 1978. Efficacy tests of some nematicides for the control of nematodes on black pepper. Pemberitaan L.P.T.I. 30, 1–10.

Nadkarni, K.M., 1976. Indian Materia Medica. Popular Prakashan, Bombay, India. pp. 971–972.

Nair, K.P.P., 1984a. Towards a better approach to soil testing based on the buffer power concept. In: Proceedings of the Sixth International Colloquium for the Optimization of Plant Nutrition, vol. **4**, September 2–8, (Prevel Pierre-Martin, Ed.), Montpellier, France, pp. 1221–1228.

Nair, K.P.P., 1984b. Zinc buffer power as an important criterion for a dependable assessment of plant uptake. Plant Soil 81, 209–215.

Nair, K.P.P., 1989. Comments on Phosphorus supply characteristics of 33 soils as influenced by seven rates of phosphorus addition. Soil Sci. Soc. Am. J. 53, 984.

Nair, K.P.P., 1992. Measuring P buffer power to improve routine soil testing for phosphate. *Eur. J. Agron.* 1 (2), 79–84.

Nair, P.M., 1993. Spice irradiation—Present and future scenario. In: Narayanan, C.S, Sankarikutty, B., Nirmala Menon, A., Ravindran, P.N., Sasikumar, B. (Eds.), Post Harvest Technology of Spices, Indian Soc. Spices (pp. 28–33). Indian Institute of Spices Research, Calicut, Kerala State, India.

Nair, K.P.P., 1996. The Buffering Power of Plant Nutrients and Effects on Availability. *Adv. Agron.* 57, 237–287.

Nair, K.P.P., 2002. Sustaining crop production in the developing world through the Nutrient Buffer Power Concept. In: Proceedings of the 17th World Soil Science Congress, Bangkok, Thailand, August 14–21, vol. 2, p. 652.

Nair, K.P.P., 2003. Measuring Zn buffer power to predict Zn availability. *J. Plant Nutr.* (in Press).

Nair, M.R.G.K., Das, N.M., Menon, M.R., 1966. On the occurrence of the burrowing nematode, *Radopholus similis* (Cobb 1893) Thorne 1949, on banana in Kerala. Indian J. Ent. 28, 553–554.

Nair, M.K., Gopalasundaram, P., 1993. Nutrient management in tree based cropping system. In: Mahatim Singh, Mishra, M.K. (Eds.), Potassium for Plantation Crops (pp. 200–211). Potash Research Institute of India, Gurgaon, India.

Nair, P.K.U., Mammootty, K.P., Sasikumaran, S., Pillay, V.S., 1993. *Phytophthora* foot rot of black pepper (*Piper nigrum* L.). A management study with organic amendments. Indian Cocoa Arecanut Spices J. 17, 1–2.

Nair, K.P.P., Mengel, K., 1984. Importance of phosphate buffer power for phosphate uptake by rye. Soil Sci. Soc. Am. J. 48 (92–95).

Nair, K.P.P., Nand, Ram., Sharma, P.K., 1984. Quantitative relationship between zinc transport and plant uptake. Plant Soil 81, 217–220.

Nair, K.P.P., Sadanandan, A.K., Hamza, S., Jose, Abraham., 1997. The Importance of Potassium Buffer Power in the Growth and Yield of Cardamom. *J. Plant Nutr.* 20 (7 & 8), 987–997.

Nair, P.K.U., Sasikumaran, S., 1991. Effect of some fungicides on quick wilt (foot rot) disease of black pepper. Indian Cocoa Arecanut Spices J. 14, 95–96.

Nakatani, N., Inatani, R., Ohta, H., Nishioka, A., 1986. Chemical constituents of pepper and application to food preservation. Naturally occurring antioxidant. Environ. Health Perspect. 67, 135–147.

Nambiar, K.K.N., 1978. Diseases of pepper in India. In: Nair, M.K, Haridasan, M. (Eds.), Proceedings of the National Seminar on Pepper (pp. 11–14). Central Plantation Crops Research Institute, Calicut, Kerala State, India.

Nambiar, K.K.N., Sarma, Y.R., 1977. Wilt diseases of black pepper. J. Plantation Crops 5, 92–103.

Nambiar, K.K.N., Sarma, Y.R., 1980. Factors associated with slow wilt of pepper Proceedings of the PLACROSYM-II 1979. Indian Society for Plantation Crops, Kasaragod, Kerala State, India. pp. 348–358.

Nandakumar, C., Ragunath, P., Visalakshi, A., and Mohandas, N. (1987). Management of pollu beetle on pepper by adjusting the spray schedules. In: Abstracts of National Symposium Integrated Pest Control: Progress and Perspectives, Trivandrum, Kerala State, India, p. 50.

Nazeem, P. A., Joseph, L., Thampi, M. S., Sujatha, R., and Nair, G. S. (1993). *In vitro* culture system for indirect organogenesis for black pepper (*Piper nigrum* L.) In: Golden Jubilee Symposium on Horticultural Research-Changing Scenario, 24–28 May, Bangalore, India, p. 250.

Ng, S.C., 1993. Uses of pepper. In: Ibrahim, M.Y., Bong, C.F.J., Ipor, I.B. (Eds.), The Pepper Industry: Problems and Prospects (pp. 347–353). Universiti Pertanian Malaysia, Sarawak, Malaysia.

Nigam, M.D., Handa, K.L., 1964. Constituents of essential oils. Indian Perfumer. 8, 15–17.

Nirmal Babu, K., Geetha, S.P., Manjula, C., Sajina, A., Minoo, D., Samsudeen, K., Ravindran, P.N., Peter, K.V., 1996b. Biotechnology-its role in conservation of genetic resources of spices. In: Das, M.R., Mundayoor, S. (Eds.), Biotechnology for Development (pp. 198–212). State Committee on Science, Technology and Environment, Trivandrum, Kerala State, India.

Nirmal Babu, K., Naik, V.G., Ravidran, P.N., 1993a. Two new taxa of Piper (*Piperaceae*) from Kerala, India with a note on their origin and inter-relationships. J. Spices Aromatic Crops 2, 26–33.

Nirmal Babu, K., Rema, J., Lukose, R., Ravindran, P. N., Johnson George, K., Sasikumar, B., and Peter, K. V., 1993b. Spices Biotechnology at National Research Centre for Spices, Technical Report II, Calicut, Kerala State, India.

Nybe, E.V., Nair, P.C.S., 1986. Nutrient deficiency in black pepper (*Piper nigrum* L.) I. Nitrogen, Phosphorus and Potassium. Agric. Res. J. Kerala 24, 132–150.

Nye, P.H., 1979. Diffusion of ions and uncharged solutes in soils and clays. *Adv. Agron.* 31, 225–272.

Pangburn, R.M., Jennings, W.G., Noelting, C.F., 1970. Preliminary examination of odour quality of black pepper oil. Flavour Ind. 1, 763.

Parmar, V.S., Jain, S.C., Bisht, K.S., Jain, R., Taneja, P., Jha, A., Tyagi, O.D., Prasad, A.K., Wengel, J., Olsen, C.E., Boll, P.M., 1977. Phytochemistry of the genus *Piper*. Phytochemistry 46, 597–673.

Parry, J.W., 1969. Spices, vol. 1. Chemical Publishing Co. Inc., New York, USA.

Paulus, A.D., Eng, L., Teo, C.H., Sim, S.L., 1993. Screening black pepper genotypes and *Piper* spp. for resistance to root knot nematode. In: Ibrahim, M.Y., Bong, C.F.J., Ipor, I.P. (Eds.), The Black Pepper Industry: Problems and Prospects (pp. 132–139). Universiti Pertanian Malaysia, Sarawak, Malaysia.

Perez, C., Anesini, Antibacterial activity of elementary plants against *Staphylococcus aureus* growth. Am. J. Chin. Med. 22, 169–174.

Peter, K.V., 1996a. Emerging potential for global spices trade. The Hindu Businessline, 21 December 1996.

Peter, K.V., 1996. Spices research and development-an updated overview. Employment News 21, 1.

Philip, V.J., Joseph, D., Trigs, G.S., Dickinson, N.M., 1992. Micropropagation of black pepper (*Piper nigrum* L.) through shoot tip cultures. Plant Cell Rep. 12, 41–44.

Pillai, V.S., 1992. Management of improved cultivars of black pepper. In: Sharma, Y.R., Devasahayam, S., Anandaraj, M. (Eds.), Black Pepper and Cardamom: Problems and Prospects (pp. 17–19). Indian Society of Spices, Calicut, Kerala State, India.

Pillai, V.S., Ali, A.B., Chandy, K.C., 1982. Effect of 3-IBA on root initiation and development in stem cuttings of pepper (*Piper nigrum* L.). Indian Cocoa Arecanut Spices J. 6 (1), 7–9.

Pillai, V.S., Chandy, K.C., Sasikumaran, S., Nambiar, P.K.V., 1979. Response of Panniyur-1 variety to nitrogen and lime application. Indian Cocoa Arecanut Spices J. 3 (2), 35–38.

Pillai, V.S., Sasikumaran, S., 1976. A note on the chemical composition of pepper plant (*Piper nigrum* L.). Arecanut Spices Bull. 8, 13–17.

Pillai, V.S., Sasikumaran, S., Nambiar, P.K.V., 1987. N, P and K requirement of black pepper. Agric. Res. J. Kerala 25 (1), 74–80.

Prabhakaran, P.V., 1994. Assessment of losses due to pests, diseases and drought in black pepper. J. Indian Soc. Agric. Stat. 47, 59–60.

Pradeep, K.U., Geervani, P., 1994. Influence of spices on protein utilisation of winged bean and horsegram. Plant Foods Hum. Nutr. 46, 187–193.

Premkumar, T., Banerjee, S.K., Devasahayam, S., Koya, K.M.A., 1986. Effect of different insecticides on the control of pollu beetle *Longitarsus nigripennis* Mots.: a major pest of black pepper *Piper nigrum* L. Entomon. 11, 219–221.

Premkumar, T., Devasahayam, S., Koya, K.M.A., 1994. Pests on spice cropsChadha, K.L.Rethinam, P. Advances in Horticulture: Plantation and Spice Crops, Part 2, vol. 10. Malhotra Publishing House, New Delhi, India.

Premkumar, T., Nair, M.R.G.K., 1987. Reduction in damage by *Longitarsus nigripennis* to berries of black pepper by insecticides. Madras Agric. J. 74, 326–327.

Premkumar, T., Nair, M.R.G.K., 1988. Effect of some planting conditions on infestation of black pepper by *Longitarsus nigripennis* Mots. Indian Cocoa Arecanut Spices J. 10, 83–84.

Pruthi, J.S., 1992. Advances in sun/solar drying and dehydration of pepper (*Piper nigrum* L.). Int. Pepper News Bull. 16 (2), 6–17.

Pruthi, J.S., 1993. Major Spices of India: Crop Management and Post-Harvest Technology. Indian Council of Agricultural Research, New Delhi, India. pp. 44–105.

Pruthi, J.S., 1997. Diversification in pepper utilization. Part I & II. Green pepper products. Int. Pepper News Bull. 14 (4), 5–9. 14(5), 6–9.

Pruthi, J.S., Bhat, A.V., Kutty, S.K., Varkey, A.G., Gopalakrishnan, M., 1976. Preservation of fresh green pepper by canning, bottling and other methods International Seminar on Pepper. Spices Export Promotion Council of India, Cochin, Kerala State, India. pp. 64–72.

Purseglove, J.W., Brown, E.G., Green, C.L., Robbins, S.R.J., 1981. Spices, vol. 1. Longman, London. pp. 1–99.

Rahiman, B.A., 1981. Biosystematic studies in varieties and spices of *Piper* occurring in Karnataka region. Ph.D. Thesis, University of Mysore, Karnataka State, India.

Raj, H.G., 1978. A comparison of the systems of cultivation of Black pepper *Piper nigrum* L. in Malaysia and Indonesia. The Directorate of Extension Education, Kerala Agricultural University, Trichur, Kerala State, India, pp. 65–75.

Raj Mohan, K., 1985. Standardization of tissue culture techniques in important horticultural crops. Ph.D. Thesis, Kerala Agricultural University, Thrissur, Kerala State, India.

Ramachandran, N., 1990. Bioefficacy of systemic fungicides against Phytophthora infections in black pepper (*Piper nigrum* L.), Ph.D. Thesis, University of Calicut, Calicut, Kerala State, India, p. 138.

Ramachandran, N., Sarma, Y.R., 1985. Efficacy of three systemic fungicides in controlling Phytophthora infections in black pepper. Indian Phytopath. 38, 160–162.

Ramachandran, N., Sarma, Y.R., Anandaraj, M., 1988a. Sensitivity of Phytophthora species affecting different plantation crops in Kerala to Metalaxyl. *Indian Phytopath.* 41, 438–442.

Ramachandran, N., Sarma, Y.R., Anandaraj, M., Abraham, J., 1988b. Effect of climatic factors on *Phytophthora* leaf infection in black pepper grown in arecanut–black pepper mixed cropping system. J. Plantation Crops 16, 110–118.

Ramachandran, N., Sarma, Y.R., Anandaraj, M., 1990. Vertical progression and spread of *Phytophthora* leaf infection in black pepper in arecanut–black pepper mixed cropping system. Indian Phytopath. 43, 414–419.

Ramachandran, N., Sarma, Y.R., Anandaraj, M., 1991. Management of *Phytophthora* infections in black pepper. In: Sarma, Y.R., Premkumar, T. (Eds.), Diseases of Black Pepper (pp. 158–174). National Research Centre for Spices, Calicut, Kerala State, India.

Ramachandran, N., Sarma, Y.R., Nambiar, K.K.N., 1986. Spatial distribution of *Phytophthora palmivora* MF4 in the root zones of *Piper nigrum*. *Indian Phytopath.* 39, 414–417.

Ramadasan, A., 1987. Canopy development and yield of adult pepper vines in relation to light interception. Indian Cocoa Arecanut Spices J. 11, 43–44.

Ramana, K.V., 1992. Final Report of the project: Role of Nematodes in the incidence of Slow Decline (Slow Wilt Disease) of Black Pepper and Screening Pepper Germplasm against Nematodes, National Research Centre for Spices, Calicut, Kerala State, India, p. 149.

Ramana, K.V., 1994. Efficacy of *Paecilomyces lilacinus* (Thom.) Samson in suppressing nematode infestations in black pepper (*Piper nigrum* L.). J. Spices Aromatic Crops 3, 130–134.

Ramana, K.V., Eapen, S.J., 1998. Plant parasitic nematodes associated with spices and condiments. In: Trivedi, P.C. (Ed.), Nematode Diseases in Plants (pp. 217–251). C. B. S. Publishers and Distributors, New Delhi, India.

Ramana, K.V., Mohandas, C., 1986. Reaction of black pepper germplasm to root knot nematode *Meloidogyne incognita*. Indian J. Nematol. 16, 138–139.

Ramana, K.V., Mohandas, C., Eapen, S.J., 1994. Plant parasitic nematodes and slow decline disease of black pepper. Tech. Bull. National Research Centre for Spices, Calicut, Kerala State, India, p. 14.

Ramana, K.V., Mohandas, C., Ravindran, P.N., 1987. Reaction of black pepper germplasm to the burrowing nematode (*Radopholus similis*). J. Plantation Crops 15, 65–66.

Ramanathan, P.K., Rao, P.N.S., 1974. Equipment needs of spice industry. In: Proceedings of the Symposium on Development and Prospects of Spice Industry in India, Central Food Technology Research Institute, Mysore, Karnataka State, India, pp. 82–84.

Ramesh, C.R., 1982. Root infection and population density of VA mycorrhizal fungi in a coconut based multi-storied cropping. In: Bavappa, K.V.A., Iyer, R.D, Das, P.K,

Sethuraj, M.R, Ranganathan, V, Nambiar, K.K.N, Chacko, M.J (Eds.), Proceedings of the Plantation Crops Symposium (PLACRO-SYM V) (pp. 548–554). Central Plantation Crops Research Institute, Kasaragod, Kerala State, India.

Rao, C.S.S., Nigam, S.S., 1976. Antimicrobial activity of some Indian essential oils. Indian Drugs 14, 62.

Ravindran, P. N., 1991. Studies on black pepper and some of its wild relatives. Ph.D. Thesis, University of Calicut, Calicut, Kerala State, India.

Ravindran, P.N., Nair, M.K., Nair, R.A., 1987. Two taxa of *Piper* (Piperaceae) from Silent Valley Forests, Kerala. J. Econ. Tax. Bot. 10, 167–169.

Ravindran, P.N., Nair, M.K., Nair, R.A., Nirmal Babu, K., Chandran, K., 1990. Ecological and taxonomical notes on Piper spp. from Silent Valley Forests, Kerala. J. Bombay Nat. His. Soc. 87, 421–426.

Ravindran, P.N., Nirmal Babu, K., 1994. Genetic resources of black pepper. In: Chadha, K.L., Rethinam, P. (Eds.), Advances in Horticulture (pp. 99–120). Malhotra Publications Co., New Delhi, India.

Ravindran, P.N., Peter, K.V., 1995. Biodiversity of major spices and their conservation in India. In: Arora, R.K., Rao, V.R. (Eds.), Proceedings of the South Asia Coordinators Meeting on Plant Genetic Resources (pp. 123–134). IPGRI, New Delhi, India.

Reddy, B.N., Sadanandan, A.K., Sivaraman, K., Abraham, J., 1992. Effects of planting density on black pepper on yield and nutrient availability in soil. J. Plantation Crops 20 (Suppl), 10–13.

Reen, R.K., Jamwal, D.S., Taneja, S.C., Koul, J.L., Dubey, P.K., Wiebel, F.J., Singh, J., 1993. Impairment of UDP-GDH and glucuronidation activities in liver and small intestine of rat and guinea pig *in vitro* by piperine. Biochem. Pharmacy 46, 229–238.

Rehiman, P.V.A., Nambiar, P.K.V., 1967. Insecticidal control of pollu beetle (*Longitarsus nigripennis*) on pepper. Madras Agric. J. 54, 39–40.

Revankar, G.D., Sen, D.P., 1974. Antioxidant effects of a spice mixture on sardine oil. J. Food Sci. Technol. (India) 11, 31–32.

Rheede, H. Van., 1678. Hortus Indicus Malabaricus, vol. 7, pp. 23–31. Amstelodami, pp. 12–16.

Richard, H.M., Jennings, W.G., 1971. Volatile composition of black pepper. *J. Food Sci.* 36, 584–589.

Richard, H.M., Russel, G.F., Jennings, W.G., 1971. The volatile components of black pepper varieties. *J. Chromatogr. Sci.* 9, 560–566.

Risfaheri Hidayat, T., 1996. Study on decorticating of pepper berries by soaking in boiling water method. Int. Pepper News Bull. 20 (1), 17–20.

Rosengarten Jr., F., 1973. The Book of Spices. Pyramid Books, New York, USA.

Ruiz, P., 1794. Cited by Trelease and Yuncker (1950).

Russel, G.F., Else, J., 1973. Volatile compositional differences between cultivars of black pepper (*Piper nigrum*). *J. Assoc. off. Anal. Chem.* 56, 344–351.

Sadanandan, A. K., 1990. Response of pepper to applied potassium (Abs.). National Seminar on Potassium on Plantation Crops, Bangalore, India, p. 149.

Sadanandan, A.K., 1994. Nutrition of black pepper. In: Chadha, K.L., Rethinam, P. (Eds.), Advances in Horticulture. Vol. 9: Plantation and Spice Crops (Part I), New Delhi (pp. 423–456).

Sadanandan, A.K., Abraham, J., Anandaraj, M., 1992. Impact of high production technology on productivity and disease incidence of black pepperNambiar, K.K.N.Iyer, R.D.Rao, E.V.V.B. Proceedings of the PLACROSYM-IX, 20. Indian Society of Plantation Crops, Kasaragod, Kerala State, India.

Sadanandan, A. K., and Hamza, S., 1996. Studies on nutritional requirement of bush pepper for yield and quality (Abs.). In: Proceedings of the PLACROSYM-XII, RRII, Kottayam, Kerala State, India, p. 48.

Saito, Y., Asari, T., 1976. Studies on the antioxidant properties of spices. I. Total tocopherol content in spices. J. Jpn. Soc. Food Nutr. 29, 289–292.

Sajina, A., Minoo, D., Geetha, S.P., Samsudeen, K., Rema, J., Nirmal Babu, K., Ravindran, P.N., Peter, K.V., 1996. Production of synthetic seeds in spices. In: Edison, S., Ramana, K.V., Sasikumar, B., Nirmal Babu, K., Eapen, S.J. (Eds.), Biotechnology of Spices, Medicinal and Aromatic Plants (pp. 65–69). Indian Society for Spices, Calicut, Kerala State, India.

Salzer, U.J., 1975. Uber die Fettsaurezussamensetzung de lipoide einiger gewurze. Fette Stefen Anstrichm. 77, 446–450.

Salzer, U.J., Broker, U., Klie, H.P., Liepe, H.U., 1977. Effect of pepper and pepper constituents on the microflora of sausage products. Fleischwirtschafe 57, 2011–2121.

Samraj, J., Jose, P.C., 1966. A Phytophthora wilt of pepper (*Piper nigrum*). Sci. Cult. 32, 90–92.

Samuel, R., Prabhu, V.K.K., Narayanan, C.S., 1984. Influence of spice essential oils on the life history of *Lasioderma serricorne*. Entomon. 9, 209.

Sanchez, P.A., 1994. Tropical soil fertility research: Towards the second paradigm. In: Proceedings of the XV ISSUS Symposium, Acapulco, Mexico, July 10–16, vol. 1, pp. 89–104.

Sankarikutty, B., Padmakumari, K.P., Howa, Umma., Narayanan, C.S., 1994. Annual Report of Regional Research Laboratory, Council of Scientific and Industrial Research, Trivandrum, Kerala State, India, pp. 7–8.

Santosh, P., 1985. Cost of cultivation and marketing of pepper in Cannanore District. M.Sc. (Ag) Thesis, Kerala Agricultural University, Thrissur, Kerala State, India.

Sarkar, B.B., Choudhary, D.M., Nath, S.C., Bhattacharjee, N.R., Barna, D.T.C., 1985. Wilt disease of pepper and its control in Tripura. Indian Cocoa Arecanut Spices J. 8, 63–66.

Sarma, Y.R., Anandaraj, M., Ramana, K.V., 1992. Present status of black pepper diseases in India and their management. In: Wahid, P., Sitepu, D., Deciyanto, S., Suparman, U. (Eds.), Proceedings of the International Workshop on Black Pepper Diseases Bander Lampung, Indonesia (pp. 67–68). Institute for Spice and Medicinal Crops, Bogor, Indonesia.

Sarma, Y.R., Anandaraj, M., Venugopal, M.N., 1996. Biological control of diseases in spices. In: Anandaraj, M., Peter, K.V. (Eds.), Biological Control in Spices (pp. 1–19). Indian Institute of Spices Research, Calicut, Kerala State, India.

Sasikumar, B., Veluthambi, K., 1994. Kanamycin sensitivity of cultured tissues of *Piper nigrum* L. J. Spices Aromatic Crops 3, 158–160.

Satheesan, K.N., Rajagopalan, A., Sukumarapillai, V., Jagadeesh Kumar, T. N., Mammootty, K. P., and Unnikrishnan Nair, P. K., 1997a. Influence of irrigation levels on yield, quality and incidence of pests and diseases in black pepper (*Piper nigrum* L.) (Abs.). In: National Seminar on Water Management for Sustainable Production and Quality of Spices. Madikheri, Karnataka State, India, p. 21.

Schenk, R.U., Hildebrandt, A.C., 1972. Medium and techniques for induction and growth of monocotyledonous and dicotyledonous plant cell cultures. *Can. J. Bot.* 50, 199–204.

Schneider, B., 1993. Steam sterilization of spices. Fleischwirtschaft 73 (6), 646–649.

Schüller, H., 1969. Die CAL-Methode, eine neue Methode zur Bestimmung des pflanzen verfügbaren Phosphates in Böden. Z. *Pflanzenernähr Bodenkd.* 123, 48–63.

Scott, R., 1992. Master spice: extruded low count spices. In: Proceedings of the New Technologies Symposium Bristol, UK, pp. 6–8.

Scott, W.P., McKibben, G.H., 1978. Toxicity of black pepper extract to ball weevil. J. Econ. Ent. 71, 343–344.

Selim, H.M., 1992. Modeling the transport and retention of inorganics in soils. *Adv. Agron.* 47, 331–384.

Sen, A.R., Roy, B.R., 1974. Adulteration in spices. In: Proceedings of the National Symposium on Development and Prospects of Spice Industry in India. Central Food Technology Research Institute, Mysore, Karnataka State, India.

Sethuram, P., 1995. Cochin pepper market—a journey through history. Ind. Spices 32, 3.

Sewell, G.W.F., 1965. The effect of altered physical conditions of soil on biological control. In: Baker, K.F., Snyder, W.C. (Eds.), Ecology and Soil Borne Plant Pathogens (pp. 479–494). University of California Press, Berkeley, USA.

Sharma, R.D., Loof, P.A.A., 1974. Nematodes of cocoa region of Bahia, Brazil IV. Nematodes in the rhizosphere of pepper (*Piper nigrum* L.) and clove (*Eugenia caryophylla* Thumb). Revista Theobroma 4, 26–32.

Sharma, N.L., Nigam, N.C., Handa, K.L., 1962. Pepper husk in oil. Perfuem. Kosmet 43, 505–506.

Sheela, M.S., Venkitesan, T.S., Mohandas, N., 1993. Status of *Bacillus* spp. as biocontrol agents of root knot nematode (*Meloidogyne incognita*) infesting black pepper (*Piper nigrum* L.). J. Plantation Crops 21 (Suppl), 218–222.

Sher, S.A., Chunram, C., Pholcharoen, S., 1969. Pepper yellows disease and nematodes in Thailand. FAO Plant Prot. Bull. 17, 33.

Shin, K.H., Yun, H.S., Won, W.S., Lee, C.K., 1980. Pharmacological activity of *Piper retrofractrum*. Soul. Tacha. Sean. Young Opi. 18, 87–89. (*Chem. Abstr.* 93, 215484s).

Shuman, L.M., 1975. The effect of soil properties on zinc adsorption by soils. Soil Sci. Soc. Am. Proc. 39, 454–458.

Shwaireb, M.H., Waba, H., El-Mofty, M.M., Dutter, A., 1990. Carcinogenesis induced by black pepper and modulated by vitamin-A. *Exp. Pathol.* 40, 233–238.

Sim, E.S., 1971. Dry matter production and major nutrient contents of black pepper (*Piper nigrum* L.). Malay. Agric. J. 48, 73.

Sim, S.L., Jafar, R., Grierson, D., Power, J.B., and Davery, M.R., 1995. Applications of molecular biology and genetic manipulation to pepper (*Piper nigrum* L.) breeding: present state and prospects. In: Abstracts FAO/IAEA International Symposium on the Use of Induced Mutations and Molecular Techniques for Crop Improvement Vienna, Austria.

Singh, A., Rao, A.R., 1993. Evaluation of the modulatory influence of black pepper on the hepatic detoxification system. Cancer Lett. 72, 5–9.

Sitepu, D., Kasim, R., 1991. Black pepper diseases in Indonesia and their control strategy. In: Sarma, Y.R., Premkumar, T. (Eds.), Diseases of Black Pepper (pp. 13–28). National Research Centre for Spices, Calicut, Kerala State, India.

Siti Hajijah, A.S., 1993. Observations of root knot infestation on pepper (*Piper nigrum* L.) in Sarawak. In: Ibrahim, M.Y., Bong, C.F.J., Ipor, I.B. (Eds.), The Pepper Industry: Problems and Prospects (pp. 140–147). Universiti Pertanian Malaysia, Sarawak, Malaysia.

Siti Hajijah, A.S., Bong, C.F.J., 1993. Pepper fruit stalk as substrate for mushroom cultivation. In: Ibrahim, M.Y., Bong, C.F.J., Ipor, I.B. (Eds.), The Pepper Industry: Problems and Prospects (pp. 342–346). Universiti, Pertanian Malaysia, Sarawak, Malaysia.

Sivadasan, C.R., 1996. Import regulations in USA. Spice India 9, 2–10.

Sivakumar, C., Wahid, P.A., 1994. Effect of application of organic materials on growth and foliar nutrient contents of black pepper (*Piper nigrum* L.). J. Spices Aromatic Crops 3, 135–141.

Soil Science Society of America, 1965. Glossary of soil science terms. Soil Sci. Soc. Am. Proc. 29, 330–351, Committee Report.

Sosamma, V.K., Koshy, P.K., 1995. Effect of *Pasteuria penetrans* and *Paecilomyces lilacinus* on population build up of root knot nematode, *Meloidogyne incognita* on black pepper. Indian J. Nematol. 25, 16–17. (Abst.)

Sparks, D.L., 1987. Potassium dynamics in soils. *Adv. Soil Sci.* 6, 2–63.

Sparks, D.L., 1989. Kinetics of Soil Chemical Processes. Academic Press, San Diego, CA.

Sreeja, T.P., Eapen, S.J., Ramana, K.V., 1996. Occurrence of *Verticillium chlamydosporium* Goddard in a black pepper (*Piper nigrum* L.) garden in Kerala, India. J. Spices Aromatic Crops 5, 143–147.

Sreekumar, M. M., Sankarikutty, B., Nirmala Menon, A., Padmakumari, K. P., and Narayanan, C.S., 1993. Annual Report of Regional Research Laboratory, Council of Scientific and Industrial Research, Trivandrum, Kerala State, India, pp. 5–6.

Sridharan, K., Kalla, A.K., Singh, J., 1978. Chemical and pharmacological screening of *Piper nigrum* L. leaves. J. Res. Indian Med. Yoga Homeopath. 13, 107.

Stirling, G.R., 1991. Biological Control of Plant Parasitic Nematodes. C. A. B. International, Wallingford, UK. p. 282

Su, H.C.F., 1977. Insecticidal properties of black pepper on cotton weevil and cow-pea weevil. J. Econ. Ent. 70, 18–21.

Subramanyam, V., Sreenivasamoorthy, N., Krishnamoorthy, K., Swaminathan, M., 1957. Studies on the antibacterial effects of spices. J. Scient. Ind. Res. 16, 240.

Sundararaju, P., Koshy, P.K., Sosamma, V.K., 1979. Plant parasitic nematodes associated with spices. J. Plantation Crops 7, 15–26.

Sundararaju, P., Koshy, P. K., and Sosamma, V. K. (1980). Survey of nematodes associated with spices in Kerala and Karnataka. In: Proceedings of the PLACROSYM-II, 1979, Indian Society for Plantation Crops, Kasaragod, Kerala State, India, pp. 39–44.

Suparman, U., Zaubin, R., 1988. Effect of defoliation, IBA and Saccharose on root growth of black pepper cuttings. Industrial Crop Res. J. 1, 54–58.

Tantaoui, E.A., Beraoud, L., 1994. Inhibition of growth and aflatoxin production in *Aspergillus parasiticus* by essential oils. J. Environ. Toxicol. Pathol. Oncol. 13, 67–72.

Thomas, K.M., Menon, K.K., 1939. The present position of pollu disease of pepper in Malabar. Madras Agric. J. 17, 347–356.

Tien, J.-K., 1981. China and the pepper trade. Hemisphere. 18, 220–222.

Trelease, W., Yuncker, T.G., 1950. The *Piperaceae* of Northern South America. University of Illinois, USA.

Tripathi, S.N., Tiwari, C.N., Upadhyaya, B.N., Singh, K.R.S., 1979. Screening of hypoglycaemic action in certain indigenous drugs. J. Res. Ind. Med. Yoga Homeopath. 14, 159.

Uijil, C. den.H., 1992. New continuous steaming method for herbs and spices. Voedingsmiddelentechnologie 25 (6), 40–43.

Uma Pradeep, K., Geervani, P., Eggum, B.O., 1993. Common Indian Spices: nutrient composition, consumption and contribution to dietary values. *Plant Foods Hum. Nutr.* 44, 137–148.

Ummer, C., 1989. Indian Spices-from the leaves of history Spice Fair Commemorative Volume. Spices Board, Cochin. pp. 27–40.

Van der Vecht, J., 1950. Plant parasitic nematodes. In: Karshoven, L.G.E., van der Vecht, J. (Eds.), Diseases of Cultivated Plants in Indonesian Colonies (pp. 16–45). W. van Hoeve, I. S' gravenhage.

Variyar, P.S., Pendharkar, M.B., Banerjee, A., Bandopadhyaya, C., 1988. Blackening in green pepper berries. Phytochemistry 27, 715–717.

Varughese, J., Anuar, M.A., 1992. Etiology and control of slow wilt disease in Johore, Malaysia. In: Wahid, P., Sitepu, D., Deciyanto, S., Suparman, U. (Eds.), Proceedings of the International Workshop on Black Pepper Diseases (pp. 188–197). Research Institute for Spice and Medicinal Crops, Bogor, Indonesia.

Velayudhan, K.C., Amalraj, V.A., 1992. *Piper pseudonigrum*—a new species from Western Ghats. J. Econ. Tax. Bot. 16, 247–250.

Venkitesan, T. S., and Charles, J. S. (1980). A note on the chemical control of nematodes infesting pepper vines in Kerala. In: Proceedings of the PLACROSYM-II 1979. Indian Society for Plantation Crops, Kasaragod, Kerala State, India, pp. 27–30.

Venkitesan, T.S., Setty, K.G.H., 1978. Reaction of 27 black pepper cultivars and wild forms to the burrowing nematode *Radopholus similis*(Cobb) Thorne. J. Plantation Crops 6, 81–84.

Venkitesan, T.S., Setty, K.G.H., 1979. Control of the burrowing nematode, *Radopholus similis* on black pepper. Pesticides 13, 40–42.

Vezyuez Olivincia, W., Shah, P., Pitchumoni, C.S., 1992. The effect of red and black pepper on orocecal transit time. J. M. College Nutr. 11, 228–231.

Vijayakumar, K.R., Unni, P.N., Vamadevan, V.K., 1984. Prevention of photo-induced chlorophyll loss by the use of lime reflectant on the leaves of black pepper (*Piper nigrum* L.). Agric. For. Meteorol. 34, 17–20.

Vinod, G., 1984. Cost of cultivation and marketing of pepper in Idukki District. M.Sc. (Ag) Thesis, Kerala Agricultural University, Thrissur, Kerala State, India.

Wahid, P., 1976. Studies on yellows disease in black pepper on the island of Bangka. Pembr. LPTI 21, 64–79.

Wealth of India. 1969. *Piper nigrum* Linn (Piperaceae) Wealth of India-Raw Materials, vol. 8. Publications and Information Directorate, Council of Scientific and Industrial Research, New Delhi. pp. 99–115.

Wilkinson, H.F., Loneragan, J.F., Quirk, J.P., 1968. The movement of zinc to plant roots. Soil Sci. Soc. Am. Proc. 32, 831–833.

Winoto, S.R., 1972. Effect of *Meloidogyne* species on the growth of *Piper nigrum* L. *Malaysian Agric. Res.* 1, 86–90. Seoul Teahakkyo Saengyak Yonguso Opidukdip

Won, W.S., Lee, E.B., and Shin, K.H., 1979. CNS depressant activity of piperine. 18, 66–70. (*Chemical Abstracts*. **93**, 215482q).

Wrolstad, R.E., Jennings, W.G., 1965. Volatile constituents of black pepper III. The monoterpene hydrocarbon fraction. *J. Food Sci.* 30, 274–279.

Wuba, H., El-Mofty, M.M., Schweureb., M.H., Dutter, A., 1992. Carcinogenicity testing of some constituents of black pepper. *Exp. Pathol.* 44, 61–65.

Yuncker, T.G., 1958. The Piperaceae: a family profile. Brittonia 10, 1–17.

Zadocks, J.C., Van den Bosch, F., 1994. On the spread of plant disease: a theory of foci. Annu. Rev. Phytopath. 32, 503–521.

Zaubin, R., Robbert, Pengaruh keasaman tanah terhadap pertumbuhan tanaman lada. Pebr. Littri 33, 1987.

2 The Agronomy and Economy of Cardamom (*Elettaria cardamomum M.*): The "Queen of Spices"

2.1 Introduction

Cardamom, popularly known as the "Queen of Spices" is the second most important spice crop in the world, next to black pepper (*Piper nigrum*), which is known as the "King of Spices." The description "Queen of Spices" is apt, because cardamom has a pleasant aroma and taste and has been a highly valued spice since time immemorial. Cardamom belongs to the genus *Elettaria* and species *cardamomum* (Maton). The term *Elettaria*, which is the generic name, has its origin in the word *Elettari* (in Tamil, one of the popular South Indian languages), referring to the cardamom seeds. In the original description, it means a "particle/seed of the leaf." *Elletaria* is a large-sized perennial herbaceous rhizomatous monocot that belongs to the Zingiberaceae family. The plant is extensively grown in the hilly tracts of southern India at elevations ranging from 800 to 1500m. It grows best as an under crop, beneath forest trees, in shade and a cool climate at high elevations. The plant is grown in Sri Lanka, in Papua New Guinea (PNG), and in Tanzania, Africa. In Latin America, Guatemala is the biggest grower of cardamom. Indeed, Guatemalan cardamom is the biggest competitor to Indian cardamom in the world market.

2.1.1 Historical Background of Cardamom

Cardamom has an interesting history dating back to Vedic times, about B.C. 3000. In the ancient Indian language Sanskrit, it is referred to as "Ela." In ancient Hindu culture, sacrificial fire was a common ritual, and in ancient texts (Mahindru, 1982) cardamom was an ingredient, along with several other materials, in the sacrificial fire, solemnizing a Hindu marriage. Both the *Charaka Samhita* and the *Susrutha Samhita*, the ancient Indian Ayurvedic texts, written in the post-Vedic period (B.C. 1400–B.C. 1600) mention cardamom. However, it is not known precisely whether cardamom, referred to as Ela in these texts, is the Indian variety or the large Nepalese variety. The Assyrians and Babylonians were familiar with medicinal plants, and cardamom was among the 200 or so plants that the former dealt with (Parry, 1969). It was mentioned that Merodach-Baladan II (reigned B.C. 721–B.C. 702), the ancient king of Babylon, grew cardamom among other herbs in his garden. Surprisingly, there was no mention of cardamom in the ancient Egyptian texts, unlike their treatment of pepper. Possibly, cardamom was

Agronomy and Economy of Black Pepper and Cardamom. DOI: 10.1016/B978-0-12-391865-9.00002-5

just beginning to reach Assyria and Babylonia through the land routes. Interestingly, cardamom is cited in some of the ancient Greek and Roman texts. Spices were the symbols of royalty and luxury, and cardamom was used in the manufacture of perfumes during Greek and Roman times. In addition, cardamom was used as an aphrodisiac (Parry, 1969). Significantly, Dioscorides (40–90 A.D.), the Greek physician and author of the legendary *Materia Medica*, mentions cardamom in his work. Cardamom was widely used to aid digestion, and that was the most important reason both the Greeks and Romans imported the spice in large quantities from India. Thus, it became one of the most popular oriental spices in Greek and Roman cuisine. This achievement led to cardamom being listed as a dutiable item in Alexandria in 176 A.D.

In his *Journal of Indian Travels* (1596), Linschoten describes two types of cardamom in use in southern India: the "greater" (large) and "lesser" (small) types. This would suggest that the large cardamom found extensively in Nepal must have been finding its way to southern India through land routes, brought by travelers dating back to nearly 4000 years from today. Referring to the introduction of cardamom to Europe, Dymock writes, "When they were first introduced into Europe is doubtful, as their identity with the *Amomum* and *Cardamomum* of the Greeks and Romans cannot be proved." Linschoten writes about lesser cardamom that "it mostly is grown in Calicut and Cannanore, places on the coast of Malabar." Paludanus, a contemporary of Linschoten, wrote that, according to Avicenna, there are two kinds of cardamoms, "greater" and "lesser," and goes on to add that cardamom was unknown to Greeks personages such as Galen and Dioscorides. In his Seventh Book of *Simples*, Galen wrote, "cardamom is not so hot as Nasturtium or water cresses," "but pleasanter of savor and smell with some small bitterness." These properties were dissimilar to those of the Indian cardamom. In his First Book, Dioscorides, commenting on the cardamom brought from Armenia and Bosphorus, wrote, "we must choose that which is full, and tough in breaking, sharp and bitter of taste, and smell there of, which cause heaviness in a man's head" (Watt, 1872). Obviously, Dioscorides was writing not about Indian cardamom, but about a different plant. Such references led Paludanus (Watt, 1872) to infer that the *Amomum* and *Cardamomum* of the ancient Greeks were not the spices of India. On the whole, references to cardamom in ancient and early centuries of the Christian era and even in the middle ages are but scanty compared with references to black pepper. Even Auboyar, in his classic work on day-to-day living in ancient India (B.C. 200 to 700 A.D.), makes only a fleeting mention of cardamom (Mahindru, 1982).

The Mediterranean merchants were clearly cheated by the Arabs on the sea route through which the latter brought home spices from India. Like pepper, cardamom was no exception. Pliny thought that cardamom was grown in Arabia. This belief persisted until the discovery of the sea route to India and the landing of the Portuguese on the west coast of India. The latter event coincided with the ending of the Arab monopoly on the spice trade, and the Portuguese started shipping out pepper, cardamom, and ginger to Europe. Since the European colonizers were more interested in procuring pepper and ginger, both crops took hold in India—the former in particular along the Malabar Coast. Cardamom was relegated to the back seat, a situation that lasted from the sixteenth to the eighteenth centuries. Cardamom was considered a minor forest produce. It was only at the beginning of the nineteenth century that cardamom plantations were established, but the spice was interplanted with coffee. Still, cardamom cultivation spread rapidly in the Western Ghats, and the region south of Palakkad (the midsouthern district of Kerala) came to be known as Cardamom Hills.

The earliest written evidence that cardamom was being grown in India was in the records of the officers working for the British East India Company. The most important among these written pieces was that of Ludlow, an assistant conservator of forests. Others were the *Pharmacographia, Madras Manual*, and *Rice Manual*. A brief description of cardamom cultivation in South India was also given by Watt (1872). The system of cardamom collection from naturally growing plants continued until 1803, but demand escalated in later years, and this naturally led to the establishment of large-scale plantations in India and Sri Lanka, then known as Ceylon (Ridley, 1912). Within the entire state of Kerala, in the two erstwhile states of Travancore and Cochin, which had their own kings, cardamom was a monopoly of the respective governments. The Raja (King) of Travancore mandated that all the cardamom produced be sold to his official representative and sent to a central depot in the central Kerala town of Alleppey, which was then a state port. Here, the produce was sold by auction. The principal buyers were Muslims, and the best lot, known as "Alleppey Green," was reserved for export. In the forestland, in the state of Kerala, owned by the then British government, cardamom was considered a "miscellaneous produce," while in the neighboring Coorg district in the state of Karnataka, forestlands were leased out to private cultivators of cardamom. In Leghorn, the conservator of forests in the Madras Presidency (an early nomenclature that included four southern states, namely, Kerala, Karnataka, Madras, and Andhra, which have all become independent since then), noted that the spread of coffee eclipsed that of cardamom in many areas of "Malabar Mountains"—a reference to the Western Ghats (Watt, 1872). Cardamom cultivation is mentioned in the *Madras Manual*, which states, "In the hills of Travancore cardamom grows spontaneously in the deep shades of the forests: it resembles somewhat turmeric and ginger plants but grows to a height of 6–10 ft, and throws out the long shoots which bear the cardamom pods." The following passage describes cardamom management: "The owners of the gardens, early in the season come up from the low country east of the Ghats, cut the brushwood and burn the creepers and otherwise clear the soil for the growth of the plants as soon as the rains fall. They come back to gather the cardamom when they ripen, about October or November" (Watt, 1872). One can surmise from the writings of the British officials that a process of bleaching used to be carried out in Karnataka, and this was done by transporting cardamom to Havre, a place in the Dharma district of Karnataka, and the bleaching process used the water from a specific well, thereby enhancing the flavor of the dried product (Watt, 1872). Mollison (1900) elaborately described a bleaching method in which soapnut water was used.

2.1.2 Cardamom Production and Productivity: A Worldview

Currently, cardamom production is concentrated primarily in India and Guatemala. Cardamom was introduced to Guatemala in 1920, most likely from India or Sri Lanka, by a New York broker and was planted in the vicinity of Cobán in the department of Alta Verapaz (Lawrence, 1978). After World War II, cardamom production in Guatemala increased substantially on account of a shortage in production and high prices, and Guatemala soon became the top cardamom producer in the world. Native Guatemalans do not relish the taste of cardamom, and the entire quantity produced is exported. Today, Guatemala produces about 13,000–14,000 t of cardamom annually. Table 2.1 presents a worldview of cardamom production and productivity.

In India, the cardamom area has come down during the last two decades, from 1,05,000 ha in 1987–1988 to 69,820 ha in 1997–1998, a decrease of 33.5%. Still, production increased 190%, from 3200 t during 1987–1988 to 9290 t in 1999–2000. During the same period, productivity has risen from 47 to 173 $kg\,ha^{-1}$, an increase of 268%. Cardamom cultivation is confined primarily to three South Indian states: Kerala, Karnataka, and Tamil Nadu. Kerala has 59% of the total area cultivated and contributes 70% of the total production. Karnataka has 34% of the total area cultivated and contributes 23% to total production, while Tamil Nadu has 7% of the area and contributes the same percentage to total production. Most of the cardamom-growing areas in Kerala are located in the districts of Idukki, Palakkad, and Waynad. In Karnataka, the crop is grown in the districts of Coorg, Chickmagalur, and Hassan, and, to some extent, in North Kanara district. In Tamil Nadu, cardamom cultivation is located in certain places of Pulney and Kodai hills. On the whole, in India, cardamom is a small-landholder's crop and there are 40,000 holdings covering an area of 80,000 ha (George and John, 1998). The cardamom-growing regions of South India lie within 8° and 30° latitude and 75° and 78° longitude. The crop grows at elevations from 800 to 1500 m above mean sea level (amsl), and these areas lie on both the

Table 2.1 Cardamom Production in the World

Time Span	Percentage Share of Total			World Production (mt)[b]
	India	Guatemala	Others[a]	
1970/1971–1974/1975	65.4	21.5	13.1	4678
1975/1976–1979/1980	53.7	34.5	11.8	6628
1980–1981	42.9	48.8	8.3	10,250
1984–1985	31.9	60.3	7.8	12,220
1985/1986–1989/1990	26.5	67.5	6.0	14,392
1990/1991–1994/1995	28.4	65.6	6.0	19,470
1995/1996–1997/1998	29.8	64.2	6.0	24,953

Source: *Cardamom Statistics, 1984–1985*, Cardamom Board, Government of India, Cochin, Kerala State. Spices Statistics, 1997, Spices Board, Government of India, Cochin, Kerala State. All India Final Estimate of Cardamom, 1997/1998, Government of India, Ministry of Agriculture.

Important note: In three decades, the percentage contribution of India to total production plummeted by 54%, while Guatemala's increased by 199%. Other countries in the same period had a similar decline of 54%, like that of India; thus, Guatemala takes the leading position in cardamom production in the world.

[a]Estimates, actual figures unavailable.

[b]Metric tons.

windward and leeward sides of the Western Ghats, which act as a barrier of the monsoon trade winds, thereby determining the spatial distribution of rainfall. The rainfall pattern differs among the cardamom-growing regions located in Kerala, Karnataka, and Tamil Nadu (Nair et al., 1991). The most important factors that have contributed to the increase in cardamom productivity are the cultivation of high-yielding varieties and improved crop management. However, cardamom export from India has plummeted during the same period. In 1985–1986, cardamom export was 3272t; in 1989–1990, it hit rock bottom at 173t—a steep decrease to 5.3% of the 1989–1990 level. In one decade, from 1985–1986 to 1994–1995, export earnings came down from Indian rupees (Rs) 53.46 crores to just Rs 7.6 crores—that is, from US$11.9 million to US$1.8 million—a dramatic decrease of 85%.

2.1.2.1 Cardamom Cultivation in Other Parts of the World

The cultivation of cardamom is getting to be popular in certain parts of PNG. Cardamom grows there in virgin forestlands, and its cultivation is exclusively with private estate owners. The productivity of these estates is very high, with yield levels of 2000–2500 kg ha^{-1} (Krishna, 1997). Total production was about 313 million tons in 1985; by 1993, it had declined to about 54 million tons. Today, PNG production hovers around 68–70 million tons. In Tanzania, the crop was introduced in the beginning of the twentieth century by German immigrants and is being grown in certain parts of the country, such as Amani and East Usambaras (Lawrence, 1978). Production was as high as 760 million tons in 1973–1974, but declined to about 127 million tons in 1984–1985, a level that continues to today. Sri Lanka is another small producer of the crop, contributing about 75 million tons to world production annually.

India was the leader in world cardamom production until the early 1980s, when Guatemala came into the picture. From thereon, India's production plummeted while that of Guatemala escalated. By the turn of the last century, whereas India's production came down by as much as 54% since the beginning of 1970s, that of Guatemala increased by as much as 199%. Guatemala is the major rival to India in cardamom production. A lot of Guatemalan cardamom is smuggled into India through the Nepalese border; this has resulted in a crash in Indian cardamom prices. As of now, nearly 90% of the global cardamom trade is controlled by Guatemala (Table 2.2).

Among the many factors that adversely affected India's cardamom production are the following:

1. Continuous drought, which lasts nearly half the year, combined with indiscriminate deforestation; together, these two factors have led to dramatic changes in the ecology of the cardamom habitat. Deforestation is the major cause of the dwindling of cardamom plantations.
2. Disease and infestation by insect pests.
3. Poor crop management. For example, cardamom nutrition in India is still rooted in "textbook knowledge." Cardamom is a heavy feeder on potassium, and Indian agronomists and soil scientists have not kept abreast of advancements in crop nutrition. (The relevance of "The Nutrient Buffer Power Concept," especially with regard to potassium nutrition of cardamom, will be discussed in later sections of this chapter.)

Table 2.2 Cardamom Scenario in Guatemala

Year	Area (ha)	Production (mt)	Productivity ($kg\,ha^{-1}$)	Export (mt)
1985	32,336	7348.32	90.89	6173.50
1986	38,333	8845.20	92.33	7978.82
1987	41,418	10,591.56	102.29	11,489.69
1988	42,656	10,432.80	97.83	11,303.71
1989	43,000	11,340.00	105.49	11,076.91
1990	43,000	11,340.00	105.49	11,113.20
1991	43,000	12,201.84	113.51	13,163.47
1992	43,000	12,474.00	116.04	13,240.58
1993	47,472	12,927.60	114.57	14,442.62
1994	45,133	14,969.80	126.13	13,213.37
1995	47,472	15,603.84	131.48	13,920.98
1996	47,472	16,329.60	137.59	21,255.70
1997	119,540	16,692.48	139.64	14,020.78
Total	11,576.70			12,491.79

4. Despite the aforesaid limiting factors, cardamom production in India has increased through enhanced productivity arising from the evolution of high-yielding clones, somewhat better crop management, and increased awareness of the importance of phytosanitary measures, especially control of diseases and pests. Between 1988–1989 and 1989–1990, cardamom export from India was 787 and 180 million tons, respectively. Exports crossed the 500 million ton mark in 1991–1992 (544 million tons precisely) and touched the 550 million ton mark in 1999–2000. The unit price of cardamom increased from about Rs 125 (US\$3 kg^{-1}) in 1987–1988 to about Rs 395 (US\$9) in 1992–1993. In 1996–1997, the unit price was about Rs 384 (US\$8.9). The current unit price hovers around Rs 450 kg^{-1}, which is equivalent to about US\$10.5. The edge that Guatemala has over India is a lower cost of production. This is the reason Guatemala edges out India in world cardamom trade. India has an extensive domestic market for cardamom. Annual consumption is around 7000 million tons, and a survey indicates that it could be as high as 7300 million tons. The total value of this market is close to Rs 2200 million, which is more than US\$50 million. This is, indeed, a large internal market. In addition, apart from individual and household consumption, cardamom in India has an extensive industrial consumptive base. Cardamom-flavored biscuits, tea, and milk are some end uses for cardamom in the culinary sector, and the spice is used in medicines of herbal origin, in food mixes, and in the ubiquitous "pan masala" (the pervasive Indian "chewing gum," which leaves a pleasant flavor in the mouth). The industrial consumption of cardamom in India currently is estimated to be about 2050–2010 million tons annually. The demand in the hotel, bakery, and fast-food sector is about 1250 million tons. In the current century, total demand of cardamom in India will escalate to about 9500 million tons annually (George and John, 1998).

2.2 Cardamom Botany

Cardamom belongs to the genus *Elettaria* and species *cardamomum* (Maton). The name is derived from the root *Elettari*, which, in the popular South Indian language Tamil, means "granules of leaf." The genus consists of about seven species (Mabberley, 1987). Only *Elettaria cardamomum* (Maton) grows in India—a fact that is of economic importance. Closely related to *E. cardamomum* (Maton) is *E. ensal* (Gaertn) Abeywick. *E. major* (Thaiw.), a much larger and sturdier plant, is a native of Sri Lanka, where it is known as the Sri Lankan "wild" cardamom; its flavor and taste are inferior to those of the Indian variety. *E. longituba* (Ridl.) Holtt., a large perennial herb whose flowering panicles often grow as tall as 3 m or more, is the Malaysian variety (Holttum, 1950), whereas the native Indian variety is a low-grown one. The flowers of *E. longituba* (Ridl.) Holtt. appear singly, and the fruit is large and is not used. Seven species have been identified from Borneo (Indonesia) and have been listed by Sakai and Nagamasu (2000). The related genera are *Elettariopsis* and *Cypbostigma*, both of which occur in the Malaysia–Indonesia region.

2.2.1 Taxonomy

Cardamom belongs to the monocot family Zingiberaceae (ginger family) of the natural order Scitaminae. Genus *Elettaria* consists of seven species, distributed over India, Sri Lanka, Malaysia, and Indonesia. Among these species, only *E. cardamomum* is economically important (Holttum, 1950; Mabberley, 1987; Willis, 1967).

2.2.1.1 *Type Species:* Elettaria cardamomum *(Linn.), Maton*

Etymology

The generic epithet *Elettaria* is derived from Rheed's *Elettari*. Elathari (the modern transcription of Rheed's name) is still used for the seeds of *E. cardamomum*. (*Thari* means "granules" in the local language.) Following is the description of *Elettaria* provided by Holttum (1950):

> *Stout or fairly stout rhizome, short intervals between leaf-shoots. Leaf shoots are tall with many blade-bearing leaves, while petioles are short and inflorescences arise from rhizome close to the base of a leaf-shoot. They are long, slender, prostrate, either just at the surface of the ground or just below it (not bearing roots), protected by alternate fairly large-scale leaves, in the axils of which cicinni arise, their attachment being sometimes supra-axillary. Cincinni short, bearing a close succession of tubular bracts, each of which encloses entirely the next flower and also the next bract; the flowers in two close rows on one side of the composite axis of the shoot, all pointing in the same direction, curved and opening in succession. Calyx tubular, split about one-fourth of its length down one side, shortly three-toothed; in some species joined at the base to the corolla-tube about as long as calyx; lobes not very broad, subequal, the upper one with a concave apex. Labellum as in Amomum, with yellow median band and red stripes, sometimes so curved that it stands as a hood over the top of the flower. Staminodes none, or short and narrow.*

Filament of anther very short, broad. Anther longer than filament, stigma small, in close contact with the distal end of the pollen sac. Fruit globose or ellipsoid, thin-walled, smooth, or with longitudinal ridges when ripe.

Following is the description provided by Burtt and Smith (1983) for E. cardamomum:

Leafy shoot nearly 4 m high, petioles 2.5 cm, lamina about 1 m × 15 cm, lanceolate, acuminate, lightly pubescent or glabrous below; ligule about 1 cm long, entire. Inflorescence normally borne separately on a prostrate, erect or semierect stalk up to 40 cm long, or more in certain cases. Bracts two to three, 0.8–1.0 cm long, lanceolate, acute glabrous, rather persistent which becomes fimbriate with age. Cincinni many flowered. Bractioles about 2.5 cm long, tubular, mucronate, glabrous. Calyx about 2 cm long, 2- or obscurely 3-lobed, lobes mucronate. Corolla tube as long as calyx. Lobes 1–1.5 cm long, rounded at the apex, the dorsal tube widens. Labellum white, streaked violet, 1.5–2.1 cm at the widest part, ovate, obscurely 3-lobed, narrowed at the base. Lateral staminodes inconspicuous, subulate. Anther sessile, about 1 cm long, parallel, connective prolonged into a short, entire crest. Ovary 2–3 mm long, glabrous. Fruit is a capsule, oblong or more or less globose. The genus has only few species—the most important being E. cardamomum and E. major (E. ensal) from South India and Sri Lanka, respectively.

E. longituba Holttum (Syn. *E. longituba*) is one of the largest species of the genus grown in Malaysia. Its flowers appear singly at long intervals, and each cincinnus contains only a few flowers. It appears that the cincinnus stops flowering as soon as fruit is formed. The fruits are large, but have no commercial value (Holttum, 1950). In their studies on Bornean Zingiberaceae, Sakai and Nagamasu (2000) described five species of *Elettaria*: *E. kapitensis, E. surculosa, E. linearicrista, E. longipilosa*, and *E. brachycalyx*.

2.2.1.2 Varieties

Based on the nature of their panicles, three varieties of cardamom are recognized (Sastri, 1952; Table 2.3). The variety Malabar is characterized by a prostrate panicle and the variety Mysore possesses an erect panicle. The third variety, Vazhukka, is considered a natural hybrid between the two, and its panicle is semierect or flexuous.

Variety Malabar

Plants are medium sized and attain a height of 2–3 m on maturity. The dorsal side of leaves may be pubescent or glabrous. Panicles are prostrate and the fruits are globose (oblong shaped). This variety grows best at elevations of 600–1200 m amsl. It is less susceptible to an infestation of *Thrips*, a common cardamom pest. Malabar can thrive under conditions of low rainfall.

Variety Mysore

Plants are robust and grow up to 3–4 m in height. Leaves are lanceolate or oblong–lanceolate, glabrous on both sides. Panicles are erect and the capsules are ovoid, bold, and dark green in color. The capsules variety is better adapted to altitudes

Table 2.3 Varietal Description

Features	Malabar	Mysore	Vazhukka
Adaptability	Low elevation (600–1000 m amsl)	High elevation (900–1200 m amsl)	High elevation (900–1200 m amsl)
Tolerance to drought	Withstands long dry spell (4–6 months)	Needs well-distributed rainfall	Needs well-distributed rainfall
Plant stature	Dwarf (2–3 m)	Tall (3–5 m)	Tall (3–5 m)
Leaf	Short petiole	Long petiole	Long petiole
Panicle	Prostrate	Erect	Semierect
Bearing nature	Early, short span of flowering	Late, long span of flowering	Late, long span of flowering
Capsule color	Pale, golden yellow	Green	Green

Source: Sudarsan et al. (1991).
Note: amsl, above mean sea level.

ranging from 900 to 1200 m amsl and thrives well under an assured, well-distributed rainfall pattern.

Variety Vazhukka

This is a natural hybrid between variety Malabar and variety Mysore and exhibits characteristics that are intermediate between both of these varieties. Plants are robust, like those of variety Mysore. Leaves are deep green, oblong–lanceolate or ovate; panicles are semierect (flexous) in nature; and capsules are bold, globose, or ovoid in shape.

Two more varieties—variety Mysorensis and variety Laxiflora—have recognizable morphological characteristics.

Variety Mysorensis

A robust, tall plant that possesses either glabrous or pubescent leaves. This variety has flexous panicles. The flowers are produced in short racemes. The capsules are bold and distinctly three angled.

Variety Laxiflora

Comparatively less robust than, nor as tall as, variety Mysorensis. Leaves are glabrous with short petioles. This variety has flexuous, lax decumbent panicles. The flowers are produced in 4–40 short lax racemes. The capsules are variable, oblong–oblong fusiform.

In India, a number of other cultivars of cardamom are also recognized. In general, they can be considered as ecotypes of var. Mysore, var. Malabar, or var. Vazhukka. Most common among them are Bijapur, Kannielam, Makaraelam, Munjarabad, Nadan, and Thara.

The Sri Lankan Wild Cardamom (E. ensal Abheywickrama)

The botanical identity of both the Sri Lankan wild cardamom and the Indian varieties just described is shrouded in much confusion. Cardamom varieties have been named differently by various authors as follows:

E. cardamomum var. minus
E. cardamomum var. miniscula
E. cardamomum var. major
E. cardamomum var. majus
E. cardamomum var. minor

Ridley (1912), who set forth one of the earliest descriptions of cardamom, gives the following details: "There are two forms of varieties of the plant, viz., var. minus with narrower and less firm leaves and globose fruits from 0.5–0.1 in. long, grayish yellow or buff in color. This is confined to South India. Var. majus with shorter stems, broader leaves and oblong fruit, 1- to 2-in. long and rather narrower than the Malabar fruit, distinctly three sided, often arched and dark grayish brown when dry, the seeds larger and more numerous and less aromatic. This is the Ceylon cardamom and is peculiar to that country."

In his notes on cardamom cultivation in Ceylon, Owen (1901) mentions three varieties, which he calls the indigenous Ceylon, the Malabar, and the Mysore. The first two can easily be recognized by the color of the stem. The Malabar plant is green or whitish at the base of the leafy or aerial stem, while the base of the Ceylon plant has a pink tinge. Owen also mentions that the Mysore form is robust, that its panicles are borne perpendicularly from the bulbs, and that the fruits grow in clusters of five to seven. This form does well at high altitudes. *E. cardamomum* var. major was described earlier as *E. major* Sm. (Rees Cyclop., 39, 1819), but this name did not find favor with cardamom workers. Many subsequent authors used the terminology indiscriminately and even began mentioning var. Mysore as var. major. While studying the flora of Sri Lanka (Ceylon then), Abheywickrama (1959) coined the name *E. ensal* for the Ceylon wild cardamom (from *Zingiber ensal*, under which the plant was described by Gaertner (1791). But Burtt (1980) is of the opinion that the differences are not reasons enough to differentiate this variety into a new species. However, Bernhard et al. (1971) and Rajapakse (1979) provided chemical evidence substantiating the distinct nature of Sri Lankan wild cardamom (Photos 2.1 and 2.2).

2.2.1.3 Fruit and Seed

The cardamom fruit has great commercial value. The fruit is a capsule developed from an inferior ovary. It is more or less three sided with rounded edges. The shape and size vary. In var. Malabar, the fruits are short and broadly ovoid and dried fruits are somewhat longitudinally wrinkled. In var. Mysore, the fruits are ovoid to narrowly ellipsoid or elongate and the surface is more or less smooth. The wild Sri Lankan cardamom is much larger, elongate, angular, and distinctly three sided. The dry pericarp is about 0.5 to 1 mm thick, with a rough, woody texture. The capsule has three locules, the septa is membranous, and the placentation is axile. There are five to eight seeds in each locule, and they adhere together to form a mass. The transverse section of a pericarp

Photo 2.1 Popular cardamom variety "*Vijetha.*"

Photo 2.2 Single bunch of cardamom variety "*Kodku Suvasini.*"

shows an outer and an inner epidermis consisting of small polygonal cells and a mesocarp of thin-walled, closely packed, parenchymatous cells. Vascular bundles traverse the mesocarp; each bundle consists of a few xylem vessels, phloem, and a sclerenchymatous sheath partially surrounding the vascular elements. Many resin canal cells (oil cells) are found distributed in the mesocarp. The xylem vessels have spiral thickening. Some of the cells contain prismatic needle-shaped calcium oxalate crystals. Externally, cardamom seed has an aril composed of a few layers of thin-walled, elongated cells. In fully mature seeds, these cells contain small oil globules. The testa consists of an epidermis composed of elongated fusiform cells, about 250–1000 μ long, that, in sectional view, are nearly square and about 18μ wide and 25μ high (Wallis, 1967). A layer of small, flattened parenchyma cells is found below the epidermis. Below the parenchyma cells is a layer of large rectangular cells, about 18–120μ long and 20–45μ wide and high, that are filled with globules of volatile oil. Interior to this layer of large cells are two or three layers of small parenchymatous cells. All these layers of cells together form the outer seed coat. The layers get widened around the raphe, where the vascular strand is surrounded by large oil cells. The inner seed coat comprises two layers, with the inner layer consisting of heavily thickened polygonal cells about 15–25μ in length and breadth and 30μ high (Wallis, 1967). These cells are so thickened that only a small lumen is found at the upper end in which there is a globule of silica that nearly fills the cavity. The inner layer of the inner seed coat consists of a narrow band of thin-walled cells (Wallis, 1967). The kernel consists mostly of a perisperm and a small endosperm. The perisperm consists of thin-walled parenchymatous cells that measure about 40–100μm, each filled with cardamom grains. One or two prismatic calcium oxalate crystals occur in each cell. The endosperm consists of thin-walled, closely packed, parenchymatous cells, 20–40μm in length, that contain pale yellow-colored deposits. On iodine staining, the contents turn deep blue, showing the presence of starch, and these deposits turn red with Millons reagent, indicating the presence of proteins. The endosperm surrounds a small, almost cylindrical embryo, which is made up of thin-walled cells. Parry (1969) and Trease and Evans (1983) provide brief descriptions of the histology of cardamom seeds.

2.2.1.4 The Cardamom Powder

When cardamom seeds are powdered, a grayish-colored powder with darker brown specks, which is gritty in texture and pleasant in smell and flavor, is obtained. The diagnostic character of cardamom powder is given by Jackson and Snowdon (1990):

1. With abundant starch grains filling the cells of the periplasm, the individual starch grains are very small and angular and a hilum is not visible.
2. The sclerenchymatous layer of the testa is composed of a single layer of thick-walled cells that are dark reddish brown in a mature seed; each cell contains a module of silica.
3. Abundant fragments of the epidermis of the testa are composed of layers of yellowish-brown parenchymatous cells with moderately thickened pitted walls.
4. The oil cells of the testa consist of a single layer of large polygonal rectangular cells with slightly thickened walls and containing globules of volatile oil. This layer is found associated with the epidermis and hypodermis.
5. The parenchyma of the testa is composed of several layers of small cells that are polygonal in surface view, with dark-brown contents and slightly thickened, heavily pitted walls.
6. The abundant parenchyma of the perisperm and endosperm are composed of closely packed, thin-walled cells.
7. The fragment of the arillus is composed of thin-walled cells, elongated and irregularly fusiform in surface view.
8. Calcium oxalate crystals, prismatic in shape, are found scattered in the cells of the perisperm and other cells.
9. Infrequently, groups of xylem vessels with spiral thickening are visible; the thickening is associated with thin-walled parenchyma.

The type of cardamom can be determined by counting the number of heavily thickened sclerenchymatous cells per square millimeter of a layer and using a standard figure for each type as follows:

Mysore: 3310
Alleppey Green: 3790
Malabar: 4600

The aforementioned classification was given by Wallis (1967). The Sri Lankan wild cardamom contains 3020 sclerenchymatous cells per millimeter of layer.

2.2.1.5 Growth, Flowering, and Fruit Set in Cardamom

As time passes, tillers emerge from the axils of underground stems and vegetative buds emerge from the bases of the stems throughout the year. However, most of the vegetative buds are produced between January and March. The linear growth of tillers increases with the onset of the southwest monsoon—the principal rainfall—in India, and once the rains cease, growth slows down. The linear growth pattern of tillers is similar in all cultivars. It takes almost 10 months for a vegetative bud to develop and about a year to the emergence of panicles from newly formed tillers (Sudarshan et al., 1988). An around-the-year study of the phenology of tiller and panicle production in three varieties of cardamom was carried out by Kuruvilla et al. (1992a). Panicles emerge from the swollen bases of tillers. Generally, two to three panicles emerge from the base of a tiller. Detailed investigations of panicle production and growth and the duration of flowering have been carried out by Pattanshetty and Prasad (1976) and Parameswar (1973). Vegetative shoots acquire maturity in about 10–12 months, when they can produce reproductive buds, and the newly emerging panicles take a period of 7–8 months to complete their growth. With the onset of the monsoon, flowering in cardamom commences. The flowering pattern depends on the region's agroclimatic characteristics and the cultivars in question. Flowers appear on the panicles after 4 months, and flowering continues during the next 6 months. The panicles grow either erect (var. Mysore), prostrate, and parallel to the ground (var. Malabar) or in a semierect (flexous) manner (var. Vazhukka). Each inflorescence (panicle) possesses a long cane-like peduncle having nodes and internodes. Each node has a scale leaf in the axil from which flowers are borne on a modified helicoids cyme (cincinnus). Thus, the panicle is branched. Multiple branching of panicles occurs in certain cultivars. In such cases, the central peduncle branches further into secondary and tertiary branches, producing multibranched panicles. Branching can be present at the lower part of the main peduncle, at the top part alone, or throughout the peduncle. The panicles bear leafy bracts on nodes, and flowers are produced in clusters (cincinnus) in the axils of bracts. Earlier researchers on cardamom described the cluster of flowers as raceme, which is incorrect. Each cluster is a cincinnus (Holttum, 1950)—that is, a modified helicoid cyme. It takes approximately 90–110 days for the first flower to emerge in a fresh panicle, irrespective of the variety of plant. The cincinni and capsules are formed during their fourth and fifth months, respectively, after the initiation of the panicle (Kuruvilla et al., 1992b). Capsule formation increases until August (Table 2.4) and thereafter declines slowly. The flowers have the typical morphology of zingiberaceous flowers. Flower

Table 2.4 Development and Pattern of Growth of Panicles

Month	Malabar			Vazhukka			Mysore		
	A	B	C	A	B	C	A	B	C
January	3.76	–	–	6.77	–	–	4.83	–	–
February	6.34	–	–	9.20	–	–	7.23	–	–
March	7.91	–	–	12.50	–	–	11.0	–	–
April	11.85	–	–	13.02	–	–	16.60	–	–
May	17.52	13.63	–	22.76	11.06	–	23.50	14.93	–
June	18.80	14.93	18.56	23.73	14.70	4.20	23.66	17.13	14.06
July	18.40	15.40	34.30	22.13	15.73	8.16	21.30	16.13	21.86
August	17.63	15.93	39.23	18.86	11.36	8.13	23.66	16.76	29.70
September	18.40	13.76	31.23	16.46	9.83	3.96	24.36	16.63	19.93
October	18.52	14.71	3.81	25.07	14.40	2.60	24.96	16.15	11.15
November	18.41	14.43	1.52	25.20	14.40	0.67	25.11	16.07	3.74
December	18.90	14.44	0.33	25.87	14.20	0	25.44	16.07	0.48

Source: Kuruvilla et al. (1992a).

Note: A, panicle length (cm); B, number of cincinni; and C, number of capsules.

opening commences from 3.30 A.M. and continues until 7.30 A.M. Between 7.30 and 8.30 A.M., anther dehiscence takes place. Flowers invariably wither in a day. Normally, flowering is seen round the year on panicles produced during the same year, as well as on panicles produced in the previous year. Flowering is spread over a period of 6 months, from May to October, in India, when the majority of cardamom plantations still are in the southwest monsoon period. Almost 75% of the flowers are produced during June–August. The time required to reach the full-bloom stage from the flower bud initiation stage ranges from 25 to 35 days, and capsules mature in about 120 days from the full-bloom stage (Krishnamurthy et al., 1989a).

2.2.1.6 Palynology and Pollination Biology

The pollen grains of cardamom plants are rich in starch and, while shedding, are two celled. Moniliform refractive bodies can be seen in some pollen grains. The exine develops warty projections that are spinescent (Panchaksharappa, 1966). Pollen fertility is maximum at the full-bloom stage and low at the commencement and cessation of flowering periods (Parameshwar and Venugopal, 1974). Pollen grain size varies from 75 to 120μ in different varieties, and the grains lose their viability quickly. After 2h, only 6.5% are viable, and after 6h of storage, none (Krishnamurthy et al., 1989a). During early and later stages of flowering, pollen fertility tends to decline. The grains germinate in a 10% sucrose solution, and the addition of 200ppm boric acid enhances germination and tube growth (Parameshwar and Venugopal, 1974). Kuruvilla et al. (1992a) found that 15% sucrose and 150ppm boric acid favored pollen germination and tube growth at an ideal temperature of 15–20°C; 5–10ppm of coconut water, GA, cycocel, and ethrel enhanced pollen germination and tube growth significantly.

Cardamom plants have bisexual flowers. The pollen and stigma are so placed within the flower that, without external intervention, pollination cannot take place. Honeybees (*Apis cerana, Apis indica*, and *Apis dorsata*) visit cardamom plantations during the flowering stage to collect nectar and pollen, and they achieve over 90% pollination. The stigma remains receptive from 4 A.M. on the day of flowering, and the receptivity is maximum between 8 A.M. and 12 P.M. (Krishnamurthy et al. 1989a; Kuruvilla and Madhusoodanan, 1988). Peak pollen activity occurs around noon (Belavadi et al., 1998; Parvathi et al., 1993) and coincides with peak pollination. A detailed study of pollination in cardamom in PNG has been carried out by Stone and Willmer (1989). There, the most common foragers are *Apis mellifera* and, to some extent, *Apis sapiens*. Over time, *Apis mellifera* was seen to show changes in pollen-foraging activity. Foraging commences around 7 A.M. and peaks at around 10 A.M. By 12.30 P.M., pollen activity declines substantially, and by that time the majority of stigmas get pollinated.

Interesting observations on flower structure and pollination by honeybees have been made by Belavadi et al. (1997). In cardamom flowers, nectar is present in the corolla tube, which is 23 ± 2.08mm long (a 21.48- to 30.4-mm range) and through which the style passes. The honeybees (*Apis cerana* and *Apis dorsata*) drew nectar up to 11.45 ± 2.65 and 11.65 ± 1.85mm, respectively, despite their short tongue lengths (14.5 and 5.5mm, respectively). Controlled experiments using capillary tubes of similar dimensions showed that the depth of feeding by the two bee species corresponded to their tongue length when there was no style. When a style was introduced, the depth of feeding increased with increase in style thickness. The presence of a style inside the corolla tube helped bees to draw more nectar from the cardamom flowers. The mean number of flowers per bush that open per day is 34.5. The mean proportion of flowers per bush visited by each *Apis mellifera* and is 25%, independent of the number of flowers present on a plant and independent of the time of the day. Hence, the mean number of flowers visited is only 8.6. Pollen production per flower is reported to be 1.3 ± 0.2 mg, and this quantity gets diminished to 0.6 ± 0.2 mg after the visit of a bee, indicating that during the first foraging about 50% of the pollen is removed. Cardamom nectar contains 55–100 mmol liter^{-1} of glucose and is neutral in reaction. The amino acid concentration at 8 A.M. is 3 mM. Over time, nectar volume varies greatly. The initial volume at dawn is about 1.6 μl and by 11 A.M. this increases to about 209 μl. The increase is due to the active secretion by the nectaries at the base of the corolla tube. Nectar volume drops rapidly following foraging by a bee (Stone and Willmer, 1989). In one location in PNG, the number of visits per day by *Apis mellifera* to each flower was 31; in another location, visits per day averaged only 10.3. In the former area, fruit set was much higher. Bee-pollinated fruits were found to contain 11 seeds per capsule, on average (Chandran et al., 1983), in South India, whereas they contained 13.8 seeds per capsule, on average, in PNG. Belavadi et al. (1993, 1997, 1998), Parvathi et al. (1993), and Belavadi and Parvathi (1998) have carried out detailed studies on pollination ecology and biology of cardamom in a cardamom-cropping system in Karnataka in South India. The pollination activity there starts around 7.30 A.M. and continues until 6.30 P.M., peaking between 11 A.M. and 1 P.M. The bees appear on cardamom clumps when the temperature is around 21 °C. Individual foragers of *Avis cerana* made four to seven trips to a single patch of flowers in a day, and the number of flowers visited on each successive trip progressively increased. In a day, individual foragers visited 157–514 flowers. A flower is visited as many as 120 times on a clear, sunny day; 57 times on a cloudy, rainy day; and, on average, 20 times a day. The mean number of flowers visited by a bee at a given clump is 12.32 when mean number of flowers per clump is 30. The number of honeybee colonies required for effective pollination in cardamom has also been calculated by the aforesaid workers. For 3000 plants per hectare planted 1.8 m apart, there will be approximately 60,000 flowers per day ready for effective pollination. On the basis of the pollinator activity, a minimum of three colonies per hectare are needed, assuming that a colony will have about 5000 foragers. An isolation distance of 15 m for seed production has been suggested for seed production (Belavadi et al., 1993).

2.2.1.7 *Fruit Setting*

When ripe, cardamom fruit is globose or ellipsoid, thin walled, and smooth or with longitudinal ridges. Fruit shapes indicate varietal variations. The fruit is green colored and turns golden yellow on ripening. The seeds are white when unripe, turn brown on aging, and become black on maturity; their numbers per capsule vary between 10 and 20, depending on genotypes. A thin mucilaginous membrane (aril) covers the seeds. The extent of fruit set is highest when the atmospheric humidity is very high in the cardamom region; fruit set is scanty in summer months, even when the crop is irrigated. This difference clearly indicates that the crop thrives best in a cool, overcast climate. In general, the percentage of fruit set is high among young plants, and when plants overshoot the economic life span of the spice, fruit set declines to 50% or even less.

2.2.1.8 Physiology of Cardamom

Photosynthesis

Among the growth parameters, total leaf area (TLA) is closely associated with photosynthesis and the production of dry matter. Hence, a precise estimation of TLA and canopy density are important in estimating productivity. Korikanthimath and Rao (1993) reported a reliable method for TLA estimation based on linear measurements of intact leaves followed by appropriate regression analysis. There were varietal differences in the TLA factor. The fraction of light that the leaves absorb has a direct bearing on crop growth and canopy development. Laboratory studies on photosynthetic efficiency in cardamom (cvs. PV-19 and PR-107) indicated that efficiency was greater at low light intensities than at higher ones. A low light compensation point favors photosynthesis. Translocation pattern showed that the rhizome was the major sink, followed by the panicle and roots. Unlabeled leaves did not receive much of the labeled photosynthates from labeled leaves (Vasanthakumar et al., 1989). Kulandaivelu and Ravindran (1982) studied the photosynthetic activity of three cardamom genotypes, measured at the rate of oxygen liberated by isolated chloroplasts. Results showed a drastic reduction in photosynthetic rates in plants exposed to warm climate. As much as a 60–80% decrease in the level of total chlorophyll was noticed in all three varieties tested. The light requirement for a cardamom nursery is approximately 50% of normal (Ranjithakumari et al., 1993), and the growth and production of tillers is best at this light intensity.

Effect of Growth Regulators on Cardamom Fruit Setting

In cardamom, a high percentage of flowers is shed before they reach maturity; nearly 80% of the fruit drops (Parameshwar and Venugopal, 1974). Temperature, wind, humidity, nutritional deficiencies, physical injuries, competition for resources, soil fertility, pests and diseases, and so on affect fruit set and fruit drop (Kuttappa, 1969a).

In cardamom, growth regulators are important for proper fruit set. Table 2.5 gives the effect of NAA, GA, and 2,4–8 on fruit set and fruit weight (Krishnamurthy et al., 1989b). Significant differences are noticeable in the case of fruit set. Tissue concentration of auxins was highest 36 h after pollination at 315 $mg\,g^{-1}$ of tissue and declined further, to 80 $mg\,g^{-1}$, 30 days after pollination. The fall in auxin activity resulted in the formation of an abscission zone, producing shedding of immature capsules. The application of 40 ppm NAA or 4 ppm of 2,4–8 decreased the capsule drop and led to an increase in yield (Vasanthakumar and Mohanakumaran, 1998). Gibberellic acid (GA) at 25, 50, 100, 150, 200, 250, and 300 ppm and 2,4–8 at 2–5 ppm and 10 ppm were

Table 2.5 Growth Regulators and Fruit Set

Treatment		**Mean Fruit Set (%)**	**Fruit Weight (g)**
Control		43.20	0.80
NAA	25 ppm	60.75	0.85
	50 ppm	61.62	0.83
	75 ppm	67.04	0.71
GA	25 ppm	37.72	0.78
	50 ppm	48.73	0.83
	75 ppm	54.11	0.90
2,4-D	2.5 ppm	69.80	0.80
	5.0 ppm	65.12	0.79
	7.5 ppm	47.56	0.82
LSD ($p = 0.05$)		10.69	N.S.
LSD ($p = 0.01$)		14.18	–

Source: Krishnamurthy et al. (1989b).
Note: LSD, least significant difference; N.S., not significant.

sprayed on cardamom plants, and the response was monitored. The plants showed increased panicle length, especially with GA at 50 ppm, and the maximum fruit set was observed at 200 ppm spray (Pillai and Santha Kumari, 1965). Indoleacetic acid (IAA) and indolebutyric acid (IBA) failed to enhance fruit set (Nair and Vijayan, 1973). Siddagangaih et al. (1993) investigated the effect of chloromequat, daminozide, ethepon, and malic hydrazide (250 ppm). Daminozide (1500 ppm), chloromequat (250 ppm), and ethephon (100 ppm) significantly enhanced tiller production and other vegetative characters when applied on 7-month-old seedlings, but had little effect on other morphological characters.

Effect of Moisture Stress on Cardamom Yield

In South India, the states of Kerala and Karnataka experience drought for about 4–5 months a year. Large-scale yield losses are observed due to drought in the Idukki, Palakkad, and Waynad districts of Kerala State, where cardamom is extensively grown. Identifying cardamom genotypes that are tolerant to moisture stress is an important prerequisite for enhancing cardamom productivity in India. At the Regional Cardamom Research Station in Mudigere, in Coorg district of Karnataka State, cardamom genotypes were screened to select clones that would be tolerant to drought. Clones differ in their susceptibility to drought. Krishnamurthy et al. (1989a) investigated the chlorophyll stability index (CSI) of three prominent varieties of cardamom—Malabar, Mysore, and Vazhukka—to see whether it would be a reliable indicator of drought tolerance. For purposes of comparison, related taxa, namely,

Hedychium flavescence and *Amomum subulatum*, were also investigated. Malabar had the highest CSI (43.14) and Mysore had the lowest (24.5); Vazhukka was intermediate (31.14). CSI is expressed as a percentage. Electrolyte leakage was highest (66.65%) in Mysore and was 69.42% in Vazhukka and 43.90% in Malabar. These results clearly indicate an inverse relationship between CSI and electrolyte leakage among the varieties, pointing to the important reason that Malabar outperforms both Mysore and Vazhukka varieties.

Among the other physiological parameters, dry matter accumulation (DMA) and harvest index (HI) are important in helping breeders in their programs for cardamom improvement. DMA during drought spells over a 3-year period has been investigated by Krishnamurthy et al. (1989a). The authors investigated a number of cardamom clones and found significant variation among them. DMA during the drought period of March–June varied from 161 g per plant to 279.5 g per plant. At the end of the drought spell (June), DMA ranged from 195 g per plant to 391 g per plant. Those clones that had the highest DMA had the maximum leaf area index, another yield-determining physiological parameter. Korikanthimath and Mulge (1998) investigated the various vegetative and physiological parameters in 12 clones planted in the "Trench System," a specialized pattern of planting in which the trenches measure 1.8 m × 0.6 m. These authors measured the dry matter content of roots, rhizome, panicles, capsules, tillers, and leaves and found that it varied significantly among the clones. Total dry matter varied from 2759 to 4853 g. Dry matter in capsules varied from 13.2 to 234.3 g. The harvest index varied from 0.06 (for the native variety) to 0.091 (for a "selection" clone). The partitioning of photosynthates within the plant is governed by genetic variability. This relationship is reflected in the DMA in capsules. These investigations help the breeders to target high productivity in cardamom.

2.2.2 Crop Improvement

Crop productivity can be enhanced in cardamom by, first, the use of high-yielding genetic materials and, second, improved crop and soil management practices. The major constraints on cardamom productivity are the lack of genotypes that produce a superior yield and the onslaught of drought and devastation by insect pests and diseases. As far as the first factor is concerned, germplasms with a high yield potential, superior capsule quality, and wide adaptability are the three criteria affecting productivity. The selection of clones that possess resistance or tolerance to major pests and diseases, as well as to drought, should be the top priority for a crop-breeding program.

2.2.2.1 Germplasm

Because cardamom is a cross-pollinated crop that is propagated mostly by seeds, the plant's natural variability is fairly high. Assembling a wide range of genetic stock, which forms the basis for further breeding or selection work, is the first step in molding new varieties for use by farmers and end users. Hence, the collection, conservation, evaluation, and exploitation of existing germplasm deserve the utmost importance

Table 2.6 Spread of Cardamom Germplasms in India

Center	Germplasm Under Cultivation	Wild and Related Taxa
Regional Research Station of the Indian Institute of Spices Research (IISR), Appangala, Coorg district, Karnataka State, India, under the administrative control of the Indian Council of Agricultural Research, New Delhi, India	314	13
Indian Cardamom Research Institute, Myladumpara, Idukki district, Kerala State, India, under the administrative control of the Ministry of Commerce, Government of India	600	12
Cardamom Research Center, Pampadumpara, Idukki district, under the administrative control of the Kerala Agricultural University, Thrissur, Kerala State, India	72	15
Regional Research Station, Mudigere, Chickmagalur district, Karnataka State	236	7

in breeding strategies. In the 1950s, in India, two surveys were conducted in cardamom-growing regions: One sought to record genetic resources and wild populations (Mayne, 1951a) and the other to understand the geographical distribution and environmental impact on cardamom (Abraham and Thulasidas, 1958). These were the first organized attempts in India to catalog the cardamom crop. Thereafter, explorations for germplasm collection were carried out by six research organizations in the country and the total number of accessions presently available at different research centers is 1350 (Madhusoodanan et al., 1998, 1999; Table 2.6). Earlier documentation was based on an old descriptor (Dandin et al., 1981), and a key for the identification of various types was formulated (Sudarshan et al., 1991). During 1994, a detailed descriptor for cardamom was brought out by the International Plant Genetic Resources Institute, in Rome, Italy. Among the germplasms collected, genotypes having marker characters include terminal panicle, narrow leaves, pink-colored tillers, compound panicles, and elongated pedicel. Asexuality, cleistogamy, and female sterility are a few of the variations that have been observed. Conservation of cardamom genetic resources under in situ conditions does not exist, although natural populations occur in protected forest areas, especially the world-famous "Silent Valley Biosphere Reserve" in Kerala State, where a sizeable population of cardamom plants exists in their natural habitat. Many organizations are now undertaking ex situ conservation of cardamom. Table 2.6 details holdings in these centers.

2.2.2.2 Genetic Variability in Cardamom

The accessions of cardamom germplasm available at the Regional Research Station, Mudigere, have been classified by Krishmurthy et al. (1989a). There are 26 distinct types, based on leaf pubescence, height and color of aerial stem, panicle type, and so on. A study to assess the variability among 210 germplasms assembled from all major cardamom-growing regions at the Cardamom Research Center, Appangala, was carried out by Padmini et al. (1999). The results of the study indicated that, in general, var. Vazhukka and var. Mysore are more robust than var. Malabar. The total number of tillers, as well as the number of bearing tillers per plant, leafy stem diameter, and the number of leaves are more in var. Vazhukka and var. Mysore than in var. Malabar. The mean number of panicles per plant is higher in var. Malabar than in the other two. Plant characters, such as panicle number, nodes per panicle, internode length, and capsule length, exhibited a high coefficient of variation. Among the Malabar accessions, percentage coefficient variation was highest for the number of panicles per plant. In var. Mysore, the characters having the highest variability were panicle per plant and internode length of panicle. In var. Vazhukka, the highest coefficient of variation was recorded for panicles per plant, followed by the number of bearing tillers per plant.

Observations on natural variations in morphological and yield parameters under the cardamom-growing situation in Idukki district of Kerala State were recorded by George et al. (1981). Highest variability was observed with regard to panicle characters (Anon., 1958, 1986a,b,c, 1987). George et al. (1981) collected 180 accessions from the wild, as well as from cardamom-growing regions of the Western Ghats of Kerala State. These researchers isolated distinct morphotypes and 12 ecotypes showing heritable adaptations. They observed that var. Mysore and var. Vazhukka attained a height of nearly 6m and were more vigorous than var. Malabar. One clone had very narrow leaves, 3cm wide. In two accessions, tillers had characteristic pink and pale-green colors. In general, each tiller had two panicles, and accessions having three or four panicles per tiller were present, especially among the Munzerabad clones. Another clone, known as "Alfred clone," produced both basal and terminal panicles. Panicle length was highly variable among accessions, ranging from 30 to 200cm, with the mean being 140cm in var. Mysore and var. Vazhukka and 80cm in var. Malabar. Some accessions produced multibranched panicles. The number of flowers or fruits varied from 4 to 36 per cincinnus. Variations were noticed in fruit shape as well. Plants having multibranched panicles or compound panicles occur in small proportions in the segregating populations of certain lines. Padmini et al. (2000) investigated the variability among different types of compound panicles, which have mostly the var. Vazhukka type of inflorescence. Among the compound panicles, the proximal-branching type is more prevalent than the distal- or entire-branching type. The contribution of such branching toward yield (weight of fresh and dry capsules) varied from 12 to 41%. Branching did not influence other yield or quality characters.

2.2.2.3 Genetic Upgradation of Cardamom

Cardamom can be propagated both sexually and asexually through vegetative cuttings. Techniques such as selection, hybridization, mutation, and polyploid breeding are used as a means of genetic upgradation of the crop. Studies on certain aspects of crop improvement in cardamom have also been carried out in Sri Lanka (Melgode,

1938), Tanzania (Rijekbusch and Allen, 1971), Guatemala (Rubido, 1967), and PNG (Stone and Willmer, 1989).

2.2.2.4 Selection

There are definite breeding pathways in cardamom selection, especially with regard to clonal selection. Gopal et al. (1990, 1992) carried out extensive correlation and path analysis, which showed that the dry weight of capsules per plant was positively correlated with other polygenic characters, such as tiller height ($r = 0.88$), productive tillers per plant ($r = 0.78$), panicles per plant ($r = 0.998$), capsules per panicle ($r = 0.998$), fresh weight of capsules per plant ($r = 0.99$), length of panicles ($r = 0.87$), nodes per panicle ($r = 0.96$), and internodal length of panicle ($r = 0.63$). Number of panicles per plant, fresh weight of capsules per plant, nodes per panicle, and internodal length of the panicle showed statistically significant positive direct effects on yield. Panicles per plant showed the maximum direct effect on yield, followed by fresh weight of capsules per plant. The authors concluded that panicles per plant, fresh weight of capsules per plant, nodes per panicle, and internodal length of panicle were useful characters for the improvement of cardamom yield. Patel et al. (1997, 1998) have suggested the use of traits such as panicles per bearing tiller, panicles per clump, recovery ratio, and capsules per panicle as important criteria in the selection for cardamom yield. In a study using 12 genotypes, these researchers found that yield per clump had a significant positive correlation with capsules per panicle ($r = 0.967$), cincinni per panicle ($r = 0.645$), tillers per clump ($r = 0.639$), panicle length ($r = 0.559$), panicles per clump ($r = 0.537$), bearing tillers per clump ($r = 0.340$), vegetative buds per clump ($r = 0.309$), and recovery ratio ($r = 0.224$). A negative correlation was observed between fresh yield per clump and dry capsules per kilogram ($r = -0.486$). Patel et al. concluded that capsules and cincinni per panicle, bearing tillers and panicles per clump, panicle length, and vegetative buds per clump are the significant attributes that are primarily responsible for high yield of cardamom, so selection for yield improvement should be based on these attributes. A systematic evaluation of germplasm accessions in India during the 1980s resulted in the identification and release of some elite clonal selections (Table 2.7). The initial collection for desirable traits was made from planters' fields, as well as from wild habitats, on the basis of certain parameters. In order to isolate elite clones, germplasm collections were subjected to initial evaluation trials, followed by comparative yield trials and multilocation tests, in various agroecological situations. Only a few studies were conducted for the selection of seedlings having precocity in bearing, and the results indicate that this phenomenon has no positive bearing on yield (Madhusoodanan and Radhakrishnan, 1996). Selection for drought tolerance has also been attempted; initial results indicate that those genotypes that are tolerant to drought are low yielders.

Selection for Biotic Stress Tolerance

At the Cardamom Research Center, Appangala, Karnataka State, under the administrative control of the Indian Institute of Spices Research, Calicut, Kerala State, a survey

Table 2.7 Elite Selections of Cardamom

Selection	Plant Type	Average Yield (kg ha^{-1})	Oil (%)	Distinguishing Features
Mudigere-1	Malabar	275	8	Compact plant suited for high-density planting. Tolerant to hairy caterpillars and white grubs. Short panicle, oval and bold pale-green capsules, leaves pubescent
Mudigere-2	Malabar	476	8	Round capsule, suited to hilly zones of Karnataka State
PV-1	Malabar	260	6.8	Early maturing variety with slightly ribbed, light-green capsules. Short panicle, close cincinni, elongated capsules
CCS-1 (IISR "Suvasini")	Malabar	409	8.7	Early maturing variety suitable for high-density planting. Long panicle, oblong, bold, parrot-green capsules
ICRI-1	Malabar	656	8.3	Early maturing, profusely flowering variety. Long panicle, globose, extrabold, dark-green capsules
ICRI-2	Mysore	766	9.0	Performs well in irrigated conditions. Suitable for high-altitude planting. Long panicle, oblong, bold, parrot-green capsules
ICRI-3	Malabar	599	6.6	Early maturing, nonpubescent leaves, oblong, bold, parrot-green capsules
BRI (IISR "Avinash")	Malabar	960	6.7	Selection from hot-spot areas. Resistant to rhizome rot disease
TDK-4	Malabar	456	6.4	Suitable for low-rainfall area, highly adapted to lower Pulney hills of Tamil Nadu State
PV-2	Malabar	982	10.45	Selection from OP seedling progeny of PV-1. Early maturing, bold capsules, high dry recovery, tolerant to *Thrips*
IISR "Vijeta" (NKE-12)*	Malabar	643	7.9	Suited to highly shaded mosaic disease areas, Oblong capsule, virus resistant
Farmer's Selection "Njallani Green Gold"**	Vazhukka	–	–	Clonal selection by a cardamom grower in Idukki district of Kerala State. High yielding, bold capsules, more than 70% of cured cardamom above 7 mm in size

*Yield up to 1600 kg ha^{-1} in farmer's fields.
**Yield up to 2475 kg ha^{-1} in farmers' fields.

was conducted to collect disease escapes from the hot-spot areas of the devastating katte disease of cardamom. Collections of natural katte escape (NKE) lines from such surveys were field evaluated and subjected to artificial inoculation through the use of insect vectors. Then, the plants that did not get infected, even after continued screening, were field evaluated again in a hot-spot area. Some of the selections were found to be good yielders with good quality, comparable to those already released for cultivation. One specific collection, RR-1, gave the highest yield and was found to be resistant to the rhizome rot caused by *Phytopthora* sp. The yield is comparable to that from NKE lines. Further studies were undertaken, and improvement of the NKE lines was accomplished with the diallel crossing technique (IISR, 1997–98).

Selection for Drought Tolerance

For drought tolerance studies on cardamom selections, parameters such as relative water content, membrane leakage, stomatal resistance, and specific leaf weight have been used as important criteria among cultivars (IISR, 1997–98).

Hybridization

The popular cardamom variety Vazhukka may have originated as a natural cross between var. Malabar and var. Mysore. Since cardamom can be propagated by both sexual and vegetative methods, hybridization is a useful tool for crop improvement. Because only one species of cardamom occurs in India, crossing in cardamom is confined to intraspecific levels. On account of its perennial, cross-pollinated, heterozygous nature, the conventional methods for evolving homozygous lines in cardamom are time consuming.

Both intergeneric and intervarietal hybridizations have been carried out by various cardamom researchers. The former method was tried with the objective of transferring disease resistance. Such attempts have not borne encouraging results, except in the case of fruit set in a cross with *Alpinia neutans* (Parameswar, 1977). All other intergeneric crosses involving *Amomum, Alpinia, Hedychium*, and *Aframomum* were sterile (Krishnamurthy et al., 1989a; Madhusoodanan et al., 1990). Intervarietal and intercultivar hybridizations have been carried out to produce high-yielding heterotic recombinants. A diallel cross involving six related types having characters of early bearing, bold capsule, high yield, long panicle, leaf rot resistance, and multiple branching was carried out, and 30 cross combinations were made. All the resulting hybrids were more vigorous than the parental lines (Krishnamurthy et al., 1989a). In another study, intervarietal hybridization was carried out with different varieties of cardamom. This study resulted in cross combinations of 56 F1 hybrids. Evaluation of these hybrids led to the isolation of a few high-yielding heterotic recombinants having an average yield of 470–610 kg ha^{-1} under moderate management (Madhusoodanan et al., 1998, 1999).

Selection among Polycross Progenies

The impact of selection in a polycross progeny population was investigated by Chandrappa et al. (1998). Promising clonal selections of var. Malabar, including the prominent variety Mudigere-1, were grown in isolation, and open-pollinated varieties of these selections were evaluated. In 34% of the progenies, average yield was found to be significantly higher than the average yield of the control variety, Mudigere-1. This increase in yield varied from 1 to 149%, and promising clones were 691, 692, D11, and D19. The researchers found that improvements in cardamom yield could be achieved more effectively through a polycross breeding program. Compared with the highest yield per plant of 1663 g, expected in the progenies of Mudigere-1, one selection from polycross seeds of Mudigere-1 yielded 2360 g per plant, which is 44% higher, and another progeny from the line D 237 gave 2670 g per plant yield, a 60% higher yield. The researchers also found, on the basis of the polycross progeny test, 38% of the clones were poor in combining ability and could be rejected. They further suggested that lines with better combining ability, such as Mudigere-1, C1-691, C1-692, D-11, D-18, D-19, D-186, D-535, and D-730, had 46–149% higher yields compared with the mean yields of the polycross progeny, and the checks could be used to establish a restricted polycross nursery to isolate higher yielding selections. The authors had investigated a population of more than 3000 plants.

Polyploidy

Polyploids ($2n = 4x = 96$) were successfully induced in cardamom by treating the sprouting seeds with 0.5% aqueous solution of colchicines. The polyploid lines exhibited gigantism. Still, an increased layer of epidermal cells, a thicker cuticle, and a higher wax coating on the leaves found in the induced polyploid lines are characters generally associated with drought tolerance in nature. The meiotic behavior of induced polyploids was almost normal, and they had reasonably good fertility. In all yield characters, the tetraploids were reported to be inferior to the diploids (Anon., 1986a).

Mutation Breeding

Attempts to induce desirable mutants by using physical mutagens, namely, X-rays and γ-rays (Co60 source), and chemical mutagens (ethyl methane sulfonate and maleic hydrazide) have been carried out. Of the large number of selfed and open-pollinated progenies of M1 plants that did not get infected after repeated cycles of inoculations with katte virus vector, 12 plants were selected as tolerant to the disease (Bavappa, 1986). There are reports on sterility (Pillai and Santha Kumari, 1965) and the absence of macromutation in M1 generation and its progeny (Anon., 1987). No desirable mutant could so far be developed in cardamom.

Biotechnology

1. Micropropagation

  a. The technique of micropropagation, which eliminates systemic pathogens such as viruses, offers the best hope for rapid vegetative propagation of elite clones or varieties (Table 2.8). Replanting of senile, seedling-raised plantations with selected high-yielding clones multiplied through micropropagation can give a five- to sixfold increase in the current average productivity of cardamom. Micropropagation can be used for the following applications (Bajaj et al., 1988):

  i. Increase in the propagation rate of plants

  ii. Availability of plants throughout the seasons round the year

Table 2.8 Tissue Culture in Cardamom

Explant	Media Composition	In vitro Response	References
Vegetative bud	MS + 1 mg liter^{-1} BA 0.5 mg liter^{-1} NAA	Multiple shoots, *in vitro* rooting	Nirmal Babu et al. (1997a)
	MS + 0.5 mg liter^{-1} BA 0.5 mg liter^{-1} kinetin, 2 mg liter^{-1} biotin, 0.2 mg liter^{-1} calcium pantothenate, 5% coconut milk	Multiple shoots, *in vitro* rooting	Nadgauda et al. (1983)
Rhizome of TC plants	MS + 1 mg liter^{-1}1 2,4–8 0.1 mg liter^{-1} NAA, 7 mg liter^{-1} BA, 0.5 mg liter^{-1} kinetin	Callus	Lukose et al. (1993)
Immature panicles	MS + 0.5 mg liter^{-1} NAA 0.5 mg liter^{-1} kinetin, 1 mg liter^{-1} BA, 0.1 mg liter^{-1} calcium pantothenate, 0.1 mg liter^{-1} folic acid, 10% coconut milk	Conversion of floral primordium into vegetative buds	Kumar et al. (1985)
Callus derived from vegetative buds	MS + 10% coconut milk 2–5 mg liter^{-1} BA, MS + 1 mg liter^{-1} 2,4-D, 0.1 mg liter^{-1} NAA, 1 mg liter^{-1} BA, 0.5 mg liter^{-1} kinetin	Regeneration of plantlets organogenesis and regeneration of plantlets	Rao et al. (1982); Lukose et al. (1993)

iii. Protection against pests and diseases under controlled conditions
iv. Production of uniform clones
v. Production of uniform secondary metabolites

In cardamom, different tissue culture approaches have made use of techniques such as (1) callusing, (2) adventitious bud formation, and (3) enhanced axillary branching.

The first report of a cardamom tissue culture was published by Rao et al. (1982), who achieved regeneration of plants from callus cultures. Nadgauda et al. (1983) achieved a multiplication ratio of 1:3 when sprouted buds were cultured in MS medium supplemented with BA (0.5 mg liter^{-1}), kinetin (0.5 mg liter^{-1}), IAA (2 mg liter^{-1}), calcium pantothenate (0.1 mg liter^{-1}), biotin (0.1 mg liter^{-1}), and coconut water (5%). The plantlets were successfully rooted and grown in the field. Kumar et al. (1985) reported direct shoot formation from inflorescence primordium cultured with MS medium containing NAA, kinetin, and BAP. These researchers could get the plantlets rooted. Reghunath and Bajaj (1992) gave a detailed account of micropropagation methods in cardamom. Lukose et al. (1993) used MS medium containing 20% coconut water, 0.5 mg liter^{-1} NAA, 0.2 mg liter^{-1} IBA, 1.0 mg liter^{-1} 6-benzyladenine, and 0.2 mg liter^{-1} kinetin. The plantlets are rooted in White's basal medium containing 0.5 mg liter^{-1} NAA and are hardened in a soil–vermiculture mixture. Other reports include those of Priyadarshan and Zachariah (1986), Vatsya et al. (1987), Priyadarshan et al. (1988), Reghunath (1989), Reghunath and Gopalakrishnan (1991), Nirmal Babu et al. (1997b), and Pradip Kumar et al. (1997). Indian biotech companies, such as A.V. Thomas and Co., Ltd., and Indo American Hybrid Seeds, based mainly in South India, are involved in the commercial multiplication of cardamom.

2. Propagation by Callus Culture

Cardamom micropropagation has been discussed in detail by Reghunath and Bajaj (1992), who investigated various explants, such as shoot primordium, inflorescence primordium, immature inflorescence segments, and immature capsules, and tested serial treatments with 95% alcohol, 2–4% sodium hypochlorite, and 0.05–0.2% mercuric chloride for decontamination of explants. Both MS and SH (Schenk–Hildebrandt) media at half and full strengths were tested, along with auxins such as NAA, IAA, and 2,4-D alone and in combination. The cultures on 0.6% agar were incubated at a light intensity of 1000 lux and a 16-h photoperiod. Maximum callus formation was seen in MS medium supplemented with 4 mg liter^{-1} NAA and 1 mg liter^{-1} BAP. On subculturing in an auxin-free medium having 3 mg liter^{-1} BAP and 0.5 mg liter^{-1} kinetin, this callus started Callogenesis, with each culture producing six to nine meristemoids. Subsequently, culturing in the same medium produced shoots within 28 days. Coconut water (15%) enhanced callogenesis.

3. Propagation through Enhanced Axillary Branching Method

Seventeen media formulations using shoot primordial as explants were tested by Priyadarshan et al. (1992), who obtained the best results with MS medium fortified with IAA, BAP, kinetin, calcium pantothenate, biotin, and coconut water. Reghunath and Bajaj (1992) have outlined the method of culturing, which uses shoot and inflorescence primordial explants; the media tested were MS and SH. The SH medium was found to be better than the full- or half-MS medium, since it gave 31% more shoot dry weight. A liquid medium culture under agitation using a gyratory shaker produced 111.5% more axillary branches than explants cultured in semi-solid medium. The cultures were maintained at 23 ± 2°C, a light intensity of 3000 lux, and a 16-h photoperiod. The number of axillary branches was maximum in the medium containing 4 mg liter^{-1} of BAP and 0.5 mg liter^{-1} of NAA. Axillary branch production was enhanced by coconut water. Var. Mysore and var. Vazhukka produced more axillary branches than those produced by var. Malabar.

The following culture materials were proposed by Nirmal Babu et al. (1997a) for micropropagation:

Explants: Rhizome bits with vegetative buds

Surface sterilization: Washing in running water, followed by washing in detergent solution, followed by washing in 0.1% $HgCl_2$

Incubation: 22 ± 2°C, 14-h photoperiod, 3000 lux

Medium: a. MS + 1 mg $liter^{-1}$ NAA for rooting

b. MS + 1 mg $liter^{-1}$ BAP, 0.5 mg $liter^{-1}$ NAA, for multiplication and rooting in one step

In vitro rooting and hardening

The excised axillary shoots can be rooted in a semisolid medium of half-strength MS salt and 0.5% activated charcoal for 1 week, followed by subculturing in one-half MS medium containing 1.5 mg $liter^{-1}$ IBA under a light intensity of 3500 lux (Reghunath and Bajaj, 1992). Rooted shoots were transferred to MS one-half liquid medium containing only mineral salts and were then shifted to a greenhouse for hardening. For planting, vermiculite-fine sand (1:1) mixture was found to be the best, giving 92% establishment (Reghunath and Bajaj, 1992).

Callus Culturing and Somaclonal Variations

Callus regeneration protocols are important for generating somaclonal variations for future crop improvement. An efficient system for callus regeneration is essential for the production of a large number of somaclones, and such a system has been reported by Rao et al. (1982). The system was standardized at IISR (Ravindran et al., 1997). High variability was noticed in the case of morphological characters in somaclones in the culture vessel itself (Ravindran et al., 1997). The most striking morphological variant is a needle-leaf one having small needle-shaped leaves that multiply and root profusely in the same modified MS medium. But its rate of establishment in the nursery and field is reported to be low (Nirmal Babu, personal communication). Nirmal Babu and his colleagues have standardized a cell culture system for large-scale production of callus through somatic embryogenesis to enhance genetic variability. These somaclones are being tested for resistance to viruses and for other characters (Nirmal Babu, unpublished).

A graphic depiction of the tissue culture cycle in cardamom (after Ravindran et al., 1997) is shown in Figure 2.1.

Field-testing of Tissue-Cultured Plants

The earliest report on field-testing of tissue-cultured (TC) plants was that of Lukose et al. (1993), although Nadgauda et al. (1983) had considered the field establishment of TC plants of cardamom. Lukose et al. carried out two statistically laid out trials to evaluate TC plants together with suckers and seedlings. The first trial was conducted with Clone-37 and the second with the cultivar Mudigere-1. Variations observed among TC plants, suckers, and seedlings were nonsignificant for most of the vegetative characters, as shown by analyzing pooled data of 4 years. Yielding tillers, panicles per plant, green capsule yield, and cumulative yield were significantly greater in TC plants in both trials. The earlier differences observed in growth characters disappeared during later years. Sudarshan et al. (1997) reported the results of a large-scale evaluation carried out by the Indian Cardamom Research Institute. In one case, eight high-yielding micropropagated clones and their open-pollinated progenies were evaluated for performance at 56 locations in an area of 37.5 ha. Unfortunately, suckers were not included in the trial. Variability was observed in the clonal population for vegetative characters. The overall variability in TC plants was 4.5%, as against 3%,

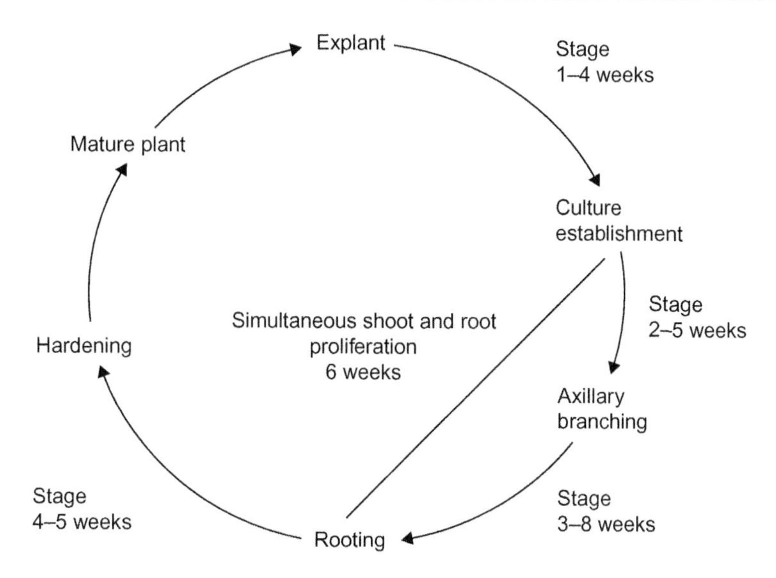

Figure 2.1 Tissue culture cycle in cardamom (Ravindran et al., 1997).

in open-pollinated seedling progenies for a given set of characters. Also, complete sterility was reported in certain clones. Microcapsules were significantly more in TC plants, accounting for a major share of variation in these plants, which was 8.4%. However, despite variations, TC plants had a 34% higher yield than open-pollinated seedlings. Among the reasons for variations were adventitious bud formation during micropropagation via axillary buds, genetic instability of adventitious meristem, and tissue culture-induced disorganization of meristems (Sudarshan et al., 1997). Chandrappa et al. (1997) tested eight TC cardamom selections against their suckers and against two local checks 92:1997a, b. Of the clones, TC 5, TC 6, and TC 7 were found promising, and they differed among themselves inasmuch as yield and other yield attributes were concerned. TC 5 was the best, recording superior values for most observations.

2.2.2.5 In Vitro *Conservation*

In vitro conservation is an alternative method for medium-term conservation. An *in vitro* gene bank will be a safe alternative in protecting the genetic resources from epidemic diseases.

Geetha et al. (1995) and Nirmal Babu et al. (1994, 1997a,b) reported the conservation of cardamom germplasm in an *in vitro* gene bank by slow growth. The two sets of researchers carried out various trials to achieve an ideal culture condition under which growth is slowed down to the minimum without affecting the physiology or genetic makeup of the plant. The slow growth is achieved by the incorporation of agents for increasing the osmotic potential of the medium, such as mannitol. Half-strength MS without growth regulators and with 10 mg liter^{-1} each of sucrose and mannitol was found to be the best medium for *in vitro* storage of cardamom under slow growth. By using this medium in screw-capped vials, the subculture interval could be extended up to 1 year or more when incubated at 22 ± 2°C at 2500 lux for a photoperiod of 10h. Low-temperature storage at 5 and 10°C was found to be lethal for cardamom, with the cultures lasting no more than 3 weeks (Geetha et al., 1995).

Additional research in biotechnology deals with the following issues:

a. *Isolation and culture of protoplast:* Protoplasts could be isolated from mesophyll tissues collected from *in vitro*-grown plantlets, achieving a yield of $35 \times 10^5 g^{-1}$ of leaf tissue on incubation in an enzyme solution containing 0.5% macerozyme R 10, 2% Onozuka cellulase R 10, and 9% mannitol for 18–20h at 25°C in darkness (Geetha et al., 2000; IISR, 1996). The yield of protoplasts from cell suspension culture was $1.5 \times 10^5 g^{-1}$ tissue when incubated in 1% macerozyne R 10 and 2% Onozuka cellulase R 10 for 24h at 25°C with gentle shaking at 53rpm in darkness. The viability of the protoplast was 75% (mesophyll) and 40% (cell suspension). On culturing, the protoplasts developed into microcalli (Geetha et al., 2000).

b. *Cryoconservation:* Cryoconservation of cardamom seed was first attempted by Chaudhary and Chandel (1995), who tried to conserve seeds at ultralow temperatures either (1) by suspending the seeds in cryovials in the vapor phase of liquid nitrogen (−150°C) by slow freezing or (2) by direct immersion in liquid nitrogen (−196°C) by fast freezing. The frozen seeds were found to possess 7.7–14.3% moisture content, which meant that they could be successfully cryopreserved. The seeds also showed more than 80% germination when tested after 1 year of storage in the vapor phase of liquid nitrogen at 150°C.

c. *Synthetic seeds:* The first report of synthetic seed production by encapsulation of shoot tips, encapsulated shoot tips of cardamom variety Malabar, isolated from multiple shoots and encapsulated in 3% w/w sodium alginate, with different gel matrices, and subsequently cultured in MS medium. Sajina et al. (1997b) reported the standardization of synthetic seed production in many species, including cardamom. Synthetic seeds have many advantages over those germinated by the normal micropropagation methods. Synthetic production is an ideal system for conservation and exchange.

2.2.2.6 Conclusions

Despite the fact that cardamom is a native of South India and has been used for many centuries, there still exist many gaps in our true understanding of this plant. Cardamom botany has been researched only superficially. The plant's developmental morphology and physiology have been neglected, and its production physiology merits a thorough investigation. Information on the origin of even a single species of cardamom and on interrelationships among the various species is practically nonexistent. Questions, such as how far Indian species are related to Sri Lankan or Malaysian species, are not at all answered. All of these issues merit further studies. In the area of crop improvement, emphasis must be placed on developing lines that are resistant to drought, a major constraint on yield. Heterosis has not been exploited, and to do so, genetically homozygous lines need to be evolved. Attempts to develop a protocol for anther culture and the production of haploids and dihaploids in cardamom are underway at IISR, Calicut. One of the most pressing areas for research is breeding for disease resistance, especially against the devastating viral diseases. Biotechnology can contribute much in this area. Yet another area of considerable importance is the molecular characterization of cardamom germplasm. Concerted efforts are required in this area. The alleviation of production constraints through conventional breeding or through molecular approaches will go a long way toward enhancing and sustaining productivity.

2.3 Cardamom Chemistry

There are three forms in which cardamom is used for flavoring: whole, decorticated seeds, and fully ground into powder. Cardamom is distilled for essential oils and solvent extracted for oleoresin. In international trade, whole cardamom is generally the item of commerce. Trade in decorticated form is small, while that in powdered form is practically negligible. The aroma and flavor of cardamom are obtained from the essential oils. As early as 1908, there were reports that cardamom contained terpinene, sabinene, limonene, 1,8-cineole, α-terpineol, α-terpinyl acetate, terpinen-4yl formate, and acetate and terpinen-4-ol (Guenther, 1975). The characteristic odor and flavor of cardamom are determined by the relative composition of the ingredients of the volatile oil (Tables 2.9 and 2.10).

Table 2.9 The Principal Components of Cardamom Volatile Oil

Component	Total Oil (%)
α-Pinene	1.5
β-Pinene	0.2
Sabinene	2.8
Myrcene	1.6
α-Phellandrene	0.2
Limonene	11.6
1,8-Cineole	36.3
Terpinene	0.7
Cymene	0.1
Terpinolene	0.5
Linalool	3.0
Linalyl acetate	2.5
Terpinen-4-ol	0.9
α-Terpineol	2.6
α-Terpinyl acetate	31.3
Citronellol	0.3
Nerol	0.5
Geraniol	0.5
Methyl eugenol	0.2
Trans-nerolidol	2.7

Table 2.10 Specification of Volatile Oil

Definition, source	Volatile oil distilled from seeds of *Elettaria cadamomum* (Linn.) Maton. Family: Zingiberaceae; cardamom grown in South India, Sri Lanka, Guatemala, Indonesia, Thailand, and South China
Physical and chemical constraints	Appearance: Colorless to very pale-yellow liquid. Odor and taste: aromatic, penetrating, somewhat camphoraceous odor, persistently pungent, strongly aromatic taste. Specific gravity: 0.917–0.947 at 25 °C; optical rotation: +22° to +44°; refractive index: 1.463–1.466 at 20 °C
Descriptive characteristics	Solubility: 70% alcohol in five volumes, occasional opalescence Benzyl alcohol: in all proportions Diethyl phthalate: in all proportions Fixed oil: in all proportions Glycerin: insoluble Mineral oil: soluble with opalescence Propylene glycol: insoluble Stability: unstable in the presence of strong alkali and strong acids; relatively stable to weak organic acids; affected by light
Containers and storage	Glass, aluminum, or suitably lined containers, filled full or tightly closed and stored in a cool place, protected from light

The dried fruit of cardamom contains steam-volatile oil, fixed fatty oil, pigments, proteins, cellulose, pentosans, sugars, starch, silica, calcium oxalate, and minerals. The major constituent of the seed is starch, up to 50%, while the crude fiber constitutes up to 31% of the fruit husk. The constituents of the spice differ among varieties and with variations in environmental conditions of growth, harvesting, drying procedures, and subsequent duration, as well as conditions of storage. The main factor that determines the quality of cardamom is the content and composition of volatile oil, which determine the fruit's flavor and aroma. Fruit color does not affect intrinsic organoleptic characteristics. However, a faded fruit color generally indicates a product stored for a longer period and, possibly, deterioration in the organoleptic characteristics through evaporation of the volatile oil (Purseglove et al., 1981).

Cardamom oil is produced commercially by steam distillation of powdered fruits. The yield and the organoleptic properties of the essential oil so obtained are dependent on many factors. Fruits from recent harvests yield more oil than oil obtained from fruits stored for a long period. To obtain full recovery of essential oil, at least 4-h extraction is essential. Industrial production of cardamom oleoresin is carried out on a relatively smaller scale. Solvent extraction yields about 10% oleoresin, and the content depends on the solvent and raw material used. Cardamom oleoresin contains about 52–58% volatile oil (Purseglove et al., 1981). Oleroresin is used to flavor the fruit and is normally dispersed in salt, flour, rusk, or dextrose before use.

The principal components of cardamom oil are given in Table 2.9. The volatile oil is extracted from the seeds, and the husks hardly give 0.2% oil. Even though the public perception about good-quality cardamom is that it comes from the greenish seed capsule, the appearance of the capsule has but little to do with the recovery of volatile oil (Sarath Kumara et al., 1985). The husk provides good protection and prevents the seeds from losing oil. Loss of oil from dehusked seeds is rapid: Seeds start losing oil the moment husk is removed, and the loss increases with storage time. Bleached cardamom tends to lose oil faster, as the husk becomes very brittle due to bleaching.

Oil from freshly separated seeds and oil from whole capsules (seeds and husk) are almost identical (Govindarajan et al., 1982a).

Steam distillation is being adopted for oil extraction by most commercial units in India and elsewhere. Cryogrinding using liquid nitrogen is ideal for preventing the loss of volatile oil during grinding. Supercritical extraction using liquid carbon dioxide is known to extract more oil, and the flavor is closer to that of natural cardamom. In oil extraction, the early fractions are rich in low-boiling terpenes and 1,8-cineole and the later fractions are rich in esters. Volatile oil content is highest 20–25 days before full maturity. The ratio of the two main components, 1,8-cineole and α-terpinyl acetate, determines the critical flavor of the oil. The volatile oil from var. Malabar represented by "Coorg Greens" is more "camphory" (smells close to the way camphor smells) in aroma because of the relatively higher content of 1,8-cineole. This oil is reported to be ideal for soft drinks. Early fractions during distillation are dominant in low-boiling monoterpenes and 1,8-cineole. Techniques are available to remove these fractions by fractional distillation so that the remaining oil will have more of α-terpinyl acetate, which contributes to the mildly herbaceous, sweet spicy flavor of the fruit, a flavor that is predominant in var. Mysore or the commercial grade, popularly known as the "Alleppey Green" (Govindarajan et al., 1982b). Eighteen export grades of Indian cardamom, as certified by Agmark (a prominent certification agency in India), were evaluated for their physical and chemical properties (Mathai, 1985). Grades with bigger and heavier capsules—"Alleppey Green Extra Bold" (AGEB) and "Coorg Green Extra Bold" (CGEB)—were inferior in their flavor constitution, compared with the medium capsule grade ("Alleppey Green Small," AGS). Chemical bleaching of the capsules reduced the amount of essential oil in them. Vasanthakumar et al. (1989) reported that cardamom at the black-seed stage, or "karimkai" (the Malayalam word meaning "black seed"), is ideal for consumption and essential oil extraction. Gopalakrishnan et al. (1989) reported that thrips-infested cardamom capsules contained a relatively higher percentage of 1,8-cineole. Nirmala Menon et al. (1999) extracted bound aroma compounds from fresh green cardamom; the free volatiles were isolated with ether:pentane (1:1) mixture and the bound compounds with methanol. The major compounds in the aglycone fraction were identified as 3-methyl-pentan-2-ol, linalool, and *cis*- and *trans*-isomers of nerolidol and farnesol. Noleau and Toulemonde (1987) reported the presence of 122 compounds in cardamom oil cultivated in Costa Rica. The identification of 122 compounds was the first-ever discovery for these authors.

2.3.1 Biosynthesis of Flavor Compounds

2.3.1.1 Sites of Synthesis

The accumulation or secretion of monoterpenes and sesquiterpenes is always associated with the presence of well-defined secretory structures such as oil cells, glandular trichomes, oil or resin ducts, or glandular epidermes. A common feature of these secretory structures is an extra cytoplasmic cavity in which the relatively toxic terpinoid cells and resins appear to be sequestered. This anatomic feature distinguishes the essential oil plants from others in which terpenes are produced as trace constituents that either volatilize inconspicuously or are rapidly metabolized. Several pieces of evidence indicate that the secretory structures are also primary sites of mono- and sesquiterpene biosynthesis (Francis and O'Connell, 1969).

2.3.1.2 Biological Function

The monoterpenes and sesquiterpenes traditionally have been regarded as functionless metabolic waste products. Yet, certain studies have shown that these compounds can play varied and important roles in mediating the interactions of plants with their environment. The monoterpenes 1,8-cineole and camphor have been shown to inhibit the germination and growth of competitors and thus act as allelopathic agents.

2.3.1.3 Early Biosynthetic Steps and Acyclic Precursors

All plants employ the general, well-known isoprenoid pathway in the synthesis of certain essential substances. The monoterpenes and sesquiterpenes are regarded as diverging at the C10 and C15 stages, respectively, in biosynthetic pathways. The isoprenoid pathway begins with the condensation of 3-acetyl-CoA in two steps to form hydroxymethyl-glutaryl-CoA, which is reduced to mevalonic acid, the precursor of all isoprenoids. A series of phosphorylations and decarboxylation with elimination of the C-3 oxygen function (as phosphate) yields isopentenyl pyrophosphate (IPP) (McCaskill and Croteau, 1995). The IPP is isomerized to dimethylallyl pyrophosphate (DMAPP), in turn leading to the synthesis of geranyl pyrophosphate (GPP) and farnesyl pyrophosphate (FPP).

A number of monoterpene cyclases have been investigated in detail, especially the one responsible for the synthesis of α-terpinene, γ-terpinene, and 1,8-cineole. Other cyclizations of interest are the cyclization of geranyl pyrophosphate to limonene and the cyclization of geranyl pyrophosphate to sabinene, the precursor of C3 oxygenated thujene-type monoterpenes. The biosynthesis of thujene monoterpenes (such as 3-thujene) involves the photooxidation of sabinene and also involves α-terpineol and terpinen-4-ol as intermediates (Croteau and Sood, 1985).

The pathways of cyclization of geranyl phosphate and farnesyl pyrophosphate to the corresponding monoterpenes and sesquiterpenes are not similar. The limited information available suggests that monoterpene and sesquiterpene cyclases are incapable of synthesizing larger and smaller analogs. Pinene biosynthesis has been extensively studied. Three monoterpene synthases (cyclases) catalyze the conversion of GPP. Pinene cyclase I converts FPP into bicyclic (+)-α-pinene, (+)-β-pinene, and monocyclic and acyclic olefins (Bramley, 1997). The biosynthesis of monoterpenes, limonene, and carvone proceeds from geranyl diphosphate, which is cyclized to (+)-limonene by monoterpene synthase. The (+)-limonene is then either stored in the essential oil ducts without further metabolism or converted by limonene-6-hydroxylase to (+)-*trans* carveol. The latter is oxidized by a dehydrogenase to (+)-carveone (Brouwmeester et al., 1998). Turner et al. (1999) demonstrated the localization of limonene synthase. Studies in peppermint (Gershenzon et al., 2000) suggested that monoterpene biosynthesis is regulated by genes, enzymes, and cell differentiation.

The biosynthesis of 1,8-cineole is suggested from linalyl pyrophosphate (Clark et al., 2000). Also known as eucalyptol, 1,8-cineole is a biosynthetic dead end in many systems, thereby allowing the accumulation of large quantities of this compound in many plants. Other than its presence in cardamom oil, 1,8-cineole is also found in essential oils of artemisia, basil, betel leaves, black pepper, carrot leaf, cinnamon bark, eucalyptus, and many other essential oil-yielding plants. Most of the processes of terpenoid biosynthesis are associated with cell organelles. Calcium and magnesium play important roles in the biosynthesis of sesquiterpenes (Preisig and Moreau, 1994). McCaskill and Croteau (1995) indicate that the cytoplasmic mevalonic acid pathway is blocked at HMG-CoA reductase and that the IPP utilized for both monoterpene and sesquiterpene biosynthesis is synthesized exclusively in the plastids.

2.3.2 Industrial Production

Industrially, cardamom oil is extracted by steam distillation. The distillation unit consists of a material-holding cage, a condenser and receiver for steam distillation, and adoptive conditions for obtaining oil of acceptable quality. Usually, lower grade capsules harvested after full maturity are used for steam distillation. Such capsules are first dehusked by shearing in a disk mill with wide distances between disks, and the seeds are separated by vibrating sieves. The dehusked seeds are further crushed to a coarse powder (Govindarajan et al., 1982a). The essential oil-containing cells in cardamom seeds are located in a single layer below the epidermis, and fine milling will result in a loss of volatile oil. Cryogrinding using liquid nitrogen is ideal for preventing this loss. Research on steam distillation revealed that nearly 100% of the volatile oil was recovered in about 1 h time. The composition of the fractions collected at 15 min shows that most are hydrocarbons and 1,8-cineole distilled over, while 25–35% of the important aroma-contributing esters were also recovered during this period. Further distillation for 2 h was required to recover remaining esters. Hence, a distillation duration of 2–3 h was essential to completely extract the volatile oils.

The value of cardamom as a food and beverage additive depends much on the aroma components, which can be recovered as volatile oil. The volatile oil has a spicy odor similar to that of eucalyptus oil. Oil yield ranges from 3 to 8% and varies with the variety of cardamom, maturity at harvest, commercial grade, freshness of the sample, whether the sample is green or bleached, and distillation efficiency.

2.3.2.1 History

Composition: Nigam et al. (1965) reported the detailed analysis of cardamom for the first time. The constituents were identified with the help of gas chromatography and infrared spectroscopy, using authentic reference compounds and published data. Ikeda et al. (1962) reported that 23.3% of the oil was hydrocarbons with limonene as a major component. Ikeda's group also reported the presence of methyl heptenone, linalool, linalyl acetate, β-terpineol, geraniol, nerol, neryl acetate, and nerolidol. Compounds present in commercial samples were identified and compared with those of the wild Sri Lankan cardamom oil (Richard et al., 1971). Govindarajan et al. (1982a) elaborated the range of concentration of major flavor constituents, their flavor description, and their effect on flavor use. Thin-layer chromatography, column chromatography, and, subsequently, gas chromatography were employed to separate oil constituents. Fractional distillation, infrared spectroscopy, mass spectrum, and nuclear magnetic resonance were adopted to identify the specific compounds. The major constituents identified were α-pinene, α-thujene, β-pinene, myrcene, α-terpinene, γ-terpinene, and penta-cymene. These were identified in the monoterpene hydrocarbon fraction of cardamom oil. Different commercial cardamom samples were compared for their chemical constituents in 1966 and 1967 (1978). Sayed et al. (1979) evaluated the oil percentage in different varieties of cardamom. Varieties Mysore and Vazhukka contained the maximum (8%). The percentage by weight of cardamom seeds in the capsules ranged from 68 to 75%. Percentage of cardamom seeds is positively correlated with volatile oil ($r = 0.436$) on a dry-seed basis, whereas percentage of husk is negatively correlated with volatile oil ($r = -0.436$).

Detailed investigations on the volatile oil revealed large differences in the 1,8-cineole content, as high as 41% in the oil of variety Malabar and as low as 26.5% in the oil of variety Mysore. Although the α-terpinyl contents were comparable, the linalool and linyl acetate were markedly higher in variety Mysore. The combination of lower 1,8-cineole with its harsh camphory note and higher linalyl acetate with its sweet, fruity floral odor results in the relatively pleasant mellow flavor in the variety Mysore, represented by the largest selling Indian cardamom grade, namely, Alleppey Green. Zachariah and Lukose (1992) and Zachariah et al. (1998) identified cardamom lines with relatively low cineole and high α-terpinyl acetate. An interesting observation is that lines Alleppey Green 221 and 223 gave a consistently higher oil yield (7.8%) and high α-terpinyl acetate content (55%). The performance of Alleppey Green 221 was consistent for about five seasons (Zachariah et al., 1998).

Previous gas chromatograms showed up to 31–33% peaks, and up to 23 compounds were identified, whereas the improved procedure gave higher resolution with more than 150 peaks. Not all peaks have been identified. All results, however, confirm the earlier observations that 1,8-cineole and α-terpinyl acetate are the major components in cardamom oil. Many investigators used techniques that were a combination of fractional distillation, column and gas chromatography, mass spectrometry, infrared spectroscopy, and nuclear magnetic resonance to identify the constituents in cardamom oil. Nirmala Menon et al. (1999) have investigated the volatiles of freshly harvested cardamom seeds by adsorption on Amberlite XAD-2 from which the free volatiles were isolated by elution with a pentane–ether mixture and glycosidically bound volatiles with methanol. Gas chromatographic–mass spectrometric analysis of the two fractions led to the identification of about 100 compounds. Among the free volatiles, the important ones are 1,8-cineole and α-terpinyl acetate. The less important ones are geraniol, α-terpineol, *p*-menth-8-en-2-ol, γ-terpinene, β-pinene, carvone oxide, and so on, while a large number of compounds were present in trace amounts. Among the aglycones, the important ones are 3-methylpentan-2-ol, α-terpineol, isosafrole, β-nerolidol, *trans*, *trans*-farnesol, *trans*, *cis*-farnesol, *cis*, *trans*-farnesol, T-muurolol, cubenol,

10-epi-cubenol, *cis*-linalol-oxide, tetrahydrolinalol. Sixty-eight compounds were identified in the volatile fraction and 61 in the glycosidically bound fraction.

2.3.2.2 Evaluation of Flavor Quality

The flavor quality of a specific food item results from the interaction of the chemical constituents contained in the food item with the taste perception of the person enjoying the item in question. As the second most important spice in the world, next only to black pepper, cardamom has as its most important component the volatile oil with its characteristic aroma, described as sweet, aromatic, spicy, camphory, and so on. Cardamom oil is rich in oxygenated compounds, all of which are potential aroma compounds. Capillary column chromatography and gas chromatography have identified 150 compounds in cardamom oil. While most of these compounds, which are alcohols, esters, and aldehydes, are commonly found in many spice oils, the dominance of the ether 1,8-cineole and the esters α-terpinyl and linalyl acetate in the composition renders the volatile oil contained in cardamom a unique one (Raghavan et al., 1982a). The bitterness compound present in cardamom is α-terpinyl, present to the extent of about 0.8–2.7%. Govindarajan et al. (1982a) described the esters: 1,8-cineole ratio (Table 2.11). In rare samples, slightly oxidized "terpinic"compounds, which are defective costituents, were observed at high dilution levels, but were overshadowed by total cardamom aroma at higher concentrations. Markedly camphory samples (lacking sweet, aromatic components) and samples that are high in defectives or oxidized terpinic, or that are resinous, oily, earthy, or bitter in flavor, are rated poor and unacceptable. The authors (Raghavan et al.) and (Govindarajan et al.) suggested that quality grading

Table 2.11 Ratios of Esters and Alcohol to 1,8-cineole in Cardamom Volatile Oils

Source	**Ratio**		
	α-Terpinyl Acetate	**α-Terpinyl + Linalyl Acetate**	**Esters + Linalool- + αTerpineol**
Alleppey Green (var. Mysore)	1.30	1.59	1.77
Alleppey Green (var. Mysore)	1.10	1.19	1.43
Kerala, Ceylon (var. Mysore)	0.83	0.91	1.03
Ceylon, commercial	1.21–1.77	–	–
Ceylon, from extract	1.67–2.40	–	–
Ceylon, green expressed (var. Mysore)	1.69	1.80	1.91
Ceylon, green from extract	2.40	2.64	2.83
Ceylon, green expressed (var. Mysore)	2.17	2.40	2.68
Coorg green (var. Malabar)	0.73	0.77	0.80

Source: Govindarajan et al. (1982).

of cardamom is possible by observing three major attributes: balance of profile, intensity or tenacity, and absence of defects. The desirable and defective notes of cardamom oil are described in Table 2.12. The general profile of the popular Alleppey Green is described in Table 2.13.

Analysis of a Japanese cardamom oil sample indicated the presence of some new compounds, including 1,4-cineole, *cis-p*-menth-2-en-1-ol, and *trans-p*-menth-2-en-1-ol,

Table 2.12 Flavor Profile of Cardamom Oil and Extracts

Desirable Notes	Defective Notes
Fresh cooling	Unbalanced
CAMPHORACEOUS	Sharp/Harsh
Green	Heavy
SWEET SPICY	Earthy
FLORAL	Oily (vegetable) oxidized
WOODY/BALSAMIC	Resinous
Herbal	Oxidized terpinic
Citrus	
Minty	
Husky	
Astringent, weakly	Bitter

Source: Govindarajan et al. (1982a).
Note: The descriptions in capital letters are the perceived dominant characteristics; the defectives are arranged in the order of increasing impact on flavor.

Table 2.13 Volatile Oil Profile of Cardamom

Origin	Commercially Distilled from Alleppey Green Varieties
Odor	**Initial Impact**
	Penetrating, slightly irritating
	Cineolic, cooling
	Camphoraceous, disinfectantlike, warm, spicy
	Sweet, very aromatic, pleasing
	Fruity, lemony, citruslike
	Persistence
	The oil rapidly "airs off" on being smelled, with its strip losing its freshness ; becomes herby, woody, with a marked musty "back-note"
	Dry Out
	No residual odor after 24 h

Source: Heath (1978).

all of them in extremely low amounts of 0.1–0.2%. Cardamom oil from Sri Lanka gave a high range of values for α-pinene plus sabinene (4.5–8.7%) and for linalool (3.6–6%) and a wider range for the principal components: 1,8-cineole (27–36.1%) and α-terpinyl acetate (38.5–47.9%) (Govindarajan et al., 1982a). Some compounds, such as α-thujene, sabinene, *p*-cymene, 2-undecanone, 2-tri-decanone, heptacosane, and *cis*- and *trans*-*p*-menth-2-en-1-ols, were rarely detected in cardamom samples. Components such as camphor, borneol, and citrals might modify the overall flavor quality of the spice, which is determined mainly by a combination of terpinyl and linyl acetate and cineole. Locations where the crop is grown alter the concentration of linalool, limonene, α-terpineol, and so on. The quality of flavor of a food depends on the interaction of the chemical constituents of the food with human taste buds, and the perception of taste by individuals depends on different attributes. A causal connection between physical and chemical characteristics of food and their sensory perception and judgment by human assessors has to be ascertained in order to establish a meaningful judgment of quality. According to many investigators, the ratio of 1,8-cineole to α-terpinyl acetate is a fairly good index of the purity and authenticity of cardamom volatile oil (Purseglove et al., 1981). The ratio is around 0.7–1.4. The Cardamom Research Center at Appangala, Coorg district in the state of Karnataka, India, under the administrative control of the Indian Institute of Spices Research, Calicut, Kerala State, India, in turn under the overall administrative control of the Indian Council of Agricultural Research at New Delhi, India, could collect many accessions with a flavor ratio of more than 1 from cardamom-growing areas. Both 1,8-cineole and α-terpinyl acetate, together with terpene alcohols (linalool, terpinen-4-ol, and α-terpineol), are important for the evaluation of aroma quality. The oils from variety Malabar exhibit the lowest flavor ratio, whereas those from variety Mysore have a high flavor ratio. Cardamom samples from Sri Lanka and Guatemala have higher ratios than those from other countries, indicating their superiority in flavor, similar to that of variety Mysore. The occurrence of components such as borneol and citral modifies the flavor quality. Pillai et al. (1984) conducted a comparative study of the 1,8-cineole and α-terpinyl acetate contents of cardamom oils derived from diverse sources (Table 2.14). Their investigation indicated that Guatemalan cardamom oil is marginally superior to Indian cardamom oil because of the former's higher content of α-terpinyl acetate content. The high concentration of 1,8-cineole makes the oil from PNG poor. The same investigators found a fair degree of concordance in the infrared spectra of oils, irrespective of their origin. These spectra provide a fingerprint of the oil as it projects the functional groups and partial structures that are present. The spectra also help in tracking the aging process of the oil.

Extraction methods such as cryogenic grinding (Gopalakrishnan et al., 1991) and supercritical extraction also influence the flavor profile. Such techniques can extract the trace compounds that are otherwise lost in other methods of extraction.

2.3.2.3 Cardamom Oleoresins and Extract

In a food, the total solvent extract or oleoresin is known to reflect the flavor quality more closely than the distilled volatile oil does. In the case of cardamom, oil more or less represents both flavor and taste. The stability of oleoresin depends on

Table 2.14 Percentage of 1,8-Cineole and Terpinyl Acetate in Volatile Oils of Cardamom Grown in Different Regions

Origin	Percentage of Oil	
	1,8-Cineole	**α-Terpinyl Acetate**
Guatemala I	36.40	31.80
Guatemala II	38.00	38.40
Guatemalayan Malabar tType	23.40	50.70
Guatemalayan I	39.08	40.26
Guatemalayan II	35.36	41.03
Synthite (commercial grade)	46.91	36.79
Mysore-type (Ceylon)	44.00	37.00
Malabar-type (Ceylon)	31.00	52.50
Mysore I	49.50	30.60
Mysore II	41.70	45.90
Mysore	41.00	30.00
Malabar I	28.00	45.50
Malabar II	43.50	45.10
Ceylon type	36.00	30.00
Alleppey I	38.80	33.30
Alleppey Green	26.50	34.50
Coorg Green	41.00	30.00
Mangalore I	56.10	23.20
Mangalore II	51.20	35.60
Papua New Guinea (PNG)	63.03	29.09
Cardamom oil (Indian origin)	36.30	31.30

Source: Pillai et al. (1984).

the changes that occur to the fat and terpenic compounds, which are usually susceptible to oxidative changes. Existing investigations point to the fact that there exists a clear difference in the flavor profile among cardamom varieties, which in turn is influenced by agroclimatic conditions, postharvest processing, and cultural practices.

2.3.2.4 Variability in Composition

Analysis of germplasm collections conserved at the Indian Institute of Spices Regional Research Station at Appangala, Coorg district, Karnataka State, India, under the administrative control of the Indian Council of Agricultural Research in New Delhi, indicated a distinct variability in oil content and concentration of the

Photo 2.3 Dried cardamom.

two important components of the oil: α-terpinyl acetate and 1,8-cineole. Selective breeding of the high-quality accessions that have a low 1,8-cineole content and high α-terpinyl acetate content, such as Appangala 221 (AG 221), will go a long way toward enhancing the total flavor quality of Indian cardamom varieties.

2.3.2.5 Pharmaceutical Properties of Cardamom Oil

Cardamom oil possesses both antibacterial and antifungal properties. The chemical composition, physicochemical properties, and antimicrobial activity of dried cardamom fruits used to assess the potential usefulness of cardamom oil as a preservative have been investigated by Badei et al. (1991a,b). The antimicrobial effect of the cardamom oil was tested against nine bacterial strains, one fungus, and one yeast, which together showed that the oil was as effective as 28.9% phenol. The minimal inhibitory concentration of the oil was $0.7\,mg\,ml^{-1}$, and it was concluded that cardamom oil could be used at a minimal inhibitory concentration range of $0.5–0.9\,mg\,ml^{-1}$ without any adverse effect whatsoever on flavor quality. Cardamom oil is effective as an antioxidant for cottonseed oil, as assessed by stability, peroxide number, refractive index, specific gravity, and rancid odor. The effect is enhanced by increasing the cardamom oil content in cottonseed from 100 to 5000 ppm. Organoleptic evaluation showed that the addition of up to 1000 ppm cardamom oil did not adversely affect the specific odor of cottonseed oil (Photo 2.3).

2.3.2.6 Fixed Oil of Cardamom Seeds

In addition to containing volatile oil, cardamom seeds contain fixed fatty oil. The composition of fatty oil has been investigated and found to contain mainly oleic and palmitic acids (Table 2.15). Gopalakrishnan et al. (1990) carried out investigations

Table 2.15 The Fixed Fatty Oil Composition of Cardamom Seed

Fixed Fatty Acid	Total Fixed Oil (%)
Oleic	42.5–44.2
Palmitic	28.4–38.0
Linoleic	2.2–15.3
Linolenic	5.8
Caproic	5.3
Stearic	3.2
Hexadecanoic	1.9
Caprylic	5.3
Capric	<0.1–3.8
Myristic	1.3–1.4
Arachidic	0.2–2.1
Hexadecanoic	1.9
Pentadecanoic	0.4
Lauric	0.2

Source: Verghese (1996).

based on nuclear magnetic resonance and mass spectroscopy and reported that the nonsaponifiable lipid fraction of cardamom consisted mainly of waxes and sterols. The waxes identified were *n*-alkanes (C21, C23, C25, C27, C29, C31, and C33). In the sterol fraction, β-sitosterol and γ-sitosterol were reported. Phytol and traces of eugenyl acetate were also identified in cardamom.

2.3.2.7 Conclusions

The cardamom plant is a wonderful gift of nature, and from a biochemical point of view, its volatile oil is so delicately constructed by kaleidoscopic permutations and combinations of terpenes, terpene alcohols, esters, and other compounds, that cardamom defies even precise and sophisticated analytical techniques. As of now, concocting "synthetic cardamom oil" that has sensory qualities identical to those associated with the components of the plant found in nature is well beyond human capabilities. It needs to be said so here because such an attempt has been made in the case of black pepper. The sensory analysis, regarded by food scientists as the touchstone of quality, is highly sensitive to concentrations ranging from 10^{-8} to 10^{-4} ppm. The superiority of the variety Alleppey Green is attributed to its superior sensory qualities. In toto, it produces a much better perception of flavor, which is not necessarily dependent on the relative concentration of any component. However, the natural quality is often lost during the extraction process, storage, and postharvest handling. The quality of the flavor can be enhanced by cryogrinding and supercritical fluid extraction. Through the indexing of genetic resources for flavor quality and incorporation into the breeding program, the superior quality genotypes can go a long way toward improving the overall flavor of cardamom. Chemical fingerprinting of the cardamom genotypes that are available in germplasm conservatories by means of infrared, gas chromatography, mass spectroscopy, or nuclear magnetic resonance spectral characters, as well as by sensory evaluation, is needed to pick up the really superior genotypes for flavor quality.

2.4 The Agronomy of Cardamom

2.4.1 Distribution

Cardamom cultivation is concentrated mostly in the evergreen forests of the Western Ghats in South India. Besides its cultivation in India, the crop is grown commercially in Guatemala and on a small scale in Tanzania, Sri Lanka, El Salvador, Vietnam, Laos, Cambodia, and PNG. Earlier, India accounted for 70% of world production of cardamom, but its output has now slid to 41%, while Guatemala contributes around 48%. Until the 1980s, the total area of cardamom under cultivation in India was 105,000 ha, a figure that has now come down to about 75,000 ha. The spice is cultivated principally in three southern states in India, namely, Kerala, Karnataka, and Tamil Nadu, which contribute approximately 60%, 31%, and 9%, respectively, of the nation's cardamom output. Cardamom is cultivated mostly under natural forest canopy, except in certain areas in Karnataka (North Karnataka, Chickmagalur, and Hassan districts) and Wayanad district in Kerala State, where it is often grown as a subsidiary crop in areca nut and coffee gardens. The important areas of cultivation in India are Uttar Kannada, Shimoga, Hassan, Chickmagalur, and Kodagu (Coorg) in Karnataka State, the northern and southern foothills of Nilgiri district in Tamil Nadu, and parts of Madurai, Salem, Tirunelveli, Annamalai, and Coimbatore districts, also in Tamil Nadu. In addition, Wayanad and Idukki districts in Kerala State, as well as the Nelliampathy hills of Palakkad district, are home to cardamom.

2.4.2 Climate

The optimum altitudinal range for cardamom is between 600 and 1500 m amsl (Anon., 1976, 1982). In South India, all cardamom plantations lie between 700 and 1300 m amsl and rarely go up to 1500 m amsl, where growth is poor. Cardamom cultivation is restricted to the Western Ghats, an extensive chain of hills parallel to the west coast of peninsular India. Variety Malabar, traditionally grown in Karnataka, can also grow at lower elevations of 500–700 m (Abraham and Thulasidas, 1958). At these elevations, vegetative growth is satisfactory but fruit production is poor. In Guatemala, cardamom is grown at varying altitudes, ranging from 900 to 1200 m amsl. Most of the plantations in southern India are at high altitudes, whereas in northern India the crop grows at both low and high elevations (George, 1990). Cardamom is highly sensitive to elevation, and the wrong choice of cultivar or an inappropriate elevation can severely affect growth and productivity. The crop is also highly prone to wind and drought damage; therefore, areas liable to be affected by such conditions are unsuitable (Mohenchandran 1984).

2.4.2.1 Temperature

The Guatemalan climate offers ideal conditions for good cardamom growth and productivity. Annual average temperature varies from 17 to 25°C in the southern part and 18–23.5°C in the northern part (George, 1990). In India, optimum growth and development are observed in the warm and humid conditions that exist at a temperature range of 10–35°C (Anon., 1976). The upper temperature limit will normally be around 31–35°C. On the eastern side of the Western Ghats, a combination of winds passing from the hinterlands of the east and low humidity leads to the desiccation of plants. In such areas, protective irrigation would be essential for the retention of humid conditions for adequate growth, panicle initiation, and capsule setting (Korikanthimath, 1991). It has been observed that the dreaded katte disease is more prevalent during summer than in the monsoon season. Cold conditions result in almost poor or no capsule setting. Hence, for healthy growth of cardamom plants, extremes of temperature and diurnal wind should be avoided.

2.4.2.2 Rainfall

In South India, cardamom is grown under a range of rainfall that varies from 1500 to 5750mm annually. The climate of the area is determined by the annual rainfall, and the year can be divided, generally, into winter, summer, and monsoon seasons. A combination of cool temperatures and relatively dry weather prevails from November to February. Hot weather prevails from March to June, marked by moderate to high temperatures and occasional showers. The southwest monsoon sets in June and continues until early September. In the more westerly areas of the hills, rains during this period are heavy and continuous, but they decrease considerably on the eastern slopes, which are characterized by strong winds, much cloud cover, and frequent light showers. After a short interval, the northeast rains commence and occasional rains continue up to December. This is a dry period in the more northerly and westerly areas, but is marked by heavy rains and overcast skies in the south and the east (Mayne, 1951b). In general, cardamom-growing areas of Karnataka State and many regions of the Idukki and Wayanad districts of Kerala State experience a dry period extending from November–December to May–June. Such a long dry period of 6–7 months is, in fact, the principal constraint on good cardamom production.

The Indian average cardamom yield is only 149 $kg\,ha^{-1}$ compared with the Guatemalan and PNG yield of 300 $kg\,ha^{-1}$. Well-distributed rainfall contributes to good yield in Guatemala (Mohenchandran 1984). In India, 70–80% of the total cardamom area is fed by rain (Charles, 1986). Following forest denudation in many parts of the Western Ghats, the normal congenial habitat for cardamom has been adversely affected, destabilizing the ideal cool, humid microclimate and the productivity of the crop.

Investigations into the effect of rainfall on cardamom productivity indicate that the distribution of rainfall is more important than either the total rainfall received or the number of rainy days. Data collected from cardamom plantations indicate that, in 10 out of the 13 plantations investigated, the highest yield was from those receiving less than 2000 mm annual rainfall. In another survey, data from 57 locations in the Coorg district of Karnataka State indicate that, in 42 cases, more than 100 kg ha^{-1} dry capsule yield was obtained when the annual rainfall was less than 2000 mm. This clearly suggests that total rainfall is not the major determinant in cardamom productivity, and even 2000 mm of rainfall that is well distributed might suffice (Ratnam and Korikanthimath, 1985; Subbarao and Korikanthimath, 1983). Most of the rainfall received during June–August would result in runoff, leading to severe soil erosion; hence, proper soil conservation measures are required to minimize soil and land degradation. Storing runoff water during rainy periods in suitable farm ponds, tanks, or embankments and recycling it during summer as protective irrigation coinciding with critical physiological stages of the crop offers great scope for avoiding total crop failure and promoting stability of yield (Cherian, 1977; Korikanthimath, 1987a).

In Guatemala, rainfall conditions are much more favorable than in India. Rainfall varies from 2000 to 5000 mm in cardamom-growing areas and is evenly distributed throughout the year, except for two peaks. Because there are no heat and drought stresses, as there are in India, cardamom yields in Guatemala are much higher than those obtained in India and, on average, stand at 300 kg ha^{-1}. A similar situation occurs in PNG as well, leading to high yields (Krishna, 1968, 1997). In India, a prolonged drought in the first 6 months of the year during the period 1985/86 to 1989/90, prolonged drought in some years had a devastating effect that led to significant crop loss, especially in exposed and partially shaded regions of Idukki district of Kerala State. India's cardamom production came down to its lowest level, 1600 million tonnes, documenting an imminent need to combat recurring drought by proper soil moisture conservation techniques, mulching, and adequate shade management, along with a need for lifesaving irrigation.

2.4.3 Management Aspects

2.4.3.1 Planting Systems

Following are the four principal planting systems (Mayne, 1951a):

a. Kodagu (Coorg district in Karnataka State) Malay system
b. North Kanara (North Kanara district of Karnataka State) system
c. Southern system
d. Mysore (Karnataka State) system.

2.4.3.2 Kodagu Malay System

The Kodagu Malay system is restricted to Coorg district in Karnataka State. Small patches of forestland, one-sixth to a quarter of a hectare in area, are cleared and planted with cardamom. Care is taken in selecting plots, which are to face north or northeast to ensure adequate lateral shade from the surrounding forest trees. Seedlings from natural regeneration are thinned out and spaces filled in, or seedlings are raised in nurseries and transplanted in shallow pits 1.5–2.5 m apart. The areas are weeded periodically, either using chemical weedicides or manually. The most commonly grown variety is Malabar. After about 15 years, the area is left to natural forest cover while cardamom cultivation is shifted to another patch of land. A somewhat similar system was followed by the Madras (now Chennai in Tamil Nadu) Forest Department, and cardamom was collected as a minor forest produce, with the areas partially cleared by selection felling.

2.4.3.3 North Kanara System

This system is followed in the districts of North Kanara, Shimoga, and parts of Chickmagalur in the state of Karnataka, where cardamom is grown as a secondary crop in areca nut gardens (*Areca catechu*, betel nut—a common nut used for chewing in India and Pakistan). Seedlings are raised in nurseries and planted in rows 1.5–1.8 m apart. About 1200 seedlings are planted per hectare. Malabar is the variety usually grown.

2.4.3.4 Southern System

The southern system is the system that is most in vogue for the commercial cultivation of cardamom and accounts for about 90% of the cardamom plantations in India. Selected areas are cleared of jungle land and all undergrowth, the overhead shade is thinned out, and cardamom seedlings are planted at regular distances and cultivated according to a regular schedule. This system has been adopted in the states of Kerala, Karnataka, and Tamil Nadu. The sizes of the holdings vary widely, but the greater part of the production comes from holdings of 2–20 ha. In most areas of Nilgiris district in Tamil Nadu, in Kerala State, and in other areas of Tamil Nadu, the principal types of cardamom grown are varieties Mysore and Vazhukka. In Karnataka State, Malabar is the variety grown exclusively.

2.4.3.5 Mysore System

Coffee is the most popular plantation crop of Karnataka State, where cardamom is grown in isolated pockets, in ravines, or in low-lying areas of coffee plantations. In such situations, cardamom is found either as a sole crop in narrow strips along the ravines or as scattered clumps interspersed with coffee plants (Photo 2.4).

2.4.4 Establishing a Cardamom Plantation

2.4.4.1 Preparation of the Main Field

Where cardamom is cultivated on a plantation scale in virgin forest, the first step consists of clearing all undergrowth and thinning out the overhead canopy in order to obtain an even density of shade. If the land slopes, it would be preferable to start clearing from the top and work downward. The shade will have to be regulated in such a way that it allows sunlight to filter through the tree canopy almost uniformly.

Photo 2.4 Cardamom plantation (mixed with coffee plants).

The bushes, shrubs, and undergrowth are cut and heaped in rows or in piles and permitted to decay. In the case of steep slopes, it is preferable to utilize the decaying debris in such a manner as to assist the checking of soil movement due to erosion. Contour terraces may be formed in cases where the land is too steep. In areca nut gardens, deep trenches and pits are dug among palms and filled up with fresh soil brought from neighboring forest. So prepared, the ground is utilized for planting. In marshy areas, adequate provisions should be made to drain off excess water by providing main and lateral drains, depending on the natural gradient of the land on which the plantation is being established.

2.4.4.2 Spacing

The variety grown and the duration of the crop determine spacing. Where cardamom is intended to grow on a regular replanting basis and for a limited period, it is obviously desirable to plant as closely as possible without unduly restricting the plants, so that early crops may be as large as possible. If a crop is meant to last 10 years, a commonly suggested crop cycle, only eight harvests are likely to be taken and at least the first two will be dependent on the number of plants per hectare. If, however, plantings are expected to remain in the field for longer periods, too close planting will lead to overcrowding and a reduction in yield as the crop ages. This is important because cardamom clumps tend to spread outward as they age, and gradually, new-shoot production will decline in the center of the plant. Korikanthimath (1983b) investigated the effect of spacing, seedling age, and their performance in relation to fertilizer rates and found that tiller number, number of leaves per plant, and plant height were significantly affected by the different treatments. Maximum tiller number (10.9 per plant) and maximum number of leaves (102.1 per plant) were seen in the case of 18-month-old seedlings planted at 2 m × 1 m spacing, and the plants were fertilized with nitrogen, phosphorus, and potassium in the ratio of 75:75:150 kg ha^{-1} and a supplemental 100 kg ha^{-1} neem (*Azadirachta indica*) cake. In a similar investigation in which spacing and fertilizer rates were considered under rain-fed conditions, the treatment differences turned out to be highly significant as regards tiller number and leaf number per plant. A spacing of 2 m × 1.5 m combined with a fertilizer schedule of 75:75:150 kg nitrogen, phosphorus, and potassium per hectare resulted in maximum tiller number and leaves per plant (Korikanthimath, 1982). Normal spacing adopted in the case of the vigorous variety Mysore is 3 m × 3 m, and for the less vigorous variety Malabar, a 2 m × 2 m spacing is adopted (Anon., 1976). In the "high-production technology" field demonstration plots, meant primarily to show to farmers, spacing at 2 m × 1 m on hill slopes along the contour and spacing at 2 m × 1.2 m on flatlands, gentle slopes, and valley bottoms yielded 500 kg dry capsules per hectare within 2 years from the date of planting (Korikanthimath and Venugopal, 1989). In a spacing trial carried out at Yercaud in Tamil Nadu, it was observed that close spacing at 1 m × 1 m and 1.5 m × 1.5 m resulted in better yield per unit area than did wider spacing at 2.5 m × 2.5 m and 2 m × 2 m. On sloping lands, it is advisable to make contour terraces in advance of the planting date, and pits may be dug along the contour for planting. Depending on the slope, a distance of 4–6 m may be provided along the slope between the contour lines. Close planting may be adopted along the contour.

2.4.4.3 Methods of Planting

The factors that determine the planting systems are the land, soil fertility, and probable period over which the plantation is expected to last. In some areas, seedlings are planted in holes, which are scooped out at the time of planting. In other areas, considerable care is taken in preparing pits for planting. Spots where pits are to be dug are marked with stakes; soil is dug out from the pits; and the pits are filled with surface soil mixed with leaf mold, compost, or cattle manure (Subbaiah, 1940). Pit size is commonly 60 cm × 60 cm × 45 cm. Some plantations use a pit size of 90 cm × 90 cm × 90 cm or 120 cm × 120 cm × 30 cm. In South and North Kanara in the state of Karnataka, pits are smaller, 45 cm × 45 cm × 45 cm (Mayne, 1951a). In the state of Kerala, varieties Mysore and Vazhukka are planted in pits of size 60 cm × 45 cm × 45 cm. Normally, pits are opened during the months of April–May, after the premonsoon showers. Pits are filled with a mixture of topsoil and compost or well-rotted farmyard manure and 100 g of rock phosphate. On sloping land, contour terraces are made sufficiently in advance of planting and the pits are dug along the contour (Anon., 1985, 1986). Most of the cardamom-growing tracts are situated on hill slopes of the Western Ghats. The undulating terrain and heavy rainfall in the region increase the problem of soil erosion and loss of plant nutrients in runoff. Thus, there is a need to conserve enough moisture while, at the same time, ensuring the safe disposal of excess rainfall. In view of these pressing demands, investigations into applying planting-fertilizer treatments under rain-fed conditions were carried out at the Cardamom Research Center at Appangala, Karnataka State, under the administrative control of

Table 2.16 Dry Cardamom Yield (kg ha^{-1}) Influenced by Planting System and Fertilizer Rates

	Nitrogen–Phosphorus–Potassium Fertilizer (kg ha^{-1}) (B)					
Planting system (A)	0:0:0	40:80:160	80:80:160	120:120:240	160:160:320	Mean
Pit	123.9	277.5	388.9	437.0	455.6	336.0
Trench	134.6	369.3	416.7	465.4	496.5	376.0
Mean	129.2	323.4	402.8	451.2	476.0	

Notes: SE/Plot: 88.0; general mean: 356.5; CV (%): 24.7 CD for (A): 57.1; CD for (B): 90.3; CD for (A) × (B): 127.7.

the Indian Council of Agriculture Research. Some of the results of the investigations, begun in 1985, are shown in Table 2.16. Korikanthimath (1989) reported greater moisture retention under the trench system of planting than with the pit system and concluded that the former was better.

Trenches may be dug to a depth of 30 cm × 45 cm wide up to any length across the slope or along the contour. The top 15 cm of soil may be removed and kept separately, while the lower 15 cm may be excavated from trenches placed below the one above. The top 15 cm of soil is filled back into the trench with cattle manure. While closing the trench, about 5 cm space may be left at the top to facilitate the application of fertilizers and mulches. Although digging trenches would be about 30–40% more expensive than digging pits, it may be worth attempting because of the benefits of soil moisture conservation and its ultimate effect on plant growth and yield. However, in low-lying areas where the danger of water stagnation is real, the pit system may be preferable.

Planting Season

Two important factors that determine the planting season are topography and pattern of rainfall. Planting is commonly done in June–July. Where the southwest monsoon is torrential, planting either is completed before July or is taken up in August–September, when the rains cease. Better crop establishment and growth are ensured through early planting compared with late planting (Mayne, 1951a). In low-lying valleys, planting should be commenced only after July, when torrential rains begin to abate (Korikanthimath, 1980). In Mudigere district of Karnataka State, better establishment and crop growth were reported when the planting was done in August (Pattanshetty and Prasad, 1972). Cardamom suckers are planted from June through August on the soil surface or 15 to 20 cm deep. Mortality of seedling is least with surface planting and when rainfall is relatively less heavy in the week following planting. Suckers planted in August survived best, with a mean mortality rate of 25%, and those planted on soil surface showed the lowest mortality rate, 17.5% (Pattanshetty et al., 1972, 1974).

Investigations into the effect of monthly planting were carried out at the Horticulture Research Station in Yercuad, Tamil Nadu, located on the eastern side of the Western Ghats along the state of Kerala. Planting was commenced from June through November during 3 years, to assess the ideal planting time at elevations of about 1300–1500 m amsl. The best establishment (87.92%) was obtained in July planting, followed by August, September, October, and November planting, with establishment rates of 77.9%, 75.4%, 63.7%, and 61.6%, respectively. June planting gave only 19.4% establishment. Obtaining good establishment requires a total rainfall of at least 322 mm, with minimum and maximum temperatures in the range of 15.5–17.5°C and 19.5–25.0°C, respectively, during the month of planting (Nanjan et al., 1981).

Planting

The general practice is to scoop a small depression in the filled soil, with the seedling placed at the center of the depression. Soil is then replaced, taking care not to disturb the roots in their normal position and pressed well around the base of the clump. Deep planting should be dispensed with, as it results in suppressing both growth of the young seedling and the emergence of new shoots, perhaps leading to decay of underground rhizomes. Seedlings are normally planted at an acute angle to the soil level, to prevent them from being broken or blown by strong winds, which follow the planting season (Anon., 1952a). Light pruning is desirable, but should be confined to longer roots of 0.3 m or more, avoiding the shorter ones. In the case of rhizome planting, the planting material can be kept in pits in a slanting manner and rhizomes covered with soil, as is done with seedling planting. Immediately following planting, the seedlings should be physically supported by stakes to prevent them from being damaged or being blown away by strong wind. A mulch cover with dry leaves is provided at the base. Crisscross staking with two stakes is the best practice to follow. Plants may be loosely tied to the stakes with dried banana sheath or jute threads, to facilitate the emergence and growth of aerial shoots. Care must be taken to offset after transplantation shock, which is due to physical causes, and the seedling must also be guarded against heavy rains. Unhealthy plants are prone to infestations of disease, and it is advisable to spray the seedlings with 1% Bordeaux mixture or any other suitable fungicide as a prophylactic measure. The newly planted area should be inspected periodically and any gaps found filled instantly if the climate is favorable.

Planting of Suckers

Using suckers to propagate cardamom consists of splitting up established clumps into sections consisting normally of at least one old and one young shoot. Planting material with 20-cm-long rhizomes results in more shoots per clump, early bearing, and large net returns than does material with short rhizomes of 2.5 cm (Pattanshetty, 1972a,b; Pattanshetty et al., 1974; Pillai, 1953). The section of rhizome is placed in a small depression in a pit that is already prepared and covered over with soil and mulch. The leafy shoots are placed almost parallel to the soil surface. A clump consists of new shoots, which arise from the rhizome. In the high ranges of Kerala State, straight planting of rhizome with stake is recommended. In those areas of Guatemala in which the dreaded katte disease is not a threat, cardamom is invariably propagated by suckers. Three suckers per pit are used to induce tillering in a short time span. Rapid growth and high yields are the essential features of Guatemala and are attributable mainly to the country's rainfall distribution pattern, complemented by fertile soils and good plantation management. In India also, farmers have realized high yields with intensive management practices of which good irrigation is an important component.

Gap Filling

Good initial establishment is crucial to raising a productive cardamom plantation. On average, 5% gaps are seen in most cardamom plantations. Healthy and disease-free seedlings or clumps can reduce seedling mortality during the establishment period. Monsoon failure during planting time is a potential hazard in which case supplemental irrigation, once a week, is a must. It is advisable to use healthy, sufficiently grown-up seedlings or, preferably, clonal materials, to fill in the gaps. May–June, when monsoon season starts, is the best time to do so. However, if this period is missed, gap filling can be extended up to August–September with proper care. The success of gap filling depends on aftercare until the gap-filled plants reach the state of earlier established plants. Regular cultural operations, consisting of regular mulching, weeding, trashing, raking or digging, irrigation, shade regulation, manuring, gap filling, and plant protection measures, must be carried out after planting in order to maintain the plants in a healthy and vigorous condition.

Mulching

Cardamom productivity is very much dependent on a proper moisture balance in the soil. In recent years in India, premonsoon showers have become quite erratic on account of climate change. As a result, cardamom plants now face drought even up to 6 months at a stretch. Mulching is used to conserve soil moisture and has been acclaimed as the most important cultural innovation for the overall improvement of soil and yield of cardamom plantations (Zachariah, 1976). Following are the advantages of mulching:

1. By minimizing surface evaporation, soil moisture is conserved.
2. When rains occur, soil does not get puddled and, because of the beating action of the raindrops, the physical condition of the soil is maintained.
3. Runoff and erosion are checked.
4. Both friability and soil structure are improved by the enhanced biotic activity under the mulch. As the biotic activity increases the number of macropores, the soil becomes more porous, in turn increasing both water percolation and moisture conservation.
5. Soil temperature equilibrium is maintained.
6. Enrichment of the soil with organic matter leads to enhanced biotic activity, thereby improving both nutrient availability and soil fertility.
7. Weed growth is controlled.
8. Roots grow better, leading to the extraction of soil moisture from deeper layers.
9. The eventual decomposition of the mulch affects soil fertility.

Soon after planting, the base of the plants is adequately mulched. Mulching, a simple cultural operation that uses dried leaves and other plant residues, should be completed before the onset of summer. Leaves shed by the shade trees come in handy for mulching, which can be done in November–December. An investigation was conducted to study the relative merit of locally available mulching materials, such as dried leaves, paddy husk, phoenix leaves, coir dust, and stratified leaf mulch, under uniform shade of coir matting and by using suckers of cv. P1 combined with two levels of irrigation, namely, 75% and 25% available moisture. The results, presented in Table 2.17, did not show any significant difference on the production of suckers due to irrigation levels. Leaf mulch and phoenix leaves were at par, statistically, but were found to be significantly superior to other mulches in the production of suckers (Raghothama, 1979).

Demulching is equally important as mulching and should be carried out in May after the premonsoon showers, in order to facilitate honeybee movement, which will ensure better pollination and capsule setting in the plantations. The practice of uncovering the panicles shortly after the commencement of flowering improves fruit set. The average number of capsules per plant is 27.4 and 2.1, respectively, in the case of exposed and covered panicles in variety Malabar (Pattanshetty and Prasad, 1974). The removal of mulch, which accumulates in the center of the clump and releases panicles beneath, not only will facilitate the movement of honeybees, but also will

Table 2.17 Effect of Different Mulch Materials on Germination of Cardamom Seeds

Treatment	Germination (%)	Leaf Spot Disease (%) 45 DAS
Paddy straw	40.8	12.0
Dry leaves of rose wood	37.4	16.2
Paddy husk	27.2	23.6
Sawdust	33.6	12.8
Wild fern	38.2	14.5
Coffee husk	21.6	25.6
Charcoal	35.2	19.2
Polythene sheet	1.8	–
Phyllanthes emblica leaf twigs	37.1	17.3
Sand	12.2	–
Control	13.3	–
SE/plot	4.3	5.9
General mean	27.1	12.8
CV (%)	15.9	46.4
LSD ($p = 0.05$)	6.2	8.6

Note: DAS, days after sowing.

provide better aeration and minimize the incidence of clump rot and rhizome rot disease.

2.4.4.4 Weed Control

Since the cardamom plant gets its nutrition from the top layers of the soil, it is crucial that frequent weed control be done in the first year of planting, in order to avoid root competition between young cardamom seedlings and weed, as both can compete for nutrients and moisture. Weeding is carried out either on the entire area covering the plants or on just the area around the plants. Both techniques are called ring weeding. The weeds are used as mulch for young plants. As many as 21 dicotyledonous weeds have been identified in cardamom estates in the Coorg district of Karnataka State. Of these, *Strobilanthes ureceolaris Gamb* is the most common. Weeds are controlled mainly by hand weeding; only in rare instances are chemicals used. Two to three rounds of weeding are essential in the first year of planting, in order to prevent the undergrowth from regenerating. Generally, the first hand weeding is done in the months of May–June, the second in August–September, and the third in December–January. When weeding is done in May–June and August–September, the weeded matter is heaped in the spaces between the rows and is used for mulching later. When weeding is done in November–December, the matter is directly utilized for mulching. Slash weeding is the most common type of weeding in cardamom plantations. Spraying Gramaxone at the rate of 1.5 ml $liter^{-1}$ twice a year, a practice that is quite economical and convenient, is also resorted to some plantations, although infrequently.

2.4.4.5 Additional Field Operations

Some other field operations are trashing, raking, digging, and earthing up. Trashing consists of removing old, drying shoots. Beginning in the second year of planting, trashing has to be continued every year. Trashing promotes better sunlight penetration and aeration, thereby enhancing tiller initiation and plant growth as well as reducing the incidence of thrips and aphid infestations. Trashing also helps honeybees pollinate the cardamom plants (Korikanthimath and Venugopal, 1989). In rainfed areas, trashing time is May, after the premonsoon showers. The trashed leaves and leafy stems may be heaped between the rows and allowed to decay or used for composting. A light raking or digging of soil around the clump up to a radius of 75 cm is done toward the end of the monsoon. The soil mulch formed around the plant base helps conserve moisture during the next weather-related period. This practice is particularly useful in rainfall areas of low rainfall. Digging not less than 25 cm deep, once in alternate years, may be done in the entire area, followed by the application of farmyard manure or any other organic manure, such as bone meal, stera meal, groundnut cake, and so on. Digging can also be done in patches; however, it is necessary to dig each year if the soil is clayey (Kuttappa, 1969b). Toward the end of the rainy season, a thin layer of fresh and fertile soil rich in organic matter may be spread at the base of the clump. This soil, which covers up to the collar region of the clump, is obtained by scraping in between the rows or collecting soil from the trenches or pits. Applied in the center of the clumps, the thin layer of soil not only will keep them intact and cover the exposed roots, but also will check the "walking" habit (radial growth) of the cardamom plant (Korikanthimath and Venugopal, 1989). Care must be taken not to heap the soil above the collar region of the clump.

2.4.4.6 Replanting in the Plantation

Decline of yield is a problem in cardamom, although it is a perennial crop. Once in 8–10 years, regular replanting has to be done to ensure high productivity. One of the main reasons for a low average cardamom yield in India is that there never has been any replanting in many old plantations (Korikanthimath et al., 1989). Clonal material from superior high-yielding varieties may be used for replanting, thereby ensuring that the yields will be high. In an investigation of the economics of replanting, Korikanthimath et al. (2000a) used the trench system to replant a cardamom plantation after a period of 10 years and maintained the replanted plantation with all the recommended inputs—fertilizers, irrigation, and so on—as per the package of practices recommended in so-called high-production technology. Planting materials used were 10-month-old seedlings from high-yielding mother plants. The replanted field gave 155 kg of dry cardamom in the second year. In the next year, a record yield of 1775 $kg\,ha^{-1}$ was obtained. In the subsequent 3 years, the dry yield obtained was 385, 560, and 870 $kg\,ha^{-1}$, respectively, which averaged 749 $kg\,ha^{-1}\,year^{-1}$. Economic analysis carried out by Korikanthimath et al. (2000b) showed a net return of Rs 203, 465 ha^{-1} (which is approximately US$4800 at the current rate of US$–Indian rupee exchange), and the benefit–cost ratio worked out to 2.78. The investigation conclusively demonstrated that replanting a cardamom field after a 10-year gap is economically advantageous.

2.4.4.7 Propagation

Cardamom can be propagated both by seeds and through vegetative means. The seedling population is variable because cardamom is a cross-pollinated crop. Hence, vegetative propagation is adopted only in the case of elite clones. Both micropropagation (tissue culture) and rhizome bits (suckers) can be used for vegetative propagation. With commercialization, micropropagation has become quite popular in cardamom production.

2.4.4.8 Propagation through Seeds

In order to obtain quality seedlings, a cardamom nursery has to be managed carefully and scientifically. This involves sowing seeds on raised beds and then transplanting them onto primary and secondary nursery beds and, finally, into the field (Cherian, 1979; Kasi and Iyengar, 1961).

2.4.4.9 Seed Selection

Seeds should be collected from high-yielding vigorous plants, with well-formed compact panicles and well-ripened capsules that are free of pest and disease infestation. The number of flowering branches formed on the panicles, percentage of fruit set, and number of seeds per capsule should be given due consideration while selecting the number of plants for seed collection (Anon., 1979; John, 1968; Ponnugangum, 1946; Siddaramaiah, 1967; Subbaiah, 1940). Apart from possessing these desirable attributes, the mother clump should have a greater number of tillers (shoots) per plant, leaves with a dark-green color, and a high percentage of fruit set. The color of the capsules should be dark green (Krishna, 1968). On average, 1 kg of fruits contain 900–1000 capsules with 10–15 seeds per capsule. Taking into consideration the percentage of germination, mortality due to diseases, and so on yields the result that, on average, 1 kg of seed capsules is required to obtain about 5000 plantable seedlings.

2.4.4.10 *Preparation of Seeds*

Seeds for sowing are collected from fully ripe capsules, preferably from the second to the third round of harvest, and are then either washed in water, sown immediately, or mixed with wood ash and dried for 2–9 days at room temperature. The first method gives better results and has been adopted widely. Following picking, seed capsules should be immersed in water and gently pressed to separate the seeds, after which they should be washed well in cold water to remove any mucilaginous coating found. After the water is drained, the seeds should be mixed with ash and surface dried in shade.

2.4.4.11 *Viability of Seeds*

Stored seeds lose viability over time, resulting in delayed germination or, sometimes, no germination at all. Seed germination was found to be 59% and 50.6% in varieties Mysore and Malabar, respectively (Korikanthimath, 1982). Germination was reduced when stored seeds were used, especially those stored in airtight containers. Seeds treated with organomercurials and stored in open bottles germinated up to 4 months later. The highest percentage of germination was 71.8%, observed when sowing was done in September (Pattanshetty and Prasad, 1973; Pattanshetty et al., 1978). In a clone of variety Malabar, germination gradually declined. Sixty days after storage, seeds sown at fortnightly intervals starting in August and going on up to October 14 showed a progressive decline in germination: 56.7%, 51.0%, 46.4%, 34.1%, 32.5%, and 29.6%.

When seeds are sown early October, when ambient temperature comes down, germination is uniform and early, and the seeds are ready for transplantation at the end of 10 months. When they are further retained in nursery beds for the next planting season, either by proper thinning or by transplantation at wider spacing in secondary nursery beds, they develop rhizomes with a large number of tillers and are ideal for field planting (Pattanshetty and Prasad, 1972). November–January has been found to be the ideal sowing time for Kerala State, and September–October for Tamil Nadu and Karnataka (Anon., 1970, 1979).

2.4.4.12 *Presowing Treatment of Seeds*

Cardamom seed possesses a hard coat, which delays germination. Investigations have been carried out to study the effect of presowing treatments on seeds to offset any delay in germination. Treating freshly extracted seeds with concentrated nitric acid or hydrochloric acid for 5 min significantly improved the germination of seeds sown during November (Pattanshetty and Prasad, 1974; Pattanshetty et al., 1978). Treating the seeds with 20% nitric acid, 25% acetic acid, and 50% hydrochloric acid for 10 min was found to produce 97.6%, 98.6%, and 91.5% germination, respectively. Korikanthimath (1982) found that a treatment with 10% nitric acid was best for enhancing germination. Ambient temperature also plays an important part in germination. Low ambient temperature in the winter in cardamom-growing areas not only reduces germination, but also delays it (Krishnamurthy et al., 1989a). Gurumurthy and Hegde (1987) found that germination is significantly correlated with maximum and minimum temperatures prevalent in the area.

2.4.4.13 Nursery Site

It is always preferable to select the nursery site on a gentle slope with easy access to a perennial source of water. The nursery area should be cleared of all existing vegetation, stumps, roots, stones, and so on. Raised beds are prepared after cultivating the land to a depth of about 30–45 cm. Usually, beds 1 m in width, of a convenient length, and raised to a height of about 30 cm are prepared for sowing the seeds. A fine layer of humus-rich forest soil is spread over the beds. Upon treatment with a 4% formaldehyde solution, the beds are found to control "damping off," another important disease of cardamom (Anon., 1985). After this treatment, the beds are covered with polythene sheets for a few days. Seeds are sown 2 weeks after treatment. Before sowing, the beds have to be flushed with water to remove any remaining formaldehyde (Photo 2.5).

Photo 2.5 Cardamom nursery.

2.4.4.14 Seed Rate and Sowing

The seed rate for raising 10-month-old seedlings is 2–5 g; for raising 18-month-old seedlings, it is10 g (Anon., 1976, 1986). Seeds are sown in lines, usually not more than 1 cm deep. Rows are spaced 15 cm apart and seeds are sown 1–2 cm apart within each row. For better and quicker germination, deep sowing should be avoided. Lindane at the rate of 60 g 5 m^{-2} is used to dust seedbeds to prevent termite attack. After sowing, a thin layer of sand or soil is spread over the beds and pressed gently with a wooden plank, and a thin mulch of leaves may be provided. Thereafter, beds are to be watered every day. Germination will start after 1 month and may continue for a month or two more. Soon after germination commences, the mulch is removed and shade is provided to protect the young seedlings from direct sunlight and rain.

2.4.4.15 Mulching of Nursery Beds

Germination is influenced by mulching (Abraham, 1958). Using locally available mulch materials, Korikanthimath (1980) carried out investigations on the effect of mulches on germination. The materials used were paddy straw, paddy husk, dry leaves of the rosewood tree, sawdust, wild fern, coffee husk, gooseberry (*Phyllanthus emblica*) leaves, sand, charcoal, and polythene sheet. Maximum germination (40%) was observed in the case of paddy straw mulch when seeds were sown in September; this treatment was statistically at par with that of rosewood tree dry leaves (37%) and wild fern (38%) (Table 2.18). Other reports show that mulching with coconut coir dust, paddy straw, or gooseberry leaves enhances germination (Korikanthimath, 1983a; Mayne, 1951a; Sulikeri and Kologi, 1978).

2.4.4.16 Secondary Nursery

In the states of Kerala and Tamil Nadu, the seedlings are transplanted to secondary nursery beds when they are about 6 months old. By contrast, in the state of Karnataka, the practice is to sow the seeds in the primary nursery and thin the seedlings out to the

Table 2.18 Mean Cardamom Yield (Gramg per Plant) under Different Shade Trees

Name of the Shade Tree	Year				Mean Yield
	1975–1976	1976–1977	1977–1978	1978–1979	
Karimaram	112	81	81	207	121
Elangi	82	62	46	126	79
Jack	66	65	52	135	79
Nandi	109	65	42	135	89
Mean yield	92	68	55	151	92

Note: SE for species = 57.5, CD; LSD $p = 0.05$=18.
SE for years = 49.4, CD; LSD $p = 0.05$=15.
CD for any 2 years for any species = 31; CD for any two species for any years = 32.

required stand, which allows them to grow in the same place. Transplanting seedlings to a secondary nursery reduces nursery-borne diseases. Korikanthimath (1982) has demonstrated that following both primary and secondary nursery practices would be needed to get vigorous seedlings that have four to five leaves within a span of 10 months and that suffer lesser disease and a lower incidence of pests. On average, 10 secondary seedbeds are required to transplant seedlings from one primary bed. Secondary nursery beds are made the same way as the primary ones. A mixture of powdered dry cattle dung and wood ash is spread over the secondary seedbeds before transplanting seedlings. In Karnataka, where seeds are sown in August–September, transplantation or thinning is done in November–January. In Kerala State and Tamil Nadu, seedlings from primary beds are transplanted to secondary beds at a spacing of 20cm × 20cm in June–July. Seedling mortality was found to be higher when transplantation was done at second leaf stage (25.4%), whereas when it was done at the fifth leaf stage, mortality was 1.1%. The number of seedlings produced was more at a wide spacing of 30cm × 30cm (11.9), compared with narrower spacing of 22.5cm × 22.5cm (9.2) or a still narrower spacing of 15cm × 15cm (7.3). However, in view of the larger area and higher expenditure involved in raising nurseries when transplantation takes place at the fifth or sixth leaf stage, a spacing of 15cm × 15cm is recommended (Korikanthimath, 1982).

2.4.4.17 *Manuring*

Generally, organic manure, such as well-decomposed compost, cattle manure, fertile top forest soil, and so on, is applied to each seedbed at the rate of 8–10kg $25\,m^{-2}$ land area, in the case of both primary and secondary nurseries. From a seedbed planted with 100 cardamom seedlings, it was found that, on average, 120g nitrogen, 20g phosphorus, 300g potassium, 50g magnesium, and 75g calcium are removed, showing that cardamom is a heavy feeder of potassium. Hence, potassium nutrition needs special attention. To facilitate uniform growth, a 250g mixture made of nine parts of nitrogen–phosphorus–potassium in the ratio of 17:17:17 and eight parts of zinc sulfate dissolved in 10 liters of water is sprayed once every 15–20 days, starting from the first month after transplantation (Anon., 1990). The Regional Research Station in Mudigere, Karnataka State, recommends a nitrogen–phosphorus–potassium mixture at the rate of 160g per seedbed 1 month after planting, to be increased by 160g every month until a maximum of 960g per bed is reached. The mixture consists of one part urea, two parts superphosphate, and one part muriate of potash (MOP) (Anon., 1979). The application of 45g nitrogen, 30g P_2O_5, and 60g K_2O per bed of 2.5m × 1m size in three equal splits at an interval of 45 days would result in better growth and a higher tiller number (Korikanthimath, 1982). The first dose of fertilizer is applied 30 days after transplantation into the secondary nursery. The application of diammonium phosphate (DAP) along with MOP was found to enhance root growth and tiller production (Anon., 1989).

2.4.4.18 *Overhead Shade*

The young seedlings must be protected from heating by the direct impact of sunlight; hence, overhead shade has to be provided. A framework erected with wooden poles and sticks upon which a sheet of nylon nursery sheet or coconut frond is spread can be used to shade the seedlings. Ideally, 50% shade has to be provided, and this aids tiller formation and growth. When the monsoon starts, the shade nets have to be removed.

2.4.4.19 Irrigation and Drainage

Nurseries should be irrigated twice daily up to 8–10 days after transplantation. Thereafter, once-daily irrigation up to 30 days after transplantation will suffice. Once the seedlings establish and put forth fresh growth, irrigation may be done on alternate days until the monsoon sets in. Flood and splash irrigation must be avoided, as it would create conditions for the onset of damping off and leaf diseases. Stagnation in the nursery should be avoided by proper drainage. In low-lying areas, central and lateral drains should be provided to prevent water from stagnating.

2.4.4.20 Weeding

Until the plants grow tall enough to smother any undergrowth of weeds, hand weeding is the best practice to keep the nursery clean, and it should be done at an interval of 20–25 days.

2.4.4.21 Earthing Up

The topsoil between the rows of cardamom seedlings would normally get washed away and deposited in pathways when rains commence. To prevent this, scraping of soil from pathways and the application of a thin layer of soil up to the collar region of the plants may be taken up 2 months after transplantation into the secondary nursery. Applying fertile soil collected from the forest, along with cattle manure, would be beneficial. Earthing up may be taken up immediately after split application of fertilizers.

2.4.4.22 Rotation and Fallow in Nursery Site

Ideally, nursery sites should be shifted once in every 2–3 years, to prevent the buildup of pathogens in the soil. If shifting is not feasible, it is advisable to grow a short-duration green-manure crop such as *Crotalaria* or *Sesbania*, which will enrich the fertility of the nursery site through the buildup of organic matter and the mobilization of plant nutrients. These green-manure crops must be plowed into the soil. Fallowing for 1 year can also be done in a partial nursery area, and when it is so done, it has the positive effect of rejuvenating the soil and also controlling any buildup of insects or other pathogens in the soil. Sunlight falling directly on fallowed land has the effect of sterilization.

When all of the preceding cultural practices are followed, seedlings will likely be ready for transplantation about 10 months after the seeds are sown. Raising seedlings in a primary nursery and subsequent transplantation into a secondary nursery is found to be more advantageous, as it facilitates better establishment and the initiation of an adequate number of suckers per plant.

2.4.4.23 Paddy Fields as Nurseries

In the state of Karnataka, small and marginal cardamom farmers in the districts of Chickmagalur, Coorg, and Hassan raise nurseries in wet paddy fields. When this is done, the most important precautionary measure is to provide good drainage to prevent water from stagnating. Beds are separated by deep channels. The wet paddy soils are heavy, and this leads to excessive moisture in the soil, which could hamper good seedling establishment and emergence (Mayne, 1951a).

2.4.4.24 Dry Nursery

The Malay system of cardamom production employs the dry nursery technique, which also is practiced in the Coorg district of Karnataka State. The forest cover provides good shade, and there is no need to irrigate as frequently as open fields would require. The technique is called the dry nursery technique because no irrigation is involved. Nursery operations are limited. After initial showers in April, dry seeds are broadcast where leaf litter is heaped. Seeds are raked into the soil and the surface is covered with leaf mold. The branches of trees are cut to regulate overhead shade. After germination, hand weeding is done. Before the monsoon fully recedes, forest soil and leaf litter are spread over the seedbeds. Seedlings withstand drought while growing. Like seedlings raised conventionally, seedlings raised by the dry nursery technique are planted in the main field after attaining sufficient growth.

2.4.4.25 Polybag Nursery

Polyethylene bags (20 cm × 20 cm) of 200–300 gauge thickness and with six to eight holes can be used to raise cardamom seedlings. The bags are first filled with a nursery mixture in the ratio 3:1:1 (forest topsoil, farmyard manure, and sand, respectively). The bags are then arranged in rows of convenient length and breadth for easy management. Seedlings at the 3–4 leaf stage can be transplanted into the bags. Subsequently, adequate space is left in between bags to facilitate better tillering. Seedlings obtained in this manner are uniform in growth and subsequent tillering and establish better in the main field. Their main advantage is that nursery growth time can be substantially shortened. An investigation carried out at the Cardamom Research Center, Appangala, Karnataka State, on the relative merits of raising seedlings in the conventional manner and in polyethylene bags showed that there was a 6% mortality of seedlings in the former, compared with just 1% in the latter, in the main field 30 days after transplantation. The high cost of raising seedlings in polyethylene bags and their subsequent transportation precludes the practice from being followed in large plantations, but the method is suitable for small or marginal cardamom planters, homestead gardens, and small estates.

2.4.4.26 Cost of Raising Seedlings

The duration of the nursery and the age of the seedlings determine the cost of operating a cardamom nursery. The rule of thumb is that one season equals 10 months and

18–22 months equal two seasons. To raise 100,000 seedlings of 10 months' age, an expenditure of about Rs 84,723 (approximately US$ 1980, based on current Indian rupee–US dollar exchange rate) is required. Usually, increasing the number of seedlings can bring down the cost of seedling production. The Spices Board of India supplies seedlings at the rate of approximately 7 US cents (Rs 3 per seedling). This translates into around Rs 4–4.50 for an 18-month-old seedling that is about 10 cents.

2.4.4.27 Age of Seedlings for Field-Testing

The factors that are considered important in selecting the age of the seedlings, for either one season (10 months) or two seasons (18–22 months) are, first, the comparative success of establishment of the seedlings in the estate (cardamom plantation) and, second, the cost of raising the seedlings. One-year-old seedlings can be good for new planting; two-year-old seedlings with a well-developed rhizome would be more suitable for gap filling. Seedlings that are 18–22 months old usually are preferred for planting in the states of Kerala and Tamil Nadu (Kasi and Iyengar, 1961).

2.4.4.28 Seed Propagation and Its Disadvantages

Since cardamom is a cross-pollinated crop, the seedling population will be highly heterogeneous and the average yield from such plantations will generally be low. Findings of a survey by Krishnamurthy et al. (1989a) indicate that only 36% of the plants are good yielders in a plantation raised from seedlings. A study of a population of 1490 plants from seedlings showed that 45% of the plants were poor yielders (less than 100 g of green capsules per plant), contributing only 12.5% of the total yield. About 36% were medium yielders, contributing 40% of the total yield. Good yielders constituted about 15% of the plants, contributing 32.1% of the total yield; in this group, plants that gave a high yield (500–900 g) were barely 4%, but they contributed 15% of the total yield. Average yield of this experimental population was 170 g green capsules per plant. The high degree of variability in yield and the high percentage of poor yielders in the seedling population necessitate selection of the elite clones and their vegetative propagation.

2.4.4.29 Vegetative Propagation

To establish plantations of high productivity, suckers of elite clones should be used. There are methods for achieving a high rate of sucker multiplication of selected high-yielding clumps (Anon., 1978). Plants raised from rhizomes come to bear earlier than those raised from seedlings. Clonal propagation is also followed in Guatemala (Anon., 1977). Tillers (suckers) and micropropagated tissues (Kumar et al., 1985) can be used to raise cardamom plants. Small-scale farmers in the states of Kerala and Tamil Nadu use suckers. In areas where the deadly viral disease known as katte is prevalent, the use of suckers should be dispensed with. The nonavailability of high-yielding selections is one of the important constraints on achieving high yield in cardamom. Various cardamom research stations in India have attempted to select high-yielding lines (Korikanthimath, 1998). In one such attempt, 80 clumps yielding more than 400 g of green capsules per clump were selected, each clump was divided into four sets, and the yield performance of 320 plants was investigated over a 6-year period. About 42% of the good yielders contributed 43% of the yield, and 17% of the very good yielders contributed 34% of the yield, indicating that the average yield of the population can be improved significantly through the use of suckers (Parameshwar et al., 1989). Such studies led to the development of clonal propagation techniques that are fairly rapid. Clonal multiplication through tissue culture has also been standardized (Kumar et al., 1985).

2.4.4.30 Rapid Clonal Nursery Technique

A quick method of proliferation of suckers was developed at the Indian Institute of Spices Research, administered at the Cardamom Research Center at Appangala, Karnataka State, in order to generate a greater number of planting units. The target was high yield in a short time span, to be achieved by resorting to high-density planting in closely spaced trenches under controlled overhead shade (Korikanthimath, 1992, 1999). Following are the steps involved in the process:

1. The method uses pest-free and disease-free plants, with bold capsules marked and part of the clump uprooted for clonal multiplication, leaving the mother clump in its original site to induce subsequent suckers for further use.
2. Each planting unit consists of one grown-up sucker and a growing young shoot.
3. The planting units are spaced 1.8 m × 0.6 m apart in trenches, an arrangement that accommodates approximately 6800 plants per hectare of clonal nursery area.
4. On average, 32–42 suckers per planting unit will be produced 12 months after planting; after 1 year, it is possible to obtain 16–21 planting units from one clump.
5. In 1 ha of clonal nursery, 1- to 1.4-lakh (100,000–140,000) planting units can be produced after 1 year.
6. A crop of 190 g per plant of dry cardamom (1759 kg ha^{-1}) was harvested within 19 months of planting (from planting date to harvest) (Korikanthimath, 1990, 1992). The system is cheap to set up and easy to manage.

2.4.5 Shade Management in Cardamom

Cardamom is highly sensitive to moisture stress and performs comparatively better in a cool, shady environment. The shade canopy provides a suitable environment by maintaining humidity and evaporation at an appropriate level (Abraham, 1965). Cardamom does not tolerate direct sunlight and also cannot do well under excess shade in which case the plant's metabolic activities become impaired because of inadequate light penetration. Shade has to be regulated in accordance with the lay of the land, moisture retention, and so on, in order to obtain 50% of filtered sunlight for growth and flowering. Following are some beneficial effects of shading in cardamom plantations:

1. A good canopy and a cool temperature protect the soil from the scorching effect of the sun. Shade also checks surface evaporation of soil moisture, thereby helping the plant to retain moisture for a longer period—an important factor in realizing good yield.

2. Shade protects plants from being scorched by the sun. On the border rows, where sunlight falls directly, pest infestations have been noticed.
3. Shading checks the high velocity of rainfall and minimizes mechanical damage to plants, such as splitting of leaves.
4. The extensive root system of shade trees prevents soil erosion and protects soil loss due to the beating action of rains, leading to improvement in the physical properties of the soil.
5. Shading maintains adequate humidity and soil moisture, which are essential for proper growth, flowering, and capsule set.
6. Shading aids the buildup of humus and organic matter and enhances soil fertility through the addition of leaf litter.
7. Shade acts as a wind break, minimizing the ill effects of gale and heavy wind.
8. One or another of the shade trees flower round the year and thus act as an alternative source of nectar to honeybees, which are the principal pollinating agents of cardamom. Forest areas ensure foraging capacity and the availability of nectar for a long period.
9. Shade trees provide a congenial microclimate for proper growth and performance of the plants and help check the growth of weeds, which otherwise would grow luxuriantly in open areas. Controlling the weeds would then be a major problem.

2.4.5.1 Ideal Shade Trees

In forests, all kinds of trees are found, and they come in handy as shade trees. In the cardamom hills of Kerala State, Karnataka State, and Tamil Nadu, trees belonging to 32 families of angiosperms constitute the major tree flora (Shankar, 1980). An ideal shade tree in cardamom plantation should possess the following characteristics:

1. A wide canopy, to minimize the number of trees required for shade.
2. No flower shedding during the pollination period so that pollination is not adversely affected.
3. Be of medium size and evergreen, retaining its foliage throughout the year.
4. Small leaves and a well-spread branching system.
5. Deep roots, to minimize competition for nutrients and water.
6. Fast growth, to provide the required shade immediately.
7. Hard heartwood, to withstand high-velocity winds.

A mixed population of medium-sized trees, which facilitates the regulation of shade and maintains more or less optimum conditions through the year, is desirable. The principal traits to be looked into while selecting shade trees are adaptability to the local climate, growth rate, and ease of establishment. Among the common trees, Balangi (*Artocarpus fraxinifolius* Wt), Nili (*Bischofia javanica* Blum), Jack (*Artocarpus heterophyllus* Lamk), Red cedar (*Cedrella toona* Roxb), Karimaram (*Diospyros ebenum Koenig*), and Karna (*Vernonia monocis* C.B. Darke) are desirable as shade trees for cardamom (Abraham, 1957; Rai, 1978). An introduction from Africa, *Maesopsis eminii*, is a very good shade tree (Korikanthimath, 1983). Another introduction is the southern silky oak (*Grevillia robusta*), and this tree is now a very popular shade tree and provides a stake for black pepper. The wood is hard and is useful as material for cabinets. The heterogeneity of shade tree species and their characteristics is the major constraint on conducting investigations into the shade requirements of cardamom. Certain studies have been carried out to evaluate the usefulness of existing shade trees and identify the most useful ones for cardamom (George et al., 1984). In this investigation, four important species—Karimaram, or ebony; Elengi (Spanish cherry, *Mimusops elengi*); Nandi, or beatrack (*Lagerstroemia lanceolata*); and Jack—were evaluated. Results indicated that cardamom plants grown under Karimaram produced significantly more and longer panicles and longer leaves, yielding 40–50% more cardamom than those plants under other shade trees (Table 2.18).

Trees that carry a crowded crown canopy are undesirable as shade trees because they hardly allow filtered sunlight. *Erythrina lithosperma* and *Erythrina indica* (Dadaps) are commonly planted by growers, especially when cardamom is planted in low-lying areas. But such trees are unsuitable, since they are shallow rooted and hence compete for nutrients and soil moisture. In addition, they act as an alternative host for nematodes.

2.4.5.2 Shade Requirements

The requirement for shade varies from place to place, depending on the lay of the land, the type of soil, the rainfall pattern, the combination of crops grown, and so on (Abraham, 1965; Korikanthimath, 1991). In Guatemala, which receives well-distributed rainfall and has a cool climate round the year, cardamom is grown practically in open areas, with either no shade or only very sparse shade (Anon., 1977). The country's rainfall and climate are major factors contributing to higher productivity in Guatemala.

Gaps in the shade canopy have almost always led to leaf scorching in Indian conditions. It appears that the performance of cardamom plants under such conditions depends on their interaction with shade, sunlight, and soil moisture (Aiyappa and Nanjappa, 1967). In an investigation on light intensity and levels of soil moisture on the growth and yield of cardamom, Sulikeri (1986) reported that, for high-density planting (9000 plants per hectare), yields were 1873 and 1928 ha^{-1} under medium (40–45%) and high (65–70%) light intensity, respectively, as against 864 under low light intensity (15–20%, Table 2.19). Heavier capsules (75.6 g per 100 capsules) were produced by plants receiving medium light intensity, compared with those receiving high light intensity (72.3 g per 100 capsules) and low light intensity (71 g per 100 capsules). The HI under medium light intensity was 0.073, as against 0.066 under low light intensity and 0.037 under high light intensity. These figures simply show that it is important to regulate light intensity, especially during the rainy season when overcast sky reduces the light intensity. It is equally important that shade trees put forth sufficient foliage and provide adequate shade by the time summer sets in. The overhead canopy should, therefore, be regulated once a year, during May–June. With the denudation of forests in the Western Ghats, the normal ecosystem is destabilized and the microclimate and rainfall pattern in the cardamom-growing tracts are vastly changed. With the onset of the dry season (November–January), a cool and humid microclimate in the plantation changes rapidly as hot air waves from the hinterlands pass across the cardamom tracts without much hindrance because of the deforestation all around cardamom pockets. As a consequence, cardamom plants face a harsh environment, resulting in poor growth and, consequently, poor yields.

Table 2.19 Dry Yield of Cardamom (kg/ha^{-1}) as Influenced by Varying Light Intensity and Soil Moisture Levels

Moisture Levels	Light Intensity			Mean Yield
	Low (5000–7000 Lux)	Medium (15,000–17,000 Lux)	High (25,000–27,000 Lux)	
Control	706.50	1217.25	645.75	856.50
Water at 25% ASM	945.90	2030.22	2521.53	1832.55
Water at 75% ASM	941.22	2373.03	2618.88	1977.51
Mean	864.54	1873.50	1928.52	1555.52
Treatment Effects	**Significance**	**LSD (p = 0.05)**	**LSD (p = 0.01)**	
Main treatment	**	366.84	521.82	
Subtreatment	**	221.49	298.26	
Interactions				
(a) Two levels of subtreatments at a fixed level of main treatment	**	383.58	510.60	
(b) Two levels of main treatments at a fixed level of subtreatment	**	492.21	685.35	

Note: ASM, available soil moisture.
** significant at LSD (p = 0.05); main treatment refers to moisture levels; subtreatment refers to light intensity.

Owing to high wind velocity, transpiration and evaporation rates are increased. Moreover, plants suffer the physical pull of the blowing of high-velocity wind. The increased evaporation and transpiration deplete soil moisture rapidly.

2.4.5.3 Pest Outbreak in Relation to Shade

Ecological upsets—especially edaphic ones—have triggered pest problems in cardamom plantations. This is a consequence of the so-called green revolution in which the indiscriminate use of chemicals, both fertilizers and pesticides, in the soil environment has led to many problems. Outbreaks of pests, once considered minor, are assuming alarming proportions in many cardamom-growing areas. Among the insect pests, root grub is seen in exposed, warm, and less shaded conditions, and the insect has emerged as a major pest of cardamom in many areas (Gopakumar et al., 1987). White flies likewise are threatening cardamom plantations in many areas. Outbreaks of locusts in Udumbanchola taluk (in the local language, Kannada, a geographical unit of land in a district) in Idukki district of Kerala State are another example of the ill effects of changes in the microclimate (Joseph, 1986).

2.4.5.4 Biorecycling

Among the plantation crops, no other has the benefit of biorecycling through the maintenance of tree growth in situ as cardamom has. Because nutrients in the soil are liable to leaching and loss, trees and other plants absorb them, and cardamom is a good example. Leaf fall affects the recycling process of these minerals to the upper soil layer, enriching soil fertility in the process.

2.4.5.5 Water Requirements and Irrigation Management

In general, cardamom is grown as a rain-fed crop, and cardamom-growing regions experience a dry spell of about 5–6 months in a year. Increased denudation of forests and deterioration in forest ecology, coupled with erratic trends of rainfall, leads to aridity effects and adversely affects cardamom production (Ratnam and Korikanthimath, 1985). Even if there is no reduction in total rainfall, the failure of premonsoon and postmonsoon showers affects the crop adversely. During the monsoon, postmonsoon, and winter months, although there is sufficient moisture in the soil, plant growth is rather slow because of the low ambient temperature. During the summer months, if adequate moisture is available, the cardamom plant puts forth luxuriant growth. Under normal conditions, panicles start emerging in January and continue to produce flowers from May onward. When postmonsoon rains fail and moisture stress precipitates, flower drop occurs and fruit set is hampered. Under severe conditions of moisture depletion, panicle tips dry up. Therefore, irrigation is necessary from January through May. In such a situation, the determination of what constitutes adequate moisture for higher yields of cardamom needs no emphasis. Raghothama (1979) used two available soil moisture (ASM) levels, 25% and 75%, to study the effect of mulches and irrigation on sucker production. There was no separate control treatment in this investigation. Irrigation at 75% ASM enhanced the performance of all growth and yield parameters, including cardamom yield, but there was no statistical difference between this treatment and 25% ASM. The differential effects of the moisture-level treatments were nullified by the effect of the mulching treatment and a diminution in the dry spell. Plants at the higher irrigation level produced more and longer panicles, a greater number of internodes, more capsules, more fruit set, and a higher capsule weight

Although cardamom requires high moisture level, it is highly sensitive to a high water table and consequent waterlogging (Sulikeri et al., 1978). For better growth, drains should be opened at regular intervals to keep the water table 30 cm below the surface. Sulikeri (1986) investigated the effect of light intensity and moisture level on the growth and yield of cardamom. Results, shown in Table 2.19, indicate that irrigation at 75% ASM resulted in a maximum yield of 1977 $kg\,ha^{-1}$, which was 8% more than that achieved by irrigation at 25% ASM (1832 $kg\,ha^{-1}$) and 31% more than that in the control treatment. Light intensity had a very positive effect, and high light intensity (25,000–27,000 lux) increased yield by 123% compared with low-intensity light at 5000–7000 lux. Also, in the control treatment, there was yield depression at the highest light intensity (25,000–27,000 lux) compared with no yield depression at low light intensity at 5000–7000 lux. In the medium light intensity level (15,000–17,000 lux), yield in the control peaked, indicating that high light intensity might, indeed, be detrimental when the other factors of production are missing.

2.4.5.6 Irrigation Methods

Among the different systems of irrigation, such as surface, subsurface, overhead, trench, and sprinkler irrigation systems, the last is the nearest to ideal in the case of cardamom plantations.

2.4.5.7 Sprinkler Irrigation

Overhead irrigation with sprinklers has many advantages. Cardamom is grown on the slopes of hills with undulating topography, and for such land, a sprinkler system can provide a uniform water supply. Since the rate of water supply can be regulated, surface loss due to runoff and evaporation, as well as conveyance loss, are greatly minimized (Anon., 1985; Bambawale, 1980; John and Mathew, 1977; Saleem, 1978; Vasanth Kumar and Sheela, 1970). A sprinkler system will also preempt puddling and leaching, which, together with runoff, are common with other irrigation systems. The humid atmosphere required for the successful growth and production of cardamom can be created by overhead sprinkling. Frequent light sprinkling can be done in soils with poor water-holding capacity. Irrigation equivalent to a rainfall of 4 cm every fortnight would be quite sufficient.

A sprinkler system should be installed only after a careful survey of the area aimed at producing an efficient and economical design. A perennial source of water is required nearby. Sprinkler systems are designed to meet specific requirements, which may vary from one plantation to another, depending on the lay of the land, the area to be irrigated, and the source of water. The pumping site should be selected in a convenient place, and the entire area should be covered with the least number of pipes. Portable units are more economical to use, but their operation costs are slightly higher compared with those of a permanent system. The main line and laterals can be made portable so that they can be moved easily from one position to another.

Vasanthakumar and Sheela (1970) conducted a field investigation on a sprinkler system for two consecutive years. The field design was a split plot, with irrigated and nonirrigated treatments as main plot and cardamom varieties Malabar, Mysore, and Vazhukka as subplots. Results, shown in Table 2.20, indicate that var. Vazhukka produced the highest number of panicles per clump (114.6) in the treatment with irrigation compared with the one without (90.8). Variety Mysore produced the least panicles in the nonirrigated treatment (51.8). All cultivars produced more flowers in sprinkler-irrigated plots (3048, 1894, and 3754 in varieties Malabar, Mysore, and Vazhukka, respectively) than in nonirrigated plots (26.4, 26.3, 27.6 in varieties Malabar, Mysore, and Vazhukka, respectively). Capsule shedding was comparatively low in the irrigated plots (14.5%, 14.9%, and 11.7%, respectively, in the three cultivars). Panicles of the irrigated plants showed faster growth, with a more pronounced effect in the variety Vazhukka (115.8 cm, as opposed to 98.4 cm in the nonirrigated plants). Even in the variety Mysore, which normally produces shorter panicles, sprinkler irrigation increased the panicle length up to 73.8 cm, whereas the nonirrigated

Table 2.20 Effect of Sprinkler Irrigation on Cardamom Yield and the Yield Components

Cultivar	Extension Panicle Growth (cm) to Total Flowers Borne			Mature Capsules (%)			Capsule Fresh Weight (g)			Capsule Dry Weight (g)			Essential Oil Content		
	I	NI	Mean	I	NI	Mean	I	NI	Mean	I	NI	Mean	I	NI	Mean
Malabar	101.3	78.5	89.9	61.3	50.6	56.0	1465	648	1061.5	353	154	253.5	8.7	7.1	7.9
Mysore	73.8	54.0	63.9	55.7	46.9	51.3	757	421	589.0	189	107	148.0	12.2	10	11.1
Vazhukka	155.8	98.4	127.1	62.5	52.7	57.6	1919	889	1404.0	446	213	329.5	9.6	7.4	8.5
Mean	110.3	76.97	59.87			50.11			1380.33		656.0	158		10.2	8.2
"F"-test			**			*		NS	**			**			**
LSD ($p = 0.05$)			19.59			5.32			3.8			67.40			0.47
Irrigation			12.42			4.43			226			52.94			0.59
Cultivars			17.56			6.27			320			74.86			0.83

Note: I, irrigation; NI, no irrigation; NS, not significant;
* Significant at 95% confidence level;
** Significant at 99% confidence level.

plants produced shorter panicles (54.0cm). The percentage of capsules that reached final maturity was significantly influenced by sprinkler irrigation in all three varieties: The percentages for Malabar, Mysore, and Vazhukkawere 61.3%, 55.8%, and 62.5%, respectively, in the irrigated plots. The corresponding figures in the nonirrigated plots were 50.6%, 46.9%, and 52.7%.

Capsule yield almost doubled in the irrigated plots compared with the nonirrigated plots. The essential oil content of capsules was greater in the irrigated plants, on a dry-weight basis. Mean values for the irrigated and nonirrigated plots were 10.2% and 8.2%, respectively; the highest in the irrigated plants was in variety Mysore (12.2%) and the lowest in the nonirrigated plants (7.1%) was in variety Malabar.

2.4.5.8 Drip Irrigation

In situations where water has become scarce, drip irrigation, like sprinkler irrigation, has become popular. The principle of a drip irrigation system is to use only enough water needed for the crop. Drip irrigation became popular in India after its great success in water-scarce Israel. It minimizes water loss from surface runoff, evaporation, and percolation to deeper layers. Since the water is applied to the base of the plant, drip irrigation is much more efficient than conventional methods of water use, such as flooding. Water economy is the greatest advantage. For cardamom, the application of 10–15 liters of water per day is sufficient. If cultivation is done on the contour, drip irrigation is easy to practice. Water from small farm ponds can be drawn into this system without pumping. Drip irrigation has 80–95 efficiency (Kurup, 1978). However, its principal limitation is the initial cost for installation. Plant spacing is also important, because, closely spaced plants, as in the State of Karnataka, would involve higher cost of installation than where spacing is wider. In Karnataka, drip irrigation costs approximately Rs 25,000 to 47,000 (US$600–1100) per hectare.

2.4.5.9 Perfospray Irrigation

In perfospray irrigation, water is sprayed under medium pressure. Aluminum or polyvinyl chloride (PVC) pipes of high density are placed 6–9cm apart, and water is pumped into the pipes at predetermined time intervals. The setup can be shifted from one place to another and has been found suitable for cardamom (Sivanappan, 1985).

2.4.5.10 Contour Furrows Irrigation

Wherever water is on the highest point, taking advantage of the natural slope of the land, contour furrows irrigation can be established. Contour furrows are opened, and water is allowed to flow through them to reach the plants. If needed, small basins can be placed around the base of the plants. The system requires no investment to set up. Water can also be stored in ponds along the slope and used for irrigation in summer months.

2.4.5.11 *Time and Frequency of Irrigation*

In a sprinkler system, to get the effect of 25 mm rainfall, the system can be operated once in a 12–15-day interval during May–June. The soil moisture level should be above 50% of the maximum water-holding capacity. A soil and/or environment induced stress period for about 45 days during December–January has been found to be quite beneficial. It would be ideal to commence irrigation during the first week of February and continue at intervals of 12–15 days until the regular monsoon commences in the first week of June.

2.4.5.12 *Water Harvesting*

Cardamom terrain often undulates, with moderate to steep slopes. Quite a number of small and fairly big streams pass through many of these areas. Runoff from cardamom watersheds can be collected in farm ponds and check dams or in underground water tapped through wells.

Harvested water can be stored in ponds and check dams by minimizing losses caused by seepage, evaporation, and recycling. Apart from improving and stabilizing yields under rain-fed cardamom cultivation, check dams, farm ponds, and dug wells reduce flood hazards and recharge groundwater. Many times, such instruments serve as percolation tanks that would substantially augment the availability of groundwater in the area.

2.4.6 *Cardamom-Based Cropping Systems*

An excellent example of a forestry–cum–cash crop combination is the cardamom plantation. Some forests are uneconomical in view of the heavy investments that are involved in building up adequate infrastructure for the extraction and transportation of timber. Where cardamom plantations can be established, such forests will turn out to be economically viable. These cropping systems are environmentally harmonious, as they will neither upset nor adversely affect the environment or the protective quality of the forest. Cardamom is, perhaps, the plantation crop that involves the least disturbance to the existing forest trees, as against the partial felling of trees required to raise coffee or black pepper.

2.4.6.1 *Sole Forest versus Cardamom Intercropped Forest: Economics and Labor Utilization*

The cultivation of cardamom beneath shade trees needs to be examined to assess the benefits derived therefrom. An evergreen forest in the Western Ghats region managed ideally (sole forestry) can yield 10 m^2 of timber per annum per hectare. The Indian yields are much below this level. At present, returns are approximately 10 times more than investment because of an increase in the cost of timber. Moreover, clearing of forests for nonforestry purposes is not permitted anymore by law. Computed over a period of 20 years, the net pecuniary return from a system as described above will be Rs 700,000 (approximately US$16,300) and the job opportunities created will be equivalent to Rs 500,000 (US$11,600), according to Joseph (1978). If the area is brought under cardamom plantation exclusively, the economic benefit, at current yield expectations and the ruling price of the produce, will be to the tune of Rs 450,00,000 (approximately US$1.05 million) and the additional creation of job opportunities to the tune of Rs 80,00,000 (approximately US$1,86,000). In other words, a cardamom–forest mix could bring about substantial economic advantage. Mixed cropping of cardamom–coffee and black pepper is called "multitier" cropping, as the plants attain different heights and utilize sunlight, soil moisture, and soil nutrients differently. (Different types of mixed or multitier crop combinations are discussed in a later section (2.4.6.2).)

2.4.6.2 Mixed Cropping System

Nutmeg-Clove-Cardamom Combination

Because of their different canopies, these three crops tap sunlight at different heights. Also, because of their varied root systems, they tap soil moisture and soil nutrients with differing degrees of efficiency. In cardamom plantations, instead of planting other forest tree species in the vacant areas, nutmeg or clove or both can be planted. The combination is both ecologically feasible and economically profitable. A good example of such a combination can be seen in Burliar at an elevation of 1680 m amsl in Tamil Nadu. This plantation was started as a nutmeg garden, and clove seedlings were planted later, in between the nutmeg plants. Subsequently, cardamom seedlings were interplanted. Nutmeg and clove started to bear in about 6–7 years time after planting. Although both of these plants require regulated shade in the early stages of growth, clove does not require much shade once it starts to yield, and clove and nutmeg can provide shade to cardamom seedlings in their early stages of growth. Cardamom plants in this combination started to yield in about 3 years time from the date of planting and gave a yield of about 150 kg ha^{-1} of dry capsules from 600 plants. Clove, on average, yielded about 1 kg in dried form from a single plant. To meet the shade requirements of cardamom, besides the nutmeg and clove, tall growing shade trees at regular intervals were retained in the garden. Sprinkler irrigation supplemented the shading effect.

Cardamom-Areca Nut (Areca catechu) Combination

Like cardamom, areca nut is a perennial crop. The long prebearing age of the areca nut, the small income from the initial harvest, the risk of pests and disease, remoteness from markets, and inadequate transport support—all of which are features of areca nut farming in the state of Karnataka primarily—led to the introduction of other crops in areca gardens (Abraham, 1956; Khader and Antony, 1968; Nagaraj, 1974). Until cocoa was introduced, cardamom was the principal crop planted in areca gardens in southern districts of Karnataka State. As regards the unit value of different crop mixtures in areca nut gardens, areca nut–cardamom is the best combination (Korikanthimath, 1990) at elevations between 700 and 1080 m amsl. This is because areca nut is not successful at higher elevations. Normally, areca nut gardens at low-lying areas on flat land with irrigation support are most nearly ideal for interplanting with cardamom. Investigations by Bhat and Leela (1968) and Bhat (1974) have shown that more than 80% of the areca nut plants have their root system confined to a radius of 75 cm from

Photo 2.6 Mixed cropping in cardamom plantation (cardamom + coconut + arecanut).

the base of the palms, spaced at 2.7 m × 2.7 m. Fourteen percent of the roots are at a radius of 25–50 cm and only 6% of the roots within a radius of 50–75 cm. Cardamom roots penetrate to a depth of only 40 cm. Although both cardamom and areca nut have most of their root systems confined to the surface, there would not be much of a competition between both plants inasmuch as moisture and nutrients are concerned. Muralidharan (1980) reported that 32.7–47.8% of incident light penetrates through the canopy of a 14-year-old areca nut garden, depending on the time of the day. In a pure areca nut crop spaced at 2.7 m × 2.7 m, this amount of light energy hits the ground and is wasted. Approximately 27% photosynthetically active radiation passes through areca nut canopy (Balasimha, 1989). Cardamom plants interplanted in areca nut gardens can utilize this otherwise wasted energy effectively (Photo 2.6).

In an areca nut–cardamom mixed-crop field, the areca nut plant should be planted in a northeast direction, spaced 2.7 m between cardamom plants and rows. September–October is the ideal time to plant areca nut (Khader and Antony, 1968). In addition, banana plants, spaced 5 m × 5 m between areca nut plants before the areca nut plants are planted, can provide enough shade to the transplanted areca nut seedlings and would also be a good supplemental source of income to the farmer. After 7 years of planting areca nut seedlings, cardamom seedlings can be planted between the areca nut plants. In old plantations, cardamom is planted in alternate rows, with a plant-to-plant spacing of 2 m. In new plantations, cardamom is planted between rows of areca nut plants. About 1250–1500 plants are accommodated in a hectare of areca nut garden.

Areca nut and cardamom share common cultural operations, such as weeding, mulching, and irrigation, thereby bringing down the total cost of cultivation. It is important to apply adequate quantities of fertilizer for the supply of plant nutrients from the time the plants establish. This ensures early bearing and sustained yielding for a long time. In the south Kannada district of Karnataka State (Sirsi and adjoining areas), cardamom replanting is taken up on a regular basis once in every 5–7 years, owing to the heavy incidence of the dreaded katte disease. Many planters pick two to three harvests and then replant. The "high-production technology" package for areca nut–cardamom mixture has given high yields of cardamom—as much as 625 kg of dry cardamom and 3750 kg of dry areca nut per hectare per year under irrigated conditions. In nonirrigated situations, the corresponding yields are 325 and 2250 $kg\,ha^{-1}\,year^{-1}$, on average (Korikanthimath, 1989). Korikanthimath et al. (2000a) investigated the microclimatic and photosynthetic characteristics in areca nut–cardamom mixture and reported higher photosynthetic and transpiration rates and more efficient carboxylation and water use. Intercellular carbon dioxide concentration and stomatal conductance were higher in cardamom than in areca nut.

Areca Nut–Tree Spices and Cardamom Combination

This combination is seen in Koppa taluk in the state of Karnataka. In it, coconut plants act as the first tier, followed by black pepper, which is trained on the coconut trees as the second tier. Cardamom and coffee (both *Arabica* and *Robusta* varieties) are planted in between the coconut rows. All of these crops are complementary to each other and help in maximizing income from a unit area.

In Belur village of Shimoga district in Karnataka State, a "spice cafeteria" has been established. Cardamom was planted between areca nut plants, followed by nutmeg at a spacing of 10 m × 10 m. Clove plants were also planted in the same garden, on the borders. The nutmeg trees have grown to a height of 8–10 m, and their yield has been established. Clove trees produced a larger yield when planted on the borders rather than in between the areca nut plants. On average, each nutmeg plant yields about 500–750 fruits. Clove trees yielded 750–1250 kg per tree. Around 275–400 $kg\,ha^{-1}$ of dry cardamom capsules was obtained. The crop combination of areca nut as a top canopy providing shade for nutmeg, cloves, and cardamom is found to be useful in increasing the income from a unit area. Cardamom is phased out in the course of time when both nutmeg and clove grow up and put forth a sufficient canopy.

2.4.6.3 Cardamom–Coffee Mixed Cropping

Cardamom can be successfully mixed with coffee. In this combination, coffee (both *Robusta* and *Arabica* species) constitutes the first tier, cardamom the second. Often, black pepper trained on the shade trees forms another component, and the shade tree canopy forms the top tier. Many experimental studies have investigated the merits of the coffee–cardamom combination (Table 2.21).

Arabica coffee (variety S. 795) planted in 1976 at a triangular spacing of 1.8 m × 1.8 m was selected to study the feasibility of mixed cropping of cardamom under rain-fed conditions. Every row of coffee was uprooted in 1981–1982, and cardamom subsequently was planted by providing a spacing of 5.4 m × 0.9 m (2058 plants per hectare). The fertilizers, as per the recommended doses, were applied separately to cardamom. Plant protection measures and cultural operations such as weeding, mulching, shade regulation, and irrigation were commonly followed for both cardamom and coffee. An average yield of 259–308 $kg\,ha^{-1}$ dry cardamom and

Table 2.21 Economics of Cardamom–Coffee (Arabica, Rain-fed Crop) Mixed Cropping System

Year	Yield (Dry, kg ha^{-1})		Gross Income		Total Expenditure	Net Profit	
	Cardamom	Coffee	Cardamom	Coffee	Total	(Cardamom+ Coffee)	
1983–1984	122.5	437.5	24,010	5486	29,496	18,750	10,746
1984–1985	300.0	175.0	54,600	2450	57,050	19,550	37,500
1985–1986	355.0	312.5	59,640	4531	64,171	21,500	42,671
Average	259.0	308.0	46,083	4156	50,239	19,933	30,305

Note: Gross income, total expenditure, and net profits are given in Rs (US$1 = approximately Rs 43).

coffee was obtained during the 3 years of the study (1983–1986). This crop combination of cardamom with arabica coffee gave a net profit of Rs 30,305per hectare (approximately US$ 700).

In another case, robusta coffee (variety Ferdinia) planted at 2.7 m × 2.7 m in 1947 was interplanted with cardamom after the removal of alternate rows of coffee in 1985 to accommodate cardamom variety Malabar (C1.37) at a spacing of 1.8 m × 1.2 m, between two rows of coffee spaced at 5.4 m × 2.7 cm. Average yields of 1907 kg ha^{-1} dry coffee (mean yield of 4 years) and 950 kg of dry cardamom capsules were obtained (Korikanthimath, 1989). These yields clearly demonstrate the high-production potential of a cardamom–coffee combination. Srinivasan et al. (1992) reported that the monetary benefit can be enhanced by more than 30% by adopting a cardamom–coffee planting system at a ratio of 1:1 in alternate rows either along or across the slopes. Thus, the combination of coffee–cardamom offers considerable financial advantage to farmers who adopt it. Additional employment potential, efficient land and water use, and effective weed control are some of the other advantages of multistoried cropping systems over monoculture.

The inclusion of either black pepper alone, as in Kerala, or a combination of coffee and areca nut, as in Karnataka, in the multiple cropping system with cardamom is highly remunerative and can give a higher benefit–cost ratio. In Karnataka State, a ratio of 3.53 was recorded in a cardamom–coconut mixed cropping system (Korikanthimath et al., 1988a). In the case of cardamom–coffee mixed cropping system, Korikanthimath et al. (1989b) found that the benefit–cost ratio was 1.94 (with a single-hedge system with arabica coffee) and 4.25 (with a double-hedge system with robusta coffee).

2.4.6.4 *Changes in the Rhizosphere due to Mixed Cropping*

It is natural that when different species of plants grow in the same soil matrix, different types of changes—physical, chemical, and microbiological—occur, and from a fertility point of view, those which occur in the rhizosphere are most important. Some of these changes have been cataloged in cardamom mixed cropping systems (Korikanthimath, unpublished). On a general comparative basis, results indicated that far more microbiological changes occur in mixed cropping systems than with monocropping. In the case of cardamom–areca nut mixture, the bacterial population registered an increase of 93% at a depth of 0–15 cm, while at a lower depth (15–30 cm) the population count was a reduced 29%. By contrast, the fungal population increased with depth, from 32% at 0–15 cm and 59% and 61% at 16–30 cm and 30–45 cm, respectively, over the figures obtained with monocropping. The actinomycetes population increased by 66% in the top 0–15 cm of soil, but thereafter decreased with depth.

When robusta coffee was mixed with cardamom, the bacterial population registered an increase of 41% at a depth of 0–15 cm, while the fungal population increased by 69% and did not show any significant changes with increased depth. There was no effect on the actinomycetes population. In the case of an arabica–cardamom–black pepper combination, the pattern in the changes of microbial or fungal population with depth was not consistent (Korikanthimath et al., 2000b). However, compared with monocropping, mixed cropping showed increases of 71%, 98%, and 52%, respectively, in the rhizosphere of arabica coffee, cardamom, and black pepper. Similar increases were noted in the mixed cropping fungal population compared with monocropping. These figures refer to the total microbial, fungal, and actinomycetes count; no efforts were made to isolate the organisms and identify them individually, which would have been far more informative.

2.4.6.5 Conclusions

The "high-production technology" for cardamom developed at the Cardamom Research Station, Appangala, Karnataka State, under the direct administrative control of the Indian Institute of Spices Research at Calicut, Kerala State, and the overall administrative control of the Indian Council of Agricultural Research has clearly shown the benefits in farmers' fields. A combination of high-quality planting material, optimum crop and soil management techniques, and the control of pests and diseases will lead to superior yields. Although production constraints do throw up challenges, alleviating these constraints, especially in the area of heat and drought stress, assumes much importance. In sum, in tracts geared toward high production and high productivity, cardamom must be considered not in isolation, but as one important component of an entire farming system.

2.4.7 Cardamom Nutrition

Traditionally, in India, cardamom is grown as an undergrowth in the dense evergreen forests of the Western Ghats of the state of Kerala, without the application of any fertilizers and without providing irrigation. Subsequently, with the development of intensive agriculture, tilling the soil became a routine practice, but it led to the soil's being depleted of its inherent fertility. The felling of forest trees compounded the problem, because the leaf litter was an important source of organic matter enriching the base soil in many ways. Heavy rainfall, a dominant characteristic of these regions, led to surface runoff, the loss of fertile topsoil, and the depletion of plant nutrients with it. Continuous cropping in the same area also leads to the same debilitating effects on the soil. These ill effects necessitate the application of supplemental crop nutrients through fertilizers, primarily chemical. Cardamom is a heavy feeder of potassium, which is one of the three principal plant nutrients. (Nitrogen and phosphorus are the other two.) Judicious agromanagement techniques and the use of high-yielding varieties are important factors influencing productivity, as manifested in the unit cost of cultivation of any crop. Among the different aspects of crop production, nutrition management is crucial because the nutrient factor decides 50% of crop productivity. And among the various factors of production, the nutrient factor is least resilient to management. This combination, then, calls for a clear understanding of the dynamics of cardamom nutrition, followed by precise estimations of the bioavailability of soil nutrients and a working schedule of dependable fertilizer practices. Accordingly, this section discusses the role of "The Nutrient Buffer Power Concept" in cardamom nutrition, with specific reference to potassium, since that is the most important of all the nutrients involved in cardamom nutrition.

2.4.7.1 The Cardamom Soils

In India, soils in which cardamom is grown come under the order Alfisol and the suborder Ustalf, which are derived from schist, granite, and gneiss, are lateritic in nature (Sadanandan et al., 1990), and are formed under alternate wet and dry conditions. The soils most favorable for the growth and development of cardamom are red lateritic loam with layers of organic debris that is present in evergreen forests, although the plant can grow on a variety of soils with only a shallow zone of humus accumulation. In general, cardamom-growing soils are fairly deep with good drainage. The clay fraction is predominantly kaolinitic; hence, potassium fixation occurs in these soils. The cardamom-growing soils of the state of Karnataka are mostly clay loam (Kulkarni et al., 1971). In Guatemala, cardamom is generally grown in rich forest soils. In the countryside, the crop is grown in the northern region in recently cleared forestlands, where soils have dolomitic limestones underlined with typical tropical clay. In the southern regions, the soil is sandy clay loam with volcanic ash deposits, which render the soils in the south more fertile than those of the north (George, 1990).

2.4.7.2 Soil Reaction

Analysis of the soil samples from several of the cardamom-growing areas shows that they are generally quite acidic, with a pH in the range of 5.0–5.5 (Zacharaiah, 1975). Nair et al. (1975) reported that the pH of the soils in the three important cardamom-growing districts of Kerala State, Idukki, Waynad, and Palakkad, ranged from 4.7 to 6.15, from 4.75 to 5.2, and from 5.2 to 5.5, respectively. They also found a slight variation in the pH with soil depth, with the surface layer having a higher pH than the subsurface. Ranganathan and Natesan (1985) reported that the pH of cardamom-growing areas of Tamil Nadu has a pH ranging from 4.7 to 7.0 (Vadiraj et al., 1998).

2.4.7.3 Cation Exchange Capacity

The cation exchange capacity (CEC) of the cardamom-growing soils varied from 8.6 c mol (p+) to 58.5 c mol(p+)kg^{-1}. The CEC of Coorg district soils in Karnataka is higher than that of the soils from Idukki district in Kerala State. In general, the CEC tends to increase with increasing altitude, a relationship attributed to the active form of humus present in high proportions compared with the total organic matter. CEC was also found to be positively correlated with soil organic matter.

2.4.7.4 Organic Carbon

The organic carbon content of the soils is low at low altitude and increases as the elevation increases. The mean organic carbon content of various cardamom-growing soils of Karnataka State was found to be 5.9% (Kulkarni et al., 1971). Nair et al. (1978) found the organic carbon content of Kerala cardamom-growing soils to be 3.3% in Palakkad district, 3.6% in Idukki district, and 4.6% in Wayanad district. They also found the organic carbon content to decrease with depth. Srinivasan (1984) found a significant positive correlation between organic carbon and total, as well as available, nitrogen. The rate of decomposition of organic matter in cardamom-growing soils is much lower than that of the other plantation crops growing soils, such as those in which tea grows, at the same elevation, because of forest tree association and a consequent lower mean annual temperature (Ranganathan and Natesan, 1985).

2.4.7.5 Soil Phosphorus

Most of the cardamom-growing soils are low to medium (less than 5–12.5 kg ha^{-1}) in available phosphorus. The percentage of soil samples falling into this category is 83 for Karnataka State, 68 for Kerala State, and 63 for Tamil Nadu, indicating that, on a relative basis, soils of Karnataka State are poorer in available phosphorus compared with both Kerala and Tamil Nadu cardamom-growing soils (Anon., 1998a). Cardamom soils also contain large quantities of iron and aluminum oxides, and their presence leads to phosphorus fixation. In laboratory studies, the overall phosphorus fixation of cardamom-growing soils ranged from 55% at 1 h to 88% after 90 days, in the lower level of soil application at 25 kg P_2O_5 ha^{-1}; at the higher rate of 50 kg P_2O_5 ha^{-1}, it ranged from 43 to 85% (Srinivasan and Mary, 1981). In field conditions, these values may not apply, as the thick mulch may prevent immediate fixation of soluble phosphorus and the organic acid content of the high levels of organic matter present would facilitate phosphate solubilization (Ranganathan and Natesan, 1985). The availability of phosphorus was found to decrease sharply with soil depth (Nair et al., 1978; Srinivasan et al., 1986).

2.4.7.6 Soil Potassium

Among the cardamom-growing soils of India, Kerala soils are rated high in available potassium (78%) while Tamil Nadu soils have 71% available potassium and Karnataka soils only 53% (Anon., 1998). These percentages are based on exchangeable potassium and may not give an accurate picture of the availability of the element. Srinivasan (1990) found the potassium-fixing capacity of cardamom-growing soils to vary from 16.9 to 32.1% for Kerala soils, 11.9 to 21.3% for Karnataka soils, and 19.0 to 23.3% for Tamil Nadu soils. Potassium availability decreased with soil depth (Nair et al., 1978; Srinivasan et al., 1986). Traditionally, NH_4OAc is used as an extractant to quantify available potassium. Nair et al. (1997) showed that recommended potassium fertilizer treatments for cardamom that are based on NH_4OAc extraction is not reliable and that a dependable method will have to be based on the potassium buffer power of soils. Studying 95 locations spread across the states of Kerala and Karnataka, they found that capsule yield is related to potassium buffer power and not to NH_4OAc-extractable potassium. On the basis of their study, these researchers have suggested a thorough overall of the potassium fertilizer schedule for cardamom, to be based on estimates of the potassium buffer power. On average, the cardamom-growing soils from the Coorg district in Karnataka State contain much less NH_4OAc-extractable potassium than HNO_3-extractable potassium, compared with soils from Idukki district of Kerala State, although the other soil parameters are quite comparable. Despite the lower NH_4OAc-extractable potassium in Coorg soils, cardamom production was nearly twice as much as in Idukki soils with high HNO_3-extractable potassium. But the Coorg soils have a high potassium buffer power, and because these soils maintain an optimum balance of potassium on cardamom plant root, they are able to meet the potassium demands much better than the Idukki soils can. These differences are manifested in the cardamom yield.

2.4.7.7 Secondary Nutrients and Micronutrients

The sulfur status of cardamom-growing soils of southern India was studied by Srinivasan et al. (2000). Results from a total of 100 samples collected (60 from Kerala State, 28 from Karnataka State, and 12 from Tamil Nadu) showed that sulfur content ranged from traces to 36 ppm in Kerala soils, traces to 27.5 ppm in Karnataka soils, and 15–36 ppm in Tamil Nadu soils. If 10 ppm is considered as the critical level, then 43.3% of Kerala soils and 50% of Karnataka soils could be categorized as deficient in sulfur while Tamil Nadu soils are not deficient in that element. Investigating the micronutrient status of cardamom-growing soils of South India, Srinivasan et al. (1993b) indicated that available iron ranged from 14.6 to 65.8 ppm, available manganese from 1.3 to 44.8 ppm, available copper from 0.66 to 32.2 ppm, available zinc from 0.01 to 2.71 ppm, available boron from 0.05 to 3.7 ppm, and available molybdenum from 0.01 to 11.1 ppm. On the basis of the critical levels prescribed for different micronutrients, it has been observed that 68% of the cardamom-growing areas are deficient in zinc, 49% deficient in boron, 28% deficient in molybdenum, and 9% deficient in manganese. None of the sampled soils showed any iron or copper deficiency. There was no deficiency of available manganese in Karnataka soils and no deficiency of available molybdenum in Tamil Nadu soils. Within the pH range of the soils investigated, there was a statistically significant negative correlation between soil pH and available iron while the relationship between both available zinc and available copper with that of soil pH was positive. Among the micronutrients investigated, only available manganese showed a statistically significant positive correlation with the organic carbon content of the soil. On the whole, these investigations indicate that, among the micronutrients, zinc is turning out to be the most problematic in cardamom-growing soils. It is perhaps relevant to mention in this context that, for precise quantification of micronutrients, especially the problematic zinc, it would be preferable to reorient soil testing for available zinc on the basis of the "buffer power concept," as in the case of another important spice crop, namely, black pepper (Nair, 2004), which grows in situations similar to those of cardamom.

2.4.7.8 Nutrient Deficiency Symptoms

Sand culture was used to characterize nutrient deficiency symptoms in cardamom (Deshpande and Kulkarni, 1973). Table 2.22 lists the symptoms observed.

2.4.7.9 Shade Trees and Soil Fertility

Cardamom is cultivated under shade trees, and the soils in general have a high fertility status that is due to the addition of leaf litter and recycling of plant nutrients (Zachariah, 1978). Nair et al. (1980) observed that shade trees bring to the soil surface the nutrients they take up from the lower layers of soil. Even though, on average, about $5.5\,t\,ha^{-1}$ of organic material are recycled as leaf litter, weeds, and pruned plant parts in a cardamom plantation in a year, the nutrients are mainly in the organic form and are available to the crop only by the process of mineralization. Because the rate of mineralization is always low, the nutrients that become available to the crop will be able to sustain only average growth with average production (Srinivasan et al., 1993a).

2.4.7.10 Plant Nutrient Composition and Uptake

Tissue Concentration

A number of investigators (Kulkarni et al., 1971; Pattanshetty, 1980; Raghothama, 1979; Ratnavele, 1968; Srivastava et al., 1968; Sulikeri, 1986; Venkatesh, 1980) have documented the tissue concentration of the entire cardamom plant and its various parts. Between the second and fifth leaves, the phosphorus and potassium content was concentrated in the leaf tip of the second leaf and all the other leaves contained lesser amounts, whereas in the case of calcium, the least content was found in the second leaf and all the others contained more (Ratnavele, 1968). The aerial stem contained more potassium than the leaves (Ratnavele, 1968). Kulkarni et al. (1971) reported that, as the plant aged, the nitrogen, phosphorus, and calcium content in the leaves increased, but there was a general reduction in magnesium and potassium content. At the bearing stage, although the major nutrients decreased in the leaf tissue, an increase was observed in the case of calcium and magnesium. A similar trend was observed in aerial stem as the plant aged. Rhizomes contained lower levels of major nutrients than did roots, whereas roots contained more of the secondary nutrients. In general, nitrogen content was maximal in leaves, followed by shoots, rhizomes, and roots (Pattanshetty, 1980; Raghothama, 1979; Venkatesh, 1980).

Table 2.22 Nutrient Deficiency Symptoms in Cardamom

Nutrient	Deficiency Symptoms	Cited by
Nitrogen	First, older leaves are affected. Leaf size is reduced. Production of suckers also is reduced, and newly formed suckers dry up after some time.	Deshpande and Kulkarni (1973)
Phosphorus	Symptoms appear 4–5 months later. Small purplish patches form on leaves, followed by premature leaf drop. Stunting appears, and sucker number is reduced.	Deshpande and Kulkarni (1973)
Potassium	Deficiency symptoms first appear in older leaves. Shoots and roots exhibit reduced of growth, and plants show browning of leaf tips, later extending downward. Finally, the entire leaf turns dark brown. Further sucker production is completely absent, and plants die after about 2 weeks from the date of the first appearance of the symptoms.	Deshpande and Kulkarni (1973)
Calcium	Deficiency symptoms appear on young leaves after 75 days. Shoots and roots exhibit reduced of growth, and further growth of aerial shoots ceases. The aerial stem thickens and shows a bulblike growth. Scattered yellow spots form on leaves and margins, which turn brown with a golden-yellow band underneath.	Deshpande and Kulkarni (1973)
Magnesium	Internodal length is reduced, and the plant appears broomlike. The top leaves become twisted, and the leaf tip drys out. Later, the whole leaf becomes pale yellow, with the mid rib turning green. The most commonly observed symptom in the nursery is white papery spots on leaf lamina. Sucker production is inhibited.	Deshpande and Kulkarni (1973)
Sulfur	First appears on young leaves. The growing leaf becomes whitish in color, after which leaves die, starting from the margins.	Deshpande and Kulkarni (1973)
Zinc	Poor plant growth, curled leaves, interveinal chlorosis.	Anon (1979)
Boron	Reduction in leaf size and cracking of leaf lamina.	

Potassium content was found to be highest in shoots, followed by rhizomes and leaves. No definitive pattern was observed with regard to phosphorus in these investigations. Sulikeri (1986) found the highest nitrogen content to be in leaves, followed by rhizomes, shoots, and roots, during the prebearing stage. By harvest time, the nutrient content had increased in all of the plant parts except the leaves, in which it was found to decrease. Phosphorus content did not vary much in different plant parts during the prebearing period, whereas at harvest the capsules contained more phosphorus than did any other plant part. During the prebearing stage, the most potassium was found in rhizomes, followed by shoots, leaves, and roots. At harvest, the same trend was noticed, except that capsules and panicles contained more potassium than did leaves and roots. Potassium content at harvest decreased in all plant parts compared with the prebearing stage, except in rhizome, where its concentration increased. Calcium was highest in leaves at the prebearing stage, whereas capsules recorded the maximum content at harvest. Magnesium content was maximal in leaves and minimal in shoots both at the prebearing stage and at harvest.

All of the preceding findings clearly show that individual nutrients have their own patterns of accumulation in the cardamom plant parts, coinciding with different growth stages of the plant. No two nutrients follow a similar pattern.

Nutrient Uptake

For the production of a kilogram of cardamom capsules, 0.122 kg of nitrogen, 0.014 kg of phosphorus, and 0.2 kg of potassium are exported by the plant from soil. For the plant as a whole, the nitrogen-to-phosphorus-to-potassium-to-calcium-to-magnesium ratio worked out to be 6:1:12:3:0.8 (Kulkarni et al., 1971). A similar ratio was reported by Venkatesh (1980). Maximum uptake of nitrogen (53%) and calcium (47.5%) was found to be by leaves, followed by shoots, whereas the reverse order was observed for phosphorus, potassium, and magnesium. Uptake by rhizome and roots was far less compared to the aerial parts of the plant (Pattanshetty, 1980). As regards uptake among the major nutrients, uptake of nitrogen was more than that of phosphorus and potassium during the prebearing stage than at harvest. At harvest, there was a considerable decrease in nitrogen uptake by leaves while the rhizomes and shoots showed an increase. In the case of phosphorus and potassium, the maximum uptake was at harvest by shoots and there was reduction in nitrogen uptake by leaves. Rhizomes occupied the second position in uptake of potassium at harvest, indicating the importance of potassium in cardamom production.

Among secondary nutrients, the maximum calcium uptake was by shoots at the prebearing and harvest stages while, in the case of magnesium, it was the leaf. Calcium uptake by shoots was greater at harvest than that at the prebearing stage. A reverse trend was observed for magnesium (Sulikeri, 1986).

From the preceding data, it can be concluded that, of all the nutrients, potassium is the most important for good cardamom production. In terms of importance, the order of the nutrients is potassium, nitrogen, and phosphorus. Soils that rate high for potassium availability, such as soils from the state of Karnataka, produce good cardamom yields (Kulkarni et al., 1971). Cardamom is extensively grown in the Western Ghats, Wayanad, and Idukki districts of Kerala State, and where the soils have been rated high for potassium availability, good cardamom yields follow.

2.4.8 Fertilizer Requirements

2.4.8.1 Scheduling Fertilizer Application

Up to the mid-1950s, cardamom grew in the forest ecosystem and not enough attention was given to scientific fertilizer scheduling. Manuring was confined to organic manures, if, at all, any was applied. As the crop gained increasing national and international stature as an important cash-earning spice crop, farmers realized the benefits of applying fertilizer and of the resultant increase in crop yield. A dose of nitrogen, P_2O_5, and K_2O in the ratio 45:45:45 kg ha^{-1} for Kerala soils, 67:34:100 kg ha^{-1} for Karnataka soils (where half of the quantity of nitrogen applied was in organic form), and 45:34:45 kg ha^{-1} for Tamil Nadu soils was suggested by de Geus (1973). By contrast, Zachariah (1978) suggested a maintenance dose of nitrogen, P_2O_5, and K_2O in the ratio 30:60:30. On the basis of further investigations, the dose of fertilizers suggested to obtain a total yield of 100 kg ha^{-1} of dry capsules was in the ratio 75:75:150 kg ha^{-1}. If the target yield was more, the fertilizer dose had to be proportionately increased. An additional dose of 0.65 kg nitrogen and P_2O_5 and 1.30 kg K_2O ha^{-1} was suggested for an increase in yield from every 2.5 kg of dry capsules over normal yield (Anon., 1976; Kologi, 1977).

Results of investigations (Srinivasan et al., 1998) indicated that a significant increase in yield and in economy of fertilizer use can be brought about through a combination of soil and foliar applications. In Karnataka, nitrogen, P_2O_5, and K_2O in the ratio 37.5:37.5:75 kg ha^{-1} through soil application supplemented with a foliar application of 2.5% urea + 0.75% single superphosphate + 1.0% MOP gave 64 kg ha^{-1} more of capsule yield compared with no fertilizer application. In Tamil Nadu, with a soil application of nitrogen, P_2O_5, and K_2O in the ratio 20:20:40 kg ha^{-1}, supplemented with a 3% urea + 1% of single superphosphate + 2% of MOP gave an increase in yield of 43 kg ha^{-1}. In irrigated plantations, the general fertilizer recommendation is to apply nitrogen, P_2O_5, and K_2O in the ratio 125:125:250 kg ha^{-1} to soil in three splits (Anon., 1997). Kumar et al. (2000) suggested applying nitrogen, P_2O_5, and K_2O in the ratio 75:75:150 kg ha^{-1} in Karnataka soils. Urea was found to be a better source of nitrogen, compared with ammonium sulfate (Deshpande et al., 1971). For cardamom-growing soils, which are rich in organic matter and acidic in reaction, Mussoori rock phosphate (a natural rock phosphate mined in North India) was found to be the ideal source of phosphorus (Nair and Zachariah, 1975). When plant density increased to 5000 plants per hectare under the trench system of planting, fertilizer has to be applied at a higher ratio, namely, 120:120:240 kg ha^{-1} of nitrogen, P_2O_5, and K_2O (Korikanthimath, 1986). This combination was subsequently revised to 150:75:300 kg ha^{-1} (Korikanthimath et al., 1998a). In low-rainfall areas of Lower Pulney tracts of Tamil Nadu, a fertilizer dose of 40:80:40 kg ha^{-1} of nitrogen, P_2O_5, and K_2O is recommended (Natarajan and Srinivasan, 1989).

Srinivasan and Bidappa (1990) investigated nutrient combinations based on phosphorus and potassium adsorption isotherms. The total phosphorus requirement (TPR) was calculated as to be TPR (kg ha^{-1}) $= (60 - x) \times 3.24$, where 60 is the desired soil fertility level for phosphorus, x denotes the soil test value for phosphorus, and 3.24 is a constant derived by taking into account soil bulk density—the weight of soil per hectare. The total potassium requirement per kilogram is $(300 - x) \times 2.64$, where 300 is the desired soil fertility level for potassium, x denotes the soil test value for potassium, and 2.64 is the constant, derived as explained previously. The recommended rates derived maintain soil phosphorus and potassium levels at 60 and 300 kg ha^{-1}, respectively, as these levels are considered optimum for cardamom.

Ranganathan and Natesan (1985) reported a beneficial application of zinc sulfate to nursery plants by incorporating the salt with the fertilizer mixture.

Sivadasan et al. (1991) found that application of 500 ppm zinc sulfate as foliar spray enhanced growth and yield, as well as the quality of the produce. Deshpande et al. (1971) found that liming cardamom soils was beneficial in correcting soil acidity and enhancing the rate of nitrification, resulting in better plant growth. However, subsequent observations (Anon., 1997; Nair et al., 1998) indicated that liming is not required as a routine practice in cardamom plantations. Table 2.23 shows the schedule of fertilizer application.

Table 2.23 Fertilizer Schedule in Cardamom and Comparison of Phosphorus Buffer Power of Central European Soils

I. Fertilizer Schedule in Cardamom[a]

Region	**Soil Application**	**Soil + Foliar Application**	**Time of Application**	
			Soil	**Foliar**
Karnataka	Nitrogen–phosphorus–potassium 75:75 (150 kg ha^{-1})	Nitrogen–phosphorus–potassium 37.5:37.5:75 (kg ha^{-1})	May–June August–September	September November/January
		Urea (2.5%)		
		Single super-phosphate (0.75%)		
		Muriate of potash (1.0%)		
Kerala	Nitrogen–phosphorus–potassium 75:75 (150 kg ha^{-1})	NPK 37.5:37.5:75 (kg ha^{-1})	May–June August–September	September November/January
		Urea (2.5%)		
		Single super-phosphate (0.75%)		
		Muriate of potash (1.0%)		
Tamil Nadu	Nitrogen–phosphorus–potassium 40:80:40 (kg ha^{-1})	NPK 20:40:20 (kg ha^{-1})	September November	June, August, November–December
		Urea (3%)		
		Single super-phosphate (1.0%)		
		Muriate of potash (2%)		

(Continued)

Table 2.23 (Continued)

II. Comparison of Phosphorus Buffer Power among Eight Widely Differing Central European Soils (Determined by Two Different Techniques)[b]

Soil	Regression Functions		*r* Values	
	(1)	(2)	(1)	(2)
Benzheimer Hof	$Y=18.8x+7.94$	$Y=0.23x+8.98$	0.912	0.995
Hungen	$Y=38.2x-1.03$	$Y=4.32+0.25x$	0.967	0.997
Oldenburg B6	$Y=49.8x+0.52$	$Y=0.72+0.26x$	0.994	0.999
Wolfersheim	$Y=70.3x+0.03$	$Y=0.11+0.27x$	0.998	0.983
Obertshausen	$Y=70.5x+2.66$	$Y=2.89+0.30x$	0.966	0.998
Oldenburg B3	$Y=73.6x+2.07$	$Y=0.61+0.31x$	0.994	0.997
Klein-Linden	$Y=75.0x+0.38$	$Y=1.81+0.32x$	0.999	0.991
Gruningen	$Y=75.4x+0.89$	$Y=3.62+0.36x$	1.000	0.996

[a]Instead of straight fertilizers, 2% each of diammonium phosphate (DAP) and muriate of potash (MOP) can be used.
[b]The *b* values in the regression functions represent the potassium buffer power of each soil. In regression function (1) (after Nair and Mengel, 1984), Y = CAL-P (Schüller's method), and in regression function (2) (after Nair, 1992), Y = the author's method. x in both refers to electroultrafiltrable phosphorus. Note the very high r values in all the cases. The soils are arranged in the sequential increase in their phosphorus buffer power.

When an application of fertilizer is routed through soil, it is necessary to apply one-third of the recommended dose (75:75:150 kg ha^{-1} and 125:125:250 kg ha^{-1} of nitrogen–phosphorus–potassium for rain-fed and irrigated conditions, respectively) during the first year of plant growth, under both rain-fed and irrigated conditions. During the second year of plant growth, the dose may be increased to one-half of the recommended dose, and fertilizer at full dose may be applied from the third year onward (Anon., 1993).

All of the preceding fertilizer recommendations are based on conventional or routine analytical techniques and not on advances in soil testing or on "The Nutrient Buffer Power Concept" developed by Nair (1996). The application of this concept in the case of the potassium requirement of cardamom led to dependable yield targeting (Nair et al., 1997). Nair (2002) has suggested a thorough reorientation of soil testing based on the "buffer power concept" for precise fertilizer recommendations in many crops, including cardamom.

Organic manures are considered essential for improving physical characteristics of soil, apart from their fertility-enriching qualities, and these manures are indispensable in cardamom cultivation whether chemical fertilizers are or are not applied. The application of organic manures, such as neem cake (1–2 kg per plant), farmyard manure, or cattle dung compost, at the rate of 5 kg per plant may be made once in a year in May–June, along with "Mussoorie rock phosphate" and MOP (Anon., 1997). Thimmarayappa et al. (2000) suggested an integrated nutrient management approach to meet the 25% requirement for nitrogen through farmyard manure and the remaining 75% through inorganic sources for sustained cardamom production over time.

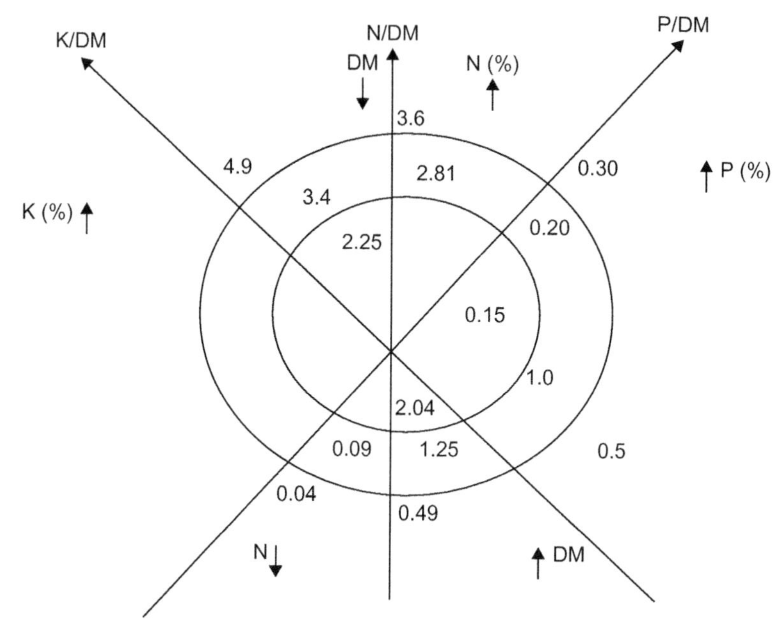

Figure 2.2 Foliar diagnostic norms for optimizing cardamom production (DRIS norms chart).

2.4.8.2 Diagnostic Recommendation Integration System

The application of the Diagnostic Recommendation Integration System (DRIS), originally detailed by De Gues (1973), has been tested in cardamom by Sadannandan et al. (2000). These researchers analyzed different nutrients in the youngest mature leaf (the fifth pair) from the terminally emerged panicle-bearing tillers, along with crop yield, and tested the efficacy of the DRIS system. Leaf nutrient composition was categorized as deficient, low, optimum, and high. As per the norms developed, leaf with 1.26–2.81% nitrogen, 0.1–0.2% phosphorus, 1.1–3.4% potassium, 0.51–1.38% calcium, 0.18–0.31% magnesium, 135–370 ppm iron, 261–480 ppm manganese, 20–45 ppm zinc, 10–46 ppm copper, and 0.28–0.84% molybdenum is considered optimum for producing cardamom yields ranging from 131 to 625 kg ha^{-1}. From this investigation, it was observed that, in order to obtain a high yield of 378 kg ha^{-1}, the indicator leaf should have mean nitrogen, phosphorus, potassium, calcium, and magnesium levels of 2.04%, 0.15%, 2.25%, 0.95%, and 0.25%, respectively. The optimum levels of micronutrients for realizing a high yield are 253 ppm iron, 371 ppm manganese, 33 ppm zinc, 28 ppm copper, and 0.56 ppm molybdenum. The principal observation made by Sadannandan et al. was that, among all the nutrient requirements they investigated, potassium was the highest for attaining maximum yield level and it should be 2.25% in the leaf. A descriptive diagnostic chart depicting all the nutrients the team of researchers investigated is shown in Figure 2.2.

2.4.8.3 Method and Time of Fertilizer Application

To decide the right time and method of fertilizer application for cardamom, the root spread should be taken into consideration. With regard to the lateral spread of roots, a full-bearing 8-year-old plant was found to have 80% of its roots in a zone of 25-cm radius, 14% in a zone of 25- to 50-cm radius, and only 6% in a zone of radius 50–75 cm. Vertically, cardamom roots penetrate only up to 40 cm deep (Khader and Sayed, 1977). Further studies conducted at the Indian Cardamom Research Institute (Nair, 1988) showed that roots of cardamom are confined to a shallow depth: Nearly 70% of roots were seen in the top 5 cm of soil. Horizontally, only 10% of the roots forage an area 120 cm away from the clump. Therefore, for maximal efficiency of applied fertilizers, it is necessary to apply them at a radius of 50 cm. Since cardamom is a surface feeder, deep placement of fertilizers is not advisable, as, for instance, in the case of cotton. Before fertilizers are applied, if the panicles spread on the ground, they are to be kept coiled, encircling the base of the plant, and the mulch should be removed. Fertilizer may be applied in a 15 cm wide circular band at a distance of 30 cm from the base of the plant. They should be incorporated into the soil with the use of a hand fork. Mulching should be followed immediately after incorporating fertilizers. Panicles may then be released and spread on the ground for the Malabar types to facilitate honeybee movement for better pollination and capsule set.

Cardamom growth is influenced by seasonal conditions, especially the rainfall pattern. Vegetative buds emerge from the bases of tillers throughout the year. However, most of the vegetative buds are produced after the rains (Sudarshan et al., 1988). It was observed that (Sudarshan et al., 1988) the linear growth of tillers increases with the onset of the southwest monsoon and the growth rate slows down with the cessation of the monsoon. The peak flowering period and the fruit-set period coincide, and nearly 70–90% of flower production occurs between May and August. Hence, for the efficient utilization of fertilizers, the time they are applied is crucial. Applying fertilizers in the month of May and later in September is found to be most beneficial (Pattanshetty and Nusrath, 1973). However, under irrigated conditions, tiller and panicle initiation are continuous processes and a split application of fertilizers is preferable.

Because cardamom grows in forest situations with shade trees all around, the plants will not make complete use of fertilizers, given the competition from the roots of other trees, herbs, and shrubs for both moisture and nutrients. Hence, a combination of soil and foliar application is preferable to resorting exclusively to soil application of fertilizers. Soil application in two rounds, first during May–June and later during August–September, and subsequent supplemental foliar applications in September, November, and January is recommended (Srinivasan et al., 1998). George (1990) reported that in Guatemala, unlike in India, almost all growers fertilize cardamom. Different fertilizer mixtures are available in the country. One such is a nitrogen–phosphorus–potassium mixture in the ratio of 15:15:15 that is marketed under the trade name "Barco Vikingo." Another mixture contains nitrogen–phosphorus–potassium–calcium–magnesium in the ratio 18:6:12:4:0.2 kg ha^{-1} with the brand name "Agrovet S.A." Fertilizer application starts from the second year after planting, when just 2 oz of the mixture is added per plant. This dose is gradually increased to 3 oz in the third year and to 4 oz from the fourth year onward. Some cardamom planters add an extra dose of 1 oz of urea to each plant over the quantity of mixture applied. The recommended dose is applied 2–4 times in a year (George, 1990). In India, when cardamom is cultivated under controlled artificial shade, a fertilizer dose of nitrogen–phosphorus–potassium in the ratio 100:25:100 kg ha^{-1} is recommended (Korkanthimath et al., 1988b).

2.5 The Role of "The Nutrient Buffer Power Concept" in Cardamom Nutrition

Historically, soil testing has been used to quantify availability of essential plant nutrients to field-grown crops. However, contemporary soil tests are based on philosophies and procedures developed several decades ago without significant changes in their general approach. For a soil test to be accurate, one needs to clearly understand the physicochemicophysiologic processes at the soil–root interface, and an understanding of the soils and plant root systems as polycationic systems is essential. It is this knowledge that leads to sound prescriptive soil-management practices in nutrient bioavailability vis-à-vis fertilizer application; because of all the factors that govern sustainability of crop production, the nutrient factor is the most important and, at the same time, least resilient to management. This section focuses on the buffering of nutrients, with specific reference to potassium, the most important major plant nutrient in cardamom nutrition, and discusses field experimental results in cardamom management.

2.5.1 The "Buffer Power" and Its Effect on Nutrient Availability

2.5.1.1 Basic Concepts

In any nutrient management approach that is sound and reproducible, one must start with a basic understanding of the chemical environment of plant roots. In doing so, the first term encountered is "soil solution" because the plant root is bathed in it and is most affected by its chemical properties. The Soil Science Society of America (1965) defines soil solution as "the aqueous liquid phase of the soil and its solutes consisting of ions dissociated from the surfaces of the soil particles and of other soluble materials." Adams (1974) has given a simple definition: "The soil solution is the aqueous component of a soil at field-moisture contents." Perhaps it is important to emphasize here that, in contemporary soil testing, "soil extraction" is often considered to be synonymous with "soil solution." Since soil extraction is supposed to simulate plant root extraction, it is apropos to consider the chemical environment of the root, although briefly, from this angle. Note that the chemical environment of roots in natural soil systems is so obviously complex that both soil scientists and plant physiologists have been unable to provide a precise definition thereof . If this complex chemical system is to be accurately quantified, thermodynamic principles will need to be used to evaluate experimental data. Even then, the limitations are obvious, as in the case of potassium, where thermodynamic investigations are quite often inapplicable under field conditions. This is because, although a quasi equilibrium in potassium exchange can be achieved in the laboratory, these conditions are seldom, if ever, attained under field conditions (Sparks, 1987). Agricultural soils are, for the most part, in a state of disequilibrium owing to both fertilizer input and nutrient uptake by plant roots. It thus appears that a universal and accurate definition of a root's chemical environment awaits the proper application of thermodynamics for the root's ambient solution (Adams, 1974)—or even kinetics, as in the case of potassium (Sparks, 1987), where thermodynamics has been found inadequate.

Soil extractions with different extractants provide a second approach to defining the root's chemical environment. This approach has been particularly successful in understanding cases such as phosphorus insolubility, soil acidity, and potassium fixation. However, it also fails to define the root's chemical environment precisely. Moreover, it suffers from other deficiencies, such as the extractants' removal of arbitrary and undetermined amounts of solid-phase electrolytes and ions (i.e., the extractants cause salts or ions to precipitate out of the soil solution) and the soil–plant interrelationship being defined in terms of the solid-phase component of the soil, even though the solid phase is essentially inert, except as it maintains thermodynamic equilibrium with the solution phase (Adams, 1974). Still, the latter aspect could be researched more to understand how solid-phase–solution-phase equilibria can be interpreted to give a newer meaning to quantifying nutrient availability, and it is in this context that the role of the plant nutrients' "buffer power" assumes crucial importance.

The close, almost linear, relationship in a low-concentration range of $<0.5\,\text{mM}$ for $NO_3^- - N$, $NH_4^- - N$, K^+, $H_2PO_4^{2-}$, and $H_2PO_4^{2-}$, which has been established by numerous solution culture experiments, can be quantitatively described by the equation

$$U = 2\pi r \alpha C_r, \tag{2.1}$$

where U is the uptake of a 1-m root segment, r denotes the root radius, C_r is the concentration of the ion at the root surface, and α is the root's absorbing power (Mengel, 1985). The metabolic rate of the root determines its absorbing power. A high root absorbing power implies that a relatively high proportion of nutrient ions coming in contact with the root surface is absorbed, and vice versa. The nutrient ion concentration at the root surface (C_r) depends on α, since a high root absorbing power tends to decrease C_r; it also depends on the rate of movement from bulk soil toward the root (Mengel, 1985). Diffusion and/or mass flow controls this movement, but it is now established that nearly 95% of the movement of nutrients such as phosphorus, potassium, zinc (among heavy metals) and, possibly, NH_4^+, is by way of diffusion. When root uptake of an ion species is less than the ion's movement towards the root, accumulation of the ion species on the root surface is bound to occur, as has been shown to be the case with Ca^{2+}, where mass flow contributes to this accumulation (Barber, 1974). The diffusive path for ions such as phosphorus and potassium, which plant roots take up at high rates, but which are in low concentration in the soil solution near the root, is the concentration gradient. In a sense, the effective diffusion coefficient, which quantifies the diffusive path, and the buffer power are analogous because the diffusive flux across the root surface is integrally related to the nutrient buffer power. This has been shown to be true in the case of phosphorus, where a highly significant positive correlation between the two was found to exist in 33 soil samples obtained from experimental sites in the United States and Canada (Kovar and Barber, 1988). However, in a routine laboratory setup, it is far easier to measure the buffer power than the effective diffusion coefficient, so this chapter will focus further on the question of how buffer power can be quantified without recourse to cumbersome analytical techniques and how its integration into routine soil test data will considerably improve the predictability of nutrient uptake.

2.5.2 Measuring the Nutrient Buffer Power and Its Importance in Affecting Nutrient Concentrations on Root Surfaces

The ability to predict the mobility of dissolved chemicals, such as fertilizers, in the soil is of considerable value in managing fertilizer applications. Soil testing, in its essence, aims to achieve this predictability. Whereas modeling the transport and retention of ions from thermodynamic (Selim, 1992), kinetic (Sparks, 1989), and mechanistic (Barber, 1984) angles could be informative, the importance of translating this information into practically feasible procedures in crop production calls for an understanding not only of the basic concepts, but of their intelligent application as well. In a dynamic state of plant growth, the concentration of any nutrient on the root surface is nearly impossible to measure, since both the nutrient in the plant tissue and the root absorbing power, which directly affects it, change quickly because of root metabolic processes. The inability of even mechanical mathematical models to accurately predict nutrient influx rates has been highlighted (Lu and Miller, 1994). Hence, if an effective soil-testing procedure that is an alternative to defining the plant root's chemical environment is to be devised for a nutrient, one must resolve the problem of quantifying the nutrient concentration on the root surface indirectly, even if it is impossible to resolve it directly, for the reasons mentioned previously.

Using Fick's first law yields

$$F = -D\left(\frac{\mathrm{d}C}{\mathrm{d}x}\right), \tag{2.2}$$

where F denotes the flux, $\mathrm{d}C/\mathrm{d}x$ is the concentration gradient across a particular section, and D is the diffusion coefficient. Nye (1972) has suggested that the formula can be applied to both ions and molecules. The negative sign on D implies net movement from high to low concentration. Although, for molecules in simple systems such as dilute solutions, D may be nearly constant over a range of concentrations, for ions in complex systems such as soils and clays, D will usually depend on the concentration of the ion and on that of other ions as well (Nye, 1979). Nye (1979) has further suggested that, although Fick's first law may be derived from thermodynamic principles in ideal systems, in a complex medium such as the soil, the preceding equation may be regarded as giving an operational definition of the diffusion coefficient. Thus, Nye (1979) defines the diffusion coefficient as

$$D = D_1 \theta f_1 \left(\frac{dC_1}{dC} \right) + D_E, \tag{2.3}$$

where D_1 is the diffusion coefficient of the solute in free solution, θ designates the fraction of the soil volume occupied by solution and gives the cross section for diffusion, f_1 is an impedance factor, C_1 specifies the concentration of solute in the soil solution, and D_E is an excess term that is zero when the ions or molecules on the solid have no surface mobility, but represents their extra contribution to the diffusion coefficient when they are mobile. D_E can generally be neglected, since only in rare instances will it play any role in the diffusion of plant nutrient ions in soil (Mengel, 1985). From the point of view of nutrient availability, dC_1/dC, which represents the concentration gradient, assumes crucial importance, as we shall see shortly.

The term dC_1/dC, where C_1 is the concentration of the nutrient ion in the soil solution and C is the concentration of the same ion species in the entire soil mass, assumes considerable significance in lending a practical meaning to nutrient availability. If we ascribe the term "capacity" or "quantity" to C and "intensity" to C_1, we have, in dC_1/dC, an integral relationship between two parameters that may crucially affect nutrient availability. Since the concentration gradient of the depletion profile of the nutrient in the zone of nutrient uptake depends on the concentration of the ion species in the entire soil mass (represented by "capacity" or "quantity") in relation to the rate at which this is lowered on the plant root surface by uptake (represented by "intensity"), it could be argued that a quantitative relationship between the two should represent the rate at which nutrient depletion and/or replenishment in the rooting zone should occur (Nair, 1984a). This relationship has been functionally quantified by Nair and Mengel (1984) for phosphorus in eight widely differing central European soils (Table 2.23), and the term dC_1/dC has been referred to as the "nutrient buffer power." Nair and Mengel (1984) used electroultrafiltration (EUF) to quantify C_1 while using an incubation and extraction technique to quantify C. For phosphorus, C was found to closely approximate the isotopically exchangeable form of the element (Keerthisinghe and Mengel, 1979), but in the experiments conducted by Nair and Mengel (1984), it was estimated by the extraction of incubated soil with an extractant that was a mixture of 0.1 M calcium lactate + 0.1 M calcium acetate +0.3 M acetic acid at pH 4.1. The extractant exchanges adsorbed phosphate and dissolved calcium phosphates except apatites; the method, known as the "CAL-method" and developed by Schuller (1969), is now widely used in central Europe. In the case of K^+ and NH_4 –N, C denotes the concentration of exchangeable and, to some extent, nonexchangeable fractions (Mengel, 1985). Since low concentrations in the range of 2.0 μM may be attained on the root surface for both phosphorus and potassium (Claassen and Barber, 1976; Claassen et al., 1981; Hendriks et al., 1981), Nair and Mengel (1984) had to use EUF to quantify C_1. Thus, the nutrient depletion around the roots, which is caused by the diffusive flux of the nutrients toward the root surface, is related to both the quantity and the intensity parameters, and a quantifiable relationship between both represents the buffer power specific to the nutrient and the soil. A growing root will at first encounter a relatively high concentration of phosphorus, one in the range of the concentration of the bulk soil solution (Nair and Mengel, 1984). As uptake continues, depletion will occur at the root surface. The depletion profile gets flatter with enhanced nutrient uptake (Claassen et al., 1981; Hendriks et al., 1981; Lewis and Quirk, 1967). But it is the capacity of the soil to replenish this depletion that ensures a supply of nutrient ions to the plant root without greatly depressing the average concentration of ions on the root surface. The nutrient's buffer power is what determines these depletion and replenishment rates. A soil with a high phosphorus buffer power implies that the phosphorus absorbed from the soil solution is rapidly replenished. In such a case, the concentration of phosphorus at the root surface decreases only slowly, and this means that the concentration remains relatively high there. In soils with a low phosphorus buffer power, the reverse is true, and the concentration of phosphorus at the root surface diminishes rapidly and remains relatively low. These observations have been borne out experimentally for phosphorus (Nair, 1992; Nair and Mengel, 1984; Table 23), Zn^{2+} (Nair, 1984a,b), potassium, and $NH_4^+ - N$ (Mengel, 1985).

2.5.3 Quantifying the Buffer Power of Soils and Testing Its Effect on Potassium Availability

The dynamics of potassium availability follows the same pattern as that of phosphorus, especially in the range of low concentrations. Beckett (1971) has used the activity ratio for K^+ and Ca^{2+} to determine potassium availability. Since interlayer potassium would play an important role in the availability of the element, it would be more logical to consider potassium buffer power in determining that availability. The ammonium acetate extractant is routinely and widely used to characterize potassium availability. The reason, however, that this approach may not be suitable for characterizing exchangeable potassium is that in routine extraction only the top layer is extracted, while interlayer potassium from which deep-rooted plants can feed is ignored (Nair et al., 1997). There is extensive evidence to substantiate this claim (Nair, 1985). The importance of interlayer potassium in the nutrition of deep-rooted perennial crops such as cardamom, the world's most valuable spice crop next to black pepper, is highlighted in the next section.

2.5.4 The Importance of Potassium Buffer Power Determination in Predicting Potassium Availability to Perennial Crops

As in the case of phosphorus, the potassium buffer power assumes great importance in predicting potassium availability, especially with regard to deep-rooted and perennial crops. The availability of exchangeable potassium has been studied; However, with perennial and deep-rooted crops, nonexchangeable and interlayer potassium play a crucial role in potassium availability. Three soil parameters that control the rate of potassium supplied to plant roots and therefore have been used to predict potassium absorption by plants are the potassium intensity in the soil solution, the potassium buffer power, and the effective diffusion coefficient (Beckett, 1971; Claassen et al., 1986; Mengel and Kirby, 1980). The potassium buffer power can be directly obtained from the quantity–intensity relationship for the element. The effective diffusion coefficient depends on, among other factors, the buffer power (Nye, 1972). Plants feed not only from exchangeable K^+, but also from nonexchangeable K^+, which consists mainly of K^+ trapped in the interlayers of nonexpanded 2:1 clay minerals (Hoagland and Martin, 1933; Schachtschabel, 1937).

Exchangeable K^+ comprises that which can be exchanged with NH_4 ion and is primarily planar K^+, the interlayer K^+ of expanded 2:1 clay minerals and some of the K^+ at the interlayer edges of nonexpanded 2:1 clay minerals. Interlayer K^+ of nonexpanded clay minerals, such as illites and interlayer, and lattice K^+ micas (present in feldspars) constitute the nonexchangeable K^+. The interlayer K^+ is of particular importance in the nutrition of deep-rooted and perennial crops, such as cardamom, as has been demonstrated by Nair et al. (1997), and also for annual crops such as ryegrass (*Lolium perenne*. cv. Taptoe) (and Uhlenbecker, 1993).

In most of the test for K^+ availability, nonexchangeable potassium is not considered. For cereals, such as wheat (*Triticum aestivum*), 80% of the K^+ extracted by the crop comes from nonexchangeable K pool. This is one of the most important reasons for the poor soil test crop response relationship with regard to potassium fertilizer applications based on such tests (Kuhlmann and Wehrmann, 1984). The contribution of nonexchangeable potassium to plant availability was assessed by 1 M HCl extraction by Schachtschabel (1961), similar to the 1 M HNO_3 extraction proposed by Pratt (1985) and McLean and Watson (1985). However, the efficiency of 1MHCl extraction to quantify plant-available potassium from the nonexchangeable pool has been disputed (Boguslawski and Lach 1971; Grimme, 1974; Kuhlmann and Wehrmann, 1984). Soils containing primarily 2:1 clay minerals, such as vermiculite and illite, have interlayer potassium in excess of crop requirements. However, the availability of interlayer potassium of nonexpanded minerals is independent of the quantity of interlayer potassium as such, but dependent on its release rate as a function of the type of K^+-bearing minerals (Sparks, 1987). The release of potassium from interlayer positions is an exchange-and-diffusion process (von Reichenbach, 1972). Although exchange depends on the cation species and their concentration near the surface of the mineral, diffusion depends largely on the expansion of the mineral and, therefore, on soil moisture. The net release of potassium will occur only if the potassium concentration of the adjacent solution is low (Mengel, 1985). While studying the release of nonexchangeable K^+ from sandy loam and loamy sand extracted with an H^+ charged ion exchanger resin, Martin and Sparks (1983) found a large release of K^+ with a K^+ concentration of about 1–2 μmole in the contact solution. This concentration may approximate the rhizosphere concentration level. Under submerged conditions, as with rice, there can be a depletion zone for potassium in the rhizosphere (Xu and Liu, 1983). Plant roots act as a sink and maintain the potassium solution concentration at low levels (Kuchenbuch and Jungk, 1984), causing further release of interlayer potassium (Mengel, K. 1985). All these considerations point to the important fact that, as far as nonchangeable interlayer potassium is concerned, a precise quantification of potassium availability hinges primarily on the

Table 2.24 Potassium Buffer Power of Cardamom-Growing Soils from Two Regions of Southern India That Grow This Crop Extensively (after Nair et al., 1997)

Region	Regression Function ($Y = a + bx$)	"r"	Crop Yield ($kg\,ha^{-1}$)
Coorg	$142.38 + 1.4443x$	0.8561^{**}	155
Idukki	$592.46 + 0.9712x$	0.5799^{*}	80

Note: Values for b in the regression functions represent the potassium buffer power of the soil. The potassium buffer power was calculated from pooled values of soil samples obtained from 94 locations covering an area of more than 20,000 ha in two cardamom-growing regions of southern India: the Coorg and Idukki districts of Karnataka and Kerala States. Yield data pertain to the same locations. The symbols (*) and (**) indicate significance at $p = 0.05$ and 0.01, respectively.

release rate, which the potassium buffer power attempts to quantify, as we shall see in the discussion that follows.

Nair et al. (1997) selected cardamom to demonstrate the importance of non-exchangeable and interlayer potassium on potassium availability vis-à-vis the potassium buffer power. Potassium buffer power curves were constructed by a two-step extraction in which 1N HNO_3 was used to determine the quantity of potassium (Wood and De Turk, 1941) and 1N NH_4OAc was used to determine its intensity. The NH_4OAc extractant is universally used to determine exchangeable potassium. The contribution of nonexchangeable potassium to plant availability has been assessed both by 1 M HCl extraction (Schachtschabel, 1961) and by extraction with 1 M HNO_3 (McLean and Watson, 1985; Pratt, 1965). Nair et al. (1997) regressed 1N HNO_3-extractable K (y) over 1N NH_4OAc-extractable potassium (x) to obtain the potassium buffer power (Table 2.24).

The data in Table 2.24 clearly indicate that the Coorg soils, which had a much higher potassium buffer power, produced a cardamom yield twice that obtained in the Idukki soils. By comparison, the 1 N NH_4OAc-extractable potassium had no significant relationship to leaf potassium (Table 2.5), and further, the integration of the computed potassium buffer power with the routine soil test potassium data (NH_4OAc-extractable potassium) remarkably improved whatever little relationship there was.

Cardamom is a heavy feeder of potassium (Sadanandan et al., 1990), and in both India, which grows most of this valuable spice crop, and other countries in Asia and Africa, potassium fertilizer needs are almost always based on the exchangeable potassium determined by 1N NH_4OAc extraction. The data in Table 2.25 unequivocally show the inability of this extraction to precisely predict potassium availability; further, the data in Table 2.26 demonstrate how the situation is remarkably improved by the integration of the potassium buffer power into the computations. Specifically, a substantial variation (302.7%) in leaf potassium is attributable to the potassium buffer power. The results shown were obtained from an extensive area (more than 20,000 ha), bolstering their significance.

The potassium buffer power in this instance integrates both exchangeable (NH_4OAc-extractable) potassium and nonexchangeable or interlayer (HNO_3-extractable) potassium, and this gives an accurate measurement of potassium depletion around the plant roots.

Table 2.25 Correlation Coefficients and Regression Functions for the Relationship between Leaf potassium (*Y*) and Exchangeable potassium (*x*, NH_4OAc-Extractable potassium) (after Nair et al., 1997)

Details	Regression Function ($Y = a + bx$)	Correlation Coefficient *r*
Leaf potassium vs exchangeable potassium	Coorg $Y = 1.2701 + 0.0004$	0.2064
Leaf potassium vs exchangeable potassium	Idukki $Y = 1.64448 + 0.000006$	−0.006

Note: The correlation coefficients refer to the leaf samples from 94 locations from which the soil samples were obtained to calculate the potassium buffer power. In cardamom, the fifth pair of leaves from the top of each panicle bearing tillers is sampled for potassium analysis (Sadanandan et al., 1993).

Table 2.26 Correlation Coefficients (*r*) for the Relationship between Leaf Potassium (Y) and Exchangeable Potassium (*x*, NH_4OAc-Extractable Potassium) for the Pooled Data (94 Locations) from Two Regions (Coorg and Idukki Districts of Karnataka and Kerala States, Respectively) without (A) and with (B) Potassium Buffer Power Integration (Nair et al., 1997)

Details	Correlation Coefficient	
	A	**B**
Leaf vs exchangeable potassium	0.2510	0.4367*

*Significant at $p = 0.01$ confidence level. Note the remarkable improvement with potassium buffer power integrated into the computations.

In an investigation (Mengel and Uhlenbecker, 1993) into potassium availability from interlayer potassium to ryegrass (*L. perenne* L. cv. Taptoe), it was observed that the rate constants (*b* values) obtained by correlating potassium released (from interlayers of clay minerals) with time spans by a modified EUF technique were closely related to potassium uptake and represented the potassium availability index from no exchangeable potassium. These rate constants, according to the authors, are of the utmost importance because (1) they provide information on the availability of nonexchangeable potassium in attaining maximum yield and (2) a set of "critical *b*" values that are useful in attaining this objective has been reported. It appears that the rate constants of Mengel and Uhlenbecker (1993) are analogous to the potassium buffer power values reported by Nair et al. (1997), because, although the techniques differ in their details, they have accomplished the same objective of precisely predicting potassium availability from the nonexchangeable pool and/or the interlayer potassium. The capability of tapping interlayer potassium varies among plant species. For instance, Steffens and Mengel (1979) found that ryegrass (*L. perenne*) could feed from interlayer potassium for a longer period without any resulting yield depression, whereas red clover (*Trifolium pratense*) could not. These authors also reported that, since *L. perenne* had a longer and deeper root system compared with *T. pratense*, the former could grow satisfactorily at relatively low

K^+ concentrations while the latter would already suffer from potassium deficiency (Steffens and Mengel, 1981). The differences in root mass, root length, and root morphology between monocots and dicots explain the better K^+ feeding capacity from interlayer K^+ of the former as opposed to the latter (Mengel, 1985).

Cotton (*Gossypium hirsutum L.*) is another deep-rooted long-duration crop on which the potassium buffer power exerts considerable influence on potassium acquisition. Brouder and Cassman (1994) used a mechanistic mathematical model to evaluate potassium uptake by cotton in a vermiculite soil and observed that, initially, the model produced substantial under- and overpredictions of whole-plant potassium accumulation. The predictions were greatly enhanced by estimating the potassium buffer power. The authors further concluded that the contribution of the fixed potassium pool to the plant available potassium pool was likely to be substantial and that this influence must be captured in estimates of the potassium soil buffer power. These studies were conducted after the cotton grown in a San Joaquin Valley, California, cotton field was observed to exhibit late season potassium deficiency while other crop species remained unaffected. In such cases, the precise estimation of the potassium buffer power will lead to far more dependable potassium fertilizer recommendations than will estimations by routine NH_4OAc extraction.

Although it has long been recognized (Schachtschabel, 1937) that the soil potassium fraction that is not exchangeable by NH_4 ions (nonexchangeable potassium) may be important for the supply of potassium to plants, it is only of late that researchers have paid more attention to this factor. The work of Sparks and Huang (1985) has critically examined the release mechanism from nonexchangeable sources and the factors controlling it. Considerable portions of initially nonexchangeable potassium can be utilized by plant roots even within a few days (Kuchenbuch and Jungk, 1984). The depletion zone, however, extends into the ambient soil for only2 mm. Hinsinger et al. (1992) embedded phlogopite in agar and observed that the interlayer potassium of this mineral was entirely lost in the close vicinity of ryegrass roots within 4 days. Since the process-limiting potassium uptake in the rhizosphere may be potassium transport through the soil rather than the release from minerals as such, some researchers have focused their attention on such transport. One example is the mechanistic mathematical model of Claassen and Barber (1976). Claassen et al. (1986)and Claassen (1990) have successfully applied the model to predict potassium depletion profiles in soil around plant roots. Meyer and Jungk (1993) have used models such as these to predict potassium uptake by test plants from exchangeable and nonexchangeable sources. They reported that 64–79% of the potassium taken up by wheat (*Triticum aestivum* L.) and sugar beet (*Beta vulgaris* L.) was derived from the rapidly released exchangeable, and 21–36% from the nonexchangeable, or less mobile, soil potassium fraction.

The buffer power describes the relationship between adsorbed potassium and the potassium concentration of the ambient solution. In simulation models, it is assumed that this relationship is linear and, hence, independent of the soil solution concentration. However, in desorption studies with soil, a sharply curved buffer relationship was found, and Meyer and Jungk (1993) have referred to it. Very near the plant roots, the soil can be subjected to a curved buffer function, since plant roots strongly reduce the soil solution concentration.

An important factor to be considered in the utilization of nonexchangeable potassium is the role of plant roots. Plant species differ in their ability to utilize nonexchangeable potassium, and this difference has been attributed to differences in root length (Mengel and Steffens, 1985). The radial distance between two single roots decreases, thereby increasing root density, and this increase would result in overlapping of the depleted soil volumes between these roots. A decrease in the rate of potassium uptake per unit root length would also ensue. In the case of the rapidly diffusing potassium fraction, which has a higher mobility than most other nutrient ions, the competition effect between roots could be intense. There is evidence to support this view, as shown by the work of Mitsios and Rowell (1987), who observed that the contribution of the nonexchangeable potassium fraction increased with a corresponding increase in root density. In addition, the differences in root hair length and density among plant species (Fohse et al., 1991) affect their ability to acquire soil potassium. The work of Meyer and Jungk (1993) has shown that potassium uptake was higher when they included root hairs in their model calculations. Since root hairs contribute not only to an increase in root absorbing surface, but also to a reduction in the distance of the diffusion from the site of potassium release to the site of potassium uptake and an increase in the potassium concentration gradient, they can be expected to exert a pronounced effect on potassium availability from the less mobile potassium fraction.

2.5.5 The Commercial Significance of the Potassium Buffer Power Determination in Potassium Fertilizer Management for Perennial Crops

The commercial significance of the potassium buffer power determination for dependable potassium fertilizer recommendation assumes great importance in those countries that are faced with importing these fertilizers at a huge cost to the national exchequer. India is a case in point. The decontrol of phosphatic and potassic fertilizer prices by the Indian government resulted in an overnight escalation of their market prices. In a situation like that, farmers become extremely wary of how they use their fields, and unless the fertilizer application is cost-effective, faith in the use of fertilizers, especially those mentioned here, would be shattered.

The potassium fertilizer recommendation for cardamom has been based exclusively on NH_4OAc extraction. An investigation by Nair et al. (1997) showed its ineffectiveness. Although the importance of the potassium buffer power in predicting potassium availability has been reported earlier in a number of research reports, the reports related mainly to annual crops, such as white clover (During and Duganzich 1979) and ryegrass (Busch, 1982); the work of Nair et al. (1997) was the first of its kind in a perennial crop.

One last point regarding the question of accurately predicting potassium availability is the role of NH_4 ion with respect to K^+ ion. One of the frequent assumptions made in predicting potassium availability in soils is that results from binary (two-ion) exchange systems can be extrapolated to ternary (three-ion) systems by using appropriate equations. The potassium–calcium exchange reactions in soils are often investigated in laboratory studies. Most of the research carried out on soil clay minerals and soils as exchanger surfaces (Argersinger et al., 1950; Gapon, 1933; Jardine and

Sparks, 1984; Sposito, 1981a,b; Sposito et al., 1981, 1983; Vanselow, 1932) is on binary exchange systems. However, field soils are at least ternary systems (Adams, 1971; Curtin and Smillie, 1983). The evaluation of soils as binary systems presupposes that these reactions can be used to predict results in ternary systems such as field soils. For this assumption to be valid, the binary exchange selectivity coefficients need to be independent of exchanger-phase composition (Lumbanraja and Evangelou, 1992). But the work of Shu-Yan and Sposito (1981) showed that it is impossible to predict exchange phase–solution phase interactions in a ternary system, such as the field soil from a binary system, as the laboratory sample. This finding highlights the importance of ternary systems. As far as potassium availability is concerned, it would be important to include NH_4^+ ion as well. The work of Lumbanraja and Evangelou (1992) has shown that K^+ adsorption to soil surfaces is suppressed in the presence of added NH_4^+ ion while the adsorption of NH_4^+ ion to the same surface is enhanced in the presence of K^+ ion. These observations point to the influence of added NH_4^+ ion on the desorption potential (chemical potential) of adsorbed potassium and vice versa (Lumbanraja and Evangelou, 1992) and would be relevant to the determination of the potassium buffer power, especially when agents containing NH_4^+ ions, such as NH_4OAc, are used in determining that power (Nair et al. 1997). The work of Lumbanraja and Evangelou (1992), however, clearly demonstrates the effect of NH_4^+ ion on potassium desorption, with an increase in K^+ desorption in the presence of added NH_4^+ ion. In its absence, it might be safe to conclude that the shape of the potassium buffer power curve will not change appreciably, even if larger quantities of potassium are removed because of cropping and therefore can be considered as a relatively constant property of soils. There is evidence to support that conclusion: while investigating the Q/I relationship – a quantitative relationship between "Quantity" (Q) and "Intensity" (I) of a specific nutrient ion in a soil system with reference to potassium uptake by wheat (*Triticum aestivum*) in calcareous vertisols and inceptisols of southwestern Spain, Jimenez and Para (1991) found that 80% of the potassium that was extracted came from the non-exchangeable potassium pool. These observations coupled with the earlier discussed observations on K dynamics, suggest that the predictions of potassium availability can be substantially enhanced by first quantifying the potassium buffer power of the soil in which the crop is intended to be grown. Admittedly, the rate-limiting steps involved in the potassium dynamics are not entirely understood (Sparks, 1987). Nonetheless, notwithstanding this limitation, if one must move forward and devise better management of potassium fertilizers in crop production, a starting point has to be made with regard to quantifying potassium availability precisely. Quantifying the potassium buffer power of soils and basing potassium fertilizer recommendations on the resulting estimates seems to be the best starting point. The investigations of Nair et al. (1997) in a crop such as cardamom, the second most important spice crop in the world, shows clearly that this can be done.

2.5.6 Conclusions

Although cardamom is a perennial crop, its growth behavior resembles that of a biennial crop in the sense that the vegetative phase (tiller production) precedes the reproductive phase in the next year, when panicle and flower initiation is followed by fruit set. Since cardamom grows in forest canopy, where shade trees also grow, competition for moisture and nutrients from the shade trees should be expected; hence, water and nutrients should be managed intelligently. Following the oil crisis, fertilizer prices have escalated drastically, and it becomes all the more important to economize and enhance efficiency. "The Nutrient Buffer Power Concept" is relevant in this regard.

2.6 Cardamom Pathology

Cardamom plant is affected by a number of pathogens of which some are fungi, others bacteria, and still others nematodes. These pathogens affect the plant in both nurseries and main plantations. To date, as many as 25 diseases caused by pathogens have been reported. On the basis of severity of damage, these diseases are categorized into major and minor. Considerable damage is caused by four major diseases in the plantations and two in the nurseries. Major diseases, such as rots, leaf blights, and nematode infestations, are often widespread and lead to crop losses, while minor diseases generally cause damage to the foliage. Unless properly managed, diseases can cause up to a 50% loss of the crop.

2.6.1 Major Diseases

Capsule rot (locally referred to as Azhukal) and rhizome rot cause the most severe damage. Leaf blight and nematode infection lead to the weakening of plants and a consequent reduction in productivity. Table 2.27 lists the major diseases.

2.6.1.1 Capsule Rot ("Azhukal" Disease)

Capsule rot, locally known as Azhukal disease. (In the South Indian state of Kerala, "Azhukal" means "rotting" in Malayalam, the language spoken there.) Azhukal is the most severe fungal disease of cardamom. Menon et al. (1972) reported it for the first time in the cardamom plantations of Idukki district in Kerala State.

Table 2.27 Major Fungal and Nematode Diseases of Cardamom

Disease	Affected Plant Parts	Causal Pathogen
Capsule rot ("Azhukal")	Capsules, leaves, panicles, young tillers	*Phytophthora meadii, Phytophthora nicotianae var. nicotianae*
Rhizome rot (clump rot)	Rhizomes, tillers, roots	*Pythium vexans, Rhizoctonia solani, Fusarium oxysporum*
Leaf blight ("Chenthal")	Leaves	*Colletotrichum gloeosporioides*
Root knot nematode	Roots, leaves	*Meloidogyne incognita*

Geographic Distribution of the Disease

Initially, symptoms of rotting are observed on the fruits or capsules, and that is the reason the disease has been named capsule rot. Subsequently, symptoms are observed in other plant parts. Azhukal is the major disease affecting cardamom, and it causes severe loss of productivity in the Idukki and Wayanad districts of Kerala State and in the Anamalai hills of Tamil Nadu (Thomas et al., 1989). The disease appears following the onset of the southwest monsoon. Capsule rot is not observed in the low-rainfall areas of Tamil Nadu. Surprisingly, although Karnataka State receives much rainfall, the disease is yet to appear there.

Symptoms and Damage

Disease symptoms develop mainly on the capsules, young leaves, panicles, and tender shoots. The first visible symptom appears as discolored, water-soaked lesions on young leaves and capsules. The lesions enlarge and the affected portions decay. Infection occurs on capsules and tender leaves simultaneously or, sometimes, first on capsules and then on foliage (Thomas et al., 1991a). When foliage is infected, water-soaked lesions appear on leaf tips or leaf margins. The lesions subsequently enlarge, and adjacent lesions coalesce to form large patches. Immature, unfurled leaf fails to unfurl. As the disease advances, the lesions on the leaves turn necrotic, the leaves decay and shrivel, and, finally, they look shredded. Infected capsules show water-soaked, discolored patches that turn brownish. Later, the infected capsules decay and drop off of the plant. A foul smell is emitted from the rotten capsules. Capsules of all ages are susceptible to infection; however, young capsules are far more prone to infection than older ones.

When favorable climate prevails, the disease is aggravated and infection extends to panicles and tender shoots. In a severe case of infection, the whole panicle or pseudostem decays completely. In such cases, the rotting extends to underground rhizomes also. The root system of the diseased plant decays, and following this, the entire plant collapses to the ground. Nair (1979) described similar symptoms and observed that the severity of the disease is uniform in the three major types of cardamom: var. Malabar, var. Mysore, and var. Vazhukka. Nambiar and Sarma (1976) investigated the disease and reported a loss in productivity of up to 30%. Subsequently, the loss has been reported to be as high as 40% (Anon., 1989a).

Causal Pathogen

Menon et al. (1972) first reported that *Phytophthora* sp. was the causal pathogen of the disease. Thankamma and Pillai (1973) identified the organism as *Phytophthora nicotianae* Brede de Haan var. *nicotianae* Waterhouse while Radha and Joseph (1974) identified it as *Phytophthora palmivora* Butler. Nambiar and Sarma (1976) reported the association of *Pythium vexans* and a *Fusarium* sp., along with *Phytophthora* sp. However, subsequent investigations (Nair, 1979) showed *Phytophthora nicotianae* var. *nicotianae* to be the causative pathogen, which was successfully isolated from all infected plant parts. *Phytophthora meadii* Mc Rae has also been widely observed to cause capsule rot disease (Anon., 1986). Host-range studies show that *Phytophthora palmivora* from coconut and rubber trees can infect cardamom (Radha and Joseph,

1974). Also, *Phytophthora palmivora* can infect coconut, cocoa, areca nut, black pepper, and rubber (Manomohanan and Abi Cheeran, 1984). *Phytophthora meadii* from cardamom can infect black pepper, cocoa, and citrus (Sastry and Hegde, 1987, 1989). Nair (1979) observed that wild colocasia plants in cardamom plantations serve as collateral hosts for *Phytophthora nicotianae* var. *nicotianae*.

Based on culture characters, sporangial morphology, sexual behavior, and pathogenic virulence, seven different isolates of *Phytophthora meadii* from different localities causing infection on capsules, leafy stems, leaves, and rizhomes have been identified (Anon., 1989a). These seven isolates fall into two groups in their requirement for optimum temperature for growth and mean sporangial dimensions. In single cultures, no oospores are formed, but when paired with A1 mating type, five of the isolates readily formed sex organs and oospores, confirming the hypothesis that most of these isolates belong to the A2 mating type. The type species of *Phytophthora meadii* from cardamom readily grows on carrot agar and sporulates; the sporangia are caduceus, ellipsoid, and papillate, and have short to medium pedicels. Although, morphologically, the seven isolates differ only slightly, all of them were found to be pathologically virulent. The pathogen *Phytophthora nicotianae* var. *nicotianae* survives in the form of chlamydospores in the soil and in plant debris, and persists up to 48 weeks in moist soil (Nair, 1979). No chlamydospore formation, however, has been observed in the case of *Phytophthora meadii*. The inability of *Phytophthora meadii* to form chlamydospores from rubber is also reported.

Epidemiology

The epidemiology of capsule rot disease has been studied by Nair (1979), who observed that a high incidence of disease is correlated with high and incessant rainfall during the southwest monsoon. The number of *Phytophthora meadii* propagules increases in soil and results in a heavy incidence of disease coinciding with high soil moisture levels (34.3–37.6%), low temperatures (20.4–21.3°C), high relative humidity (83–90.6%), and high rainfall (320–400mm annual) during the months of June–August (Nair and Menon, 1980). The presence of a high level of soil inoculum, thick shade in the plantation, close spacing, high soil moisture, and waterlogging, together with favorable weather conditions, such as high relative humidity, continuous rainfall, and low temperatures, predisposes the plants to *Phytophthora meadii* infection. Nair (1979) also observed that the *Phytophthora* population decreases with increasing distance from the plant base and with depth from the surface of the soil.

2.6.1.2 Disease Management

Because the outbreak of disease occurs in the monsoon season, disease management protocols have to be in place sufficiently early—that is, prior to the onset of primary infection. During earlier years, various fungicides were extensively used to control the disease. Spraying and drenching with copper fungicides, such as 1% Bordeaux mixture and 0.2% copper oxychloride (Menon et al., 1973; Nair, 1979; Nair et al., 1982; Nambiar and Sarma, 1974), has been recommended as the disease control measure. Inhibition of the fungus *in vitro* has been reported following treatments with organomercurials (Wilson et al., 1974). Nair (1979) observed an 86% reduction in soil population levels of *Phytophthora* when drenched with 1% Bordeaux mixture or 100 ppm Dexon (Bay-5072). Alagianagalingam and Kandaswamy (1981) observed that the disease could be controlled by spraying the plants with 0.2% Dexon (Bay-5072) at the rate of $4\,kg\,ha^{-1}$. Although a number of fungicides have been reported to control the disease, disease control in the field has often been a challenging task. Factors hindering satisfactory disease control include lack of adequate phytosanitation, effective and timely application schedules, the high cost or nonavailability of fungicides, and the continuous rainfall that makes spraying a difficult operation and reduces its efficacy when the fungicide is sprayed.

Thomas et al. (1989, 1991a) evaluated a number of contact and systemic fungicides under field conditions and concluded that two to three rounds of sprays, including one round of prophylactic spray, with 1% Bordeaux mixture or 0.3% Aliette (Fosetyl-aluminum) after proper phytosanitation effectively controlled the spread of the disease.

2.6.1.3 Biological Control of Diseases

Bioagents play an important role in an ecofriendly system of disease management to fight against plant pathogens in a totally safe manner that avoids the use of expensive and hazardous chemical fungicides. Inhibition of *Phytophthora meadii* in laboratory conditions and disease suppression in cardamom nurseries have been investigated by Thomas et al. (1991b), employing *Trichoderma viride, Trichoderma harzianum, Laetisaria arvalis*, and *Bacillus subtilis*. Employing *Trichoderma viride* and *Trichoderma harzianum*, Suseela Bhai et al. (1993) achieved field control of capsule rot disease. Subsequently, (Suseela Bhai et al., 1994, 1997) developed a simple carrier-cum-multiplication medium for *Trichoderma* sp. application in fields. Cardamom-growing soils in their native state harbor *Trichoderma viride* and *Trichoderma harzianum* isolates for which they have been screened and effective strains for high-biocontrol potential have been developed (Dhanapal and Thomas, 1996). Field control of capsule rot disease has become effective, environmentally safe, and economically cost-effective through the biocontrol potential of *Trichoderma* sp.

2.6.1.4 Rhizome Rot Disease

Rhizome rot disease is also known as clump rot. The onset of the disease occurs during the southwest monsoon. Park (1937) was the first to report the occurrence of the disease. Subba Rao (1938) described the disease as clump rot. Rhizome rot disease is widely distributed throughout cardamom plantations in the states of Kerala and Karnataka and also in Tamil Nadu, where heavy rainfall occurs, as in the Anamalai hills.

Symptoms of the Disease

It is during the southwest monsoon, by about the middle of June, that the disease makes its appearance. The first visible symptoms are a pale-yellow color in the foliage, and wilting and the premature death of older leaves. The collar portion of the aerial shoots becomes brittle, and the tiller breaks off with just slight physical disturbance. Rotting develops at the collar region, which becomes soft and brown. At this stage, the affected aerial shoots fall off, emitting a foul smell. Mayne (1942) reported the incidence of the disease in cardamom hills of the state of Kerala. The tender shoots or the young tillers also turn brown and rot completely. With the advancement of the disease, all the affected aerial shoots fall off from the base. The panicles and young shoots attached to the base also are affected by the rot. Rotting extends to the rhizomes and the roots. The falling off of shoots resulting from rhizome rot infection becomes severe during July–August. In severely affected areas, as much as a 20% incidence of disease is recorded.

Causal Pathogen

Subba Rao (1938) observed that cardamom rhizome rot is caused by *Rhizoctonia solani Kuhn.* and associated the pathogen with a nematode. Ramakrishnan (1949) reported *Pythium vexans* de Barry as the causal pathogen. Thomas and Vijayan (1994) reported that *Fusarium oxysporum* is also occasionally found to cause rhizome rot and root infections.

Disease Management

The disease is usually observed in areas previously affected by rhizome rot disease. Therefore, phytosanitation plays an important role in disease management. The presence of inoculum in the soil and in plant debris, overcrowding of plants, and thick shade all promote the development of the disease; therefore, any disease management schedule has to be followed with these factors in mind. The application of superphosphate at the rate of 300–400 g per plant has been recommended for controlling clump rot in cardamom plantations (Anon., 1955). Soil drenching with 1% Bordeaux mixture or 0.25% copper oxychloride or neem oil cake at the rate of 500 g per plant, followed by one round of premonsoon and two rounds of postmonsoon soil drenching with 0.25% copper oxychloride at an interval of a month has been reported to be effective in controlling the disease (Thomas and Vijayan, 1994).

Biological Control

As in the case of capsule rot, attempts to control rhizome rot disease make use of *Trichoderma* sp., namely, *Trichoderma viride* and *Trichoderma harzianum* (Thomas et al., 1991b). A formulation of *Trichoderma harzianum* in a carrier medium consisting of farmyard manure and coffee husk mixture has been developed for field application in an integrated disease management system for the control of rot diseases of cardamom (Thomas et al., 1997).

2.6.1.5 Leaf Blight Disease ("Chenthal" Disease)

Chenthal is a leaf blight disease, and the name means "shredding" in the Malayalam language of the state of Kerala. The disease was first reported in the Idukki district of the state of Kerala (George et al., 1976). Since then, the disease has been observed in many plantations. Its spread is faster in partially deforested areas and less shaded plantations. Although Chenthal was reported as a minor disease of limited spread, currently the situation is alarming because the disease is spreading to newer areas and is becoming a major problem.

Disease Symptoms and Damage

Chenthal appears mostly during the premonsoon period, and the severity increases during the summer months. Symptoms develop on the foliage as water-soaked rectangular lesions, which subsequently elongate to form parallely arranged streaks. The length of these streaks varies from a few millimeters up to 5 cm. The lesion areas become yellowish brown to orange red in color, and often the central portions become necrotic. Usually, the two youngest leaves are not attacked by the disease. As the disease advances, more and more lesions develop on older leaves, adjacent lesions coalesce, and these areas begin to dry up. Severely infected plants show a burnt appearance. George and Jayasankar (1979) reported a reduction in plant height and in panicle length and crop loss due to failure of panicle formation in severely affected plants. However, Govindaraju et al. (1996) studied the symptomatology in detail and found that Chenthal infects only the leaves, and not the plant height, panicle emergence, or crop yield.

Causal Pathogen

Chenthal was originally reported as a bacterial disease caused by *Corynebacterium* sp. (George and Jayasankar, 1977). These researchers recommended penicillin spray to control the disease. Because later investigators could neither isolate *Corynebacterium* sp. nor control the disease with penicillin sprays, the bacterial etiology became suspect and the cause of the disease remained obscure for more than a decade. Govindaraju et al. (1996) conducted detailed investigations on the symptomatology, etiology, and management strategies of Chenthal and have shown beyond a doubt that the causal pathogen is the fungus *Colletotrichum gloeosporioides* (Penz.) Penz and Sacc. The fungus closely resembles *Colletotrichum gloeosporioides*, which causes the anthracnose disease of capsule rot reported by Suseela Bhai et al. (1988). Both leaf and capsule isolates showed similar cultural and morphological characters and were cross infective to capsules and leaves, and vice versa. However, these two kinds of isolate exhibited considerable differences in their period of occurrence, type of symptoms, distribution, and spread of the disease.

Disease Management

Since the disease was considered to be caused by *Corynebacterium* sp., penicillin spray was suggested as a control measure (George and Jayasankar, 1977), but when this treatment was not effective, it was abandoned by the planters. Govindaraju et al. (1996) reported that three sprays of Carbendazim (Bavistin, 0.3%) at monthly intervals, or Mancozeb (0.3%), or copper oxychloride (0.25%) effectively controlled the spread of Chenthal disease in the cardamom plantations.

Diseases Caused by Nematodes

Heavy loss of a crop could be brought about by nematode infestation. Although as many as 20 different genera of plant-parasitic nematodes have been reported in cardamom-growing soils (Ali, 1983), the root knot nematode (*Meloidogyne incognita*), which attacks black pepper as well (Nair, 2004), causes the most damage to cardamom. The root knot nematode is widely observed in almost all cardamom-growing regions, in both the nurseries and main plantations (Ramana and Eapen, 1992), while the lesion nematode (*Pratylenchus coffeae*) and the burrowing nematode (*Radopholus similis*) are observed in mixed plantations.

Disease Symptoms and Damage

Aerial damages to plant parts, such as stunting, reduced tillering, resetting and narrowing of leaves, yellow banding of leaf blades, and drying of leaf tips or leaf margins, are noticed. Flowering is normally delayed. The dropping of immature fruits results in a reduction in yield (Anon., 1972, 1989b). Underground symptoms develop on the roots of infected plants in the form of pronounced root galling. Tender root tips show spherical–ovoid swellings. Severe infestation can result in crop losses of up to 80% (Ramana and Eapen, 1992). The nematode population is high in cardamom soils during postmonsoon period (September–January). Heavy shade in plantations, moist soil, and warm, humid weather are predisposing factors for nematodes to multiply. Nematode infestation is a chronic problem in cardamom nurseries, where the same site is repeatedly used for raising seedlings. Nematode-infested soils affect seed germination and result in severe galling of the root system, marginal yellowing and drying of leaves, stunting, and reduced tillering. The leaves become narrow, and the leaf tips show upward curling.

Nematode Control

Because infected seedlings serve as the source of inoculum, extreme care has to be taken in transplanting affected seedlings. Preferably, transplanting should be avoided, as it is highly likely to cause new infections. Pretreatment of infested nursery beds with methyl bromide at the rate of $500\,g/10\,m^2$ or soil drenching with 2% formalin is usually recommended. Solarization of nursery beds is reported to reduce nematode populations in the soil. Nematicides, such as Adicarb, Carbofuran, or Phorate, at the rate of 5 g active ingredient (a.i.) per plant twice a year has been recommended for controlling nematodes in plantations (Ali, 1984). A biocontrol schedule employing *Trichoderma viride* or *Trichoderma harzianum* isolates and also *Pacilomyces lilacinus* to control damping-off disease and nematode damage in cardamom nurseries has been put in place (Eapen and Venugopal, 1995).

2.6.2 Minor Diseases

A number of minor diseases that affect leaves, capsules, and aerial stems occur sporadically in cardamom plantations. Some of these are frequently observed in all areas, while others are restricted to specific localities. Among these diseases are various types of leaf spots and capsule spots, stem infections, and so on, caused predominantly by fungal pathogens. Details are given in Table 2.28.

Table 2.28 Minor Fungal and Bacterial Diseases in Cardamom Plantations

Disease	The Affected Plant Part	Causal Pathogen
Leaf blotch	Leaves	*Phaeodactylium alpiniae*
Phytophthora leaf blight	Leaves	*Phytophthora meadii*
Phytophthora leaf rust	Leaves	*Phakopsora elettariae* (*Uredo elettariae*)
Phytophthora leaf spot	Leaves	*Sphaceloma cardamomi, Cercospora zingiberi, Glomerella singulata, Phaeotrichoconis crotalariae Ceriospora elettariae*
Sooty mould	Leaves	*Trichosporiopsis* sp.
Stem lodging	Pseudostem (tillers)	*Fusarium oxysporum*
Anthracnose	Capsules	*Colletotrichum gloeosporioides*
Capsule tip rot	Capsules	*Rhizoctonia solani*
Fusarium capsule rot	Capsules	*Fusarium moniliforme*
Capsule canker (Vythiri spot)	Capsules	*Bacterium* (?)
Capsule ring spot	Capsules	*Marasmius* sp.
Bacterial rot	Rhizomes	*Erwinia chrysanthemi*

2.6.2.1 Leaf Blotch

Agnihothrudu (1968) reported a foliar disease in cardamom characterized by the typical blotching of leaves. The disease appears during monsoon season, from June to August, normally under heavily shaded conditions. Thick shade, continuous rainfall, and high atmospheric humidity predispose the cardamom plants to infection. Leaf blotch was thought to be a minor disease. Recently, however, it was found to spread in great severity in certain regions.

2.6.2.2 Disease Symptoms

Nair (1979) studied the symptomatology of leaf blotch in detail. During the monsoon, round, ovoid, or irregular water-soaked lesions appear on middle leaves, usually near the tips or at the midrib areas. These areas enlarge and become dark brown with a necrotic center. In moist weather, a thick, gray fungal growth is seen on the underside of the blotched areas. Following a dry period, the spread of lesions is limited.

2.6.2.3 Causal Pathogen

Leaf blotch is a fungal disease caused by *Phaeodactylium venkatesanum* (Agnihothrudu, 1969). Subsequently, the fungus was identified as *Phaeodactylium alpinae* (Sawada) (Ellis, 1971). The pathogen grows profusely on the underside of the leaves and on potato dextrose agar medium. Hyphae are hyaline, smooth, partially submerged, 6–10 μ thick, and dichotomously or often trichotomously branched with conidia formed at their tips. Conidia are solitary, hyaline with three transverse septa, smooth, and elliptical with a tapered basal end and broad apices. Conidia measure 15–25 μ × 4.7 μ. The pathogen infects and produces typical symptoms on *Alpinia* sp. and *Amomum sp.*, and it has been observed that the fungus was completely inhibited *in vitro* by 1% Bordeaux mixture, 0.1% Bavistin, or 0.15% Hinosan (Nair, 1979). Fungicidal spray with copper oxychloride or Bordeaux mixture was reported to control leaf blotch infection in the field (Ali, 1982).

2.6.2.4 Phytophthora *Leaf Blight*

In many cardamom plantations, a widespread leaf blight disease is observed during the postmonsoon season. The infection starts on the young- to middle-aged leaves in the form of large, brown elongate or ovoid patches, which soon become necrotic and dry off. These necrotic dry patches are seen mostly on leaf margins, and in severe cases the entire leaf area on one side of the midrib is found affected. The disease appears during the months of October–November and may extend up to January–February. Thick shade, the low night temperature, and fog prevailing in isolated pockets predispose the plants to leaf blight infection. The causal organism is *Phytophthora meadii*. The pathogen can easily be isolated from infected leaf portions by means of a water-floating technique. The infection is aerial, and the infected plant debris serves as the source of primary inoculum. The pathogen grows internally and, under moist and misty conditions, produces abundant sporangia, which are disseminated by wind, spreading the disease to other areas. Disease symptoms are seen only on the leaves. Although *Phytophthora* is a potential pathogen that infects all parts of cardamom, the leaf blight isolate is specific to the leaves under natural conditions. However, cross inoculations of *Phytophthora meadii* leaf isolate on capsules were found to be infective on plant parts tested under laboratory conditions, and vice versa.

2.6.2.5 Disease Management

Leaf blight can spread rapidly, and it leads to severe leaf necrosis and drying if the disease is not controlled in the initial stage. One round of foliar spray with 1% Bordeaux mixture, Aliette 0.3%, or Akomin 40 (potassium phosphonate) at 0.3% was found to limit the spread of the disease.

2.6.2.6 Leaf Rust

Thirumalachar (1943) first reported a type of rust in cardamom in the state of Karnataka. The disease appears after the monsoon during October–May, and symptoms appear on leaves in the form of numerous yellowish rust-colored pustules distributed across the leaf surface in several patches, mostly on the underside of the leaves. As the disease progresses, pustules, or uredosori, become reddish brown and protrude from the leaf surface. The mature pustules break open and release uredospores. In severe cases, infected leaves show several yellowish patches with numerous rust-colored pinhead spots distributed throughout the patches. These areas dry off as the disease advances. The rust fungus *Phakospora elettariae* (Racib.) Cummins (Syn: *Uredo elettariae* Racib.) causes the disease. Naidu (1978) reported a mycoparasite, *Darluca filum* (Biv) Cast, hyperparasitizing this rust fungus. The mycoparasite produces dark-brown to black pycnidia in large numbers, which protrude from the uredosori. The hyperparasitized uredospores shrivel and do not germinate. *Darluca filum* develops only in the advanced stage of rust development. However, it helps to prevent further secondary spread of the rust fungus. The spread of leaf rust infection can be minimized by spraying fungicide such as Mancozeb 0.2% (Dithane M-45) and Indofil M-45.

2.6.2.7 Leaf Spot Diseases

A number of leaf spot diseases caused by a variety of pathogenic fungi affect cardamom. They infect both the seedling and the mature plant. The types of leaf spots occurring in main plantations are discussed in the sections that follow.

2.6.2.8 Sphaceloma *Leaf Spot*

The occurrence of this disease in the Coorg district of Karnataka State was first reported by Muthappa (1965). Symptoms appear on leaves in the form of scattered spherical blotches measuring a few millimeters in diameter. Adjacent lesions coalesce to form large necrotic patches. The disease was reported as a major problem in Coorg district. Though present throughout the year, the disease is more abundant and severe during the postmonsoon period. This leaf spot disease is caused by *Sphaceloma cardamomi* Muthappa. Naidu (1978) reported that Ceylon and Alleppey Green cultivars in Coorg district showed resistance to *Sphaceloma* leaf spots. Cultivars having erect panicles are mostly resistant to *Sphaceloma* leaf spot, while cultivars with creeping or prostrate panicles are susceptible.

2.6.2.9 Cercospora *Leaf Spot*

Another leaf spot occurring in Coorg district was reported by Rangaswami et al. (1968). Leaf blades show the first symptoms as water-soaked linear lesions, which are rectangular and parallely arranged alongside the veins. On the upper leaf surface, lesions turn dark brown with dirty white long patches in the center. In advanced stages, the lesions become grayish brown and dry off. The disease is caused by *Cercospora zingiberi* Togshi Katsaki. The fungus produces conidiophores in clusters from many-celled dark-brown stromata. Conidiophores are simple or (rarely) branched, septate, straight or curved, and geniculate; often, they undulate at the tip and are light brown in color. The conidiophores measure $17.5–56\,\mu \times 5.23–3.5\,\mu$. Conidia are formed singly, linear, indistinctly septate with three to six septa, and mostly curved with an obtuse base $37–195\,\mu \times 1.75–2.5\,\mu$. Naidu (1978) observed that cultivars having erect (variety Mysore) panicles are relatively resistant to *Cercospora* leaf spot, compared with variety Malabar, which has prostrate panicles.

2.6.2.10 Glomella *Leaf Spot*

Nair (1979) reported the occurrence of a leaf spot disease characterized by the presence of circular–ovoid dark-brown, concentric spots on the middle leaves. This disease appears during the postmonsoon period in isolated spots. The disease is generally seen only in variety Malabar. The infection starts as small pale-yellow water-soaked lesions on leaves. The lesions may be irregular in shape, measuring 1–2 mm in size. Later, they enlarge and form a depressed central area surrounded by a dark band of tissue. Later still, alternate concentric dark and pale-brown bands develop, with a yellow halo around the entire spot. Large mature spots may coalesce and the lesions may start drying. Sometimes, lesions measure as large as about 4 cm in diameter. The fruiting bodies of the fungus are seen as dark-brown dotlike structures in the lesions.

The causal organism is identified as *Glomerella cingulata* Stoneum Spanding and Shronk. The fungus forms a grayish-white mycelial growth in potato dextrose agar medium. Later, the growth becomes dark gray with zonations. Acervuli are produced in cultures. Conidiophores are short and hyaline, and conidia are cylindrical, hyaline, aseptate, and 12–25 μ × 3–5 μ in size. Perithecia are globose dark brown, and ostiolate, and measure 85–135 μ in diameter.

2.6.2.11 Phaeotrichoconis *Leaf Spot*

Phaeotrichoconis was reported by Dhanalakshmy and Leelavathy (1976). Symptoms appeared on young and old leaves and are characterized by irregular papery white spots with brown margins on the leaf blade. Under moist conditions, the lesions enlarge and coalesce. During dry weather, the central portion of lesions dries off. The causal organism is identified as *Phaeotrichoconis crotalariae* (Salam and Rao) Subram. The pathogen grows profusely in culture and produces a yellowish-brown mycelium with numerous dark-brown sclerotia. Conidiophores are indistinguishable from hyphae, and the conidia are solitary, elongate, fusoid, straight, or slightly curved and have five to eight septate.

2.6.2.12 Ceriospora *Leaf Spot*

Yet another type of leaf spot on cardamom leaves was reported by Ponnappa and Shaw (1978). This disease, seen rarely, was observed in the Coorg district and is caused by *Ceriospora elettariae* Ponnappa and Shaw. The main symptom is the appearance of numerous circular or oval spots on the foliage, up to 8 mm in diameter, that coalesce to form larger patches. The center of the lesion is dirty white and is surrounded by light-brown, circular necrotic areas.

2.6.2.13 *Management of Leaf Spot Diseases*

Most of the leaf spot diseases just described occur sporadically in minor proportions and, as such, do not have a deleterious effect on crop yield. The spread of these diseases can be prevented by one or two rounds of spraying with common fungicides such as 1% Bordeaux mixture and 0.25% Mancozeb.

2.6.2.14 Sooty Mold

A sooty mold infection on leaves of cardamom growing under the shade tree *Cedrella toona* Roxb was reported by Nair (1979). Symptoms appear during January–February when the shade trees are in blossom. Infection starts as a minutely scattered dark mycelial growth on the upper leaf surface. This growth spreads rapidly to cover the entire lamina; in severe cases, it extends to the petioles and leafy stems, which later are covered with the black mycelial growth. In advanced stages, leaves tear off at margins along the veins and dry prematurely. The sooty mold fungus is identified as *Trichosporiopsis* sp.

2.6.2.15 Stem Lodging

In the cardamom plantations of Idukki district in Kerala State, a relatively new disease affecting leafy stem has been found. The same disease has appeared in the lower Pulney area of Tamil Nadu (Dhanapal and Joseph, 1996). The disease attacks middle portions of the tillers in the form of pale discolored patches, which lead to a sort of dry rotting. The leafy stem is weakened at this portion and leads to partial breakage. The partially broken tillers bend downward and hang from the point of infection. Where the infection occurs at the lower part of the tillers, the tillers lodge. In such tillers, leaves and leaf sheaths soon dry up. The disease is caused by *Fusarium oxysporum* and usually appears during the postmonsoon period.

2.6.2.16 Anthracnose

Anthracnose occurring on cardamom capsules was reported as a minor disease in certain localities of cardamom cultivation (Suseela Bhai, Thomas, and Naidu, 1988). Symptoms appear on capsules as reddish-brown round or oval spots 1–2 mm in diameter, often with a soft, depressed center. The lesions vary in number and size and, in rare cases, coalesce to form large lesions. Often, less than 2% of plants with the disease alone is noticed. But in Anamalai areas of Tamil Nadu, as high as a 10–28% incidence was noticed.

2.6.2.17 Causal Pathogen

Colletotrichum gloeosporioides (Penz) Penz and Sacc has been shown to be the causative pathogen of anthracnose disease (Suseela Bhai et al., 1988). The fungus grows profusely in potato dextrose agar medium, producing a dark, gray-colored dense mycelium. Setae are dark brown, conidia abundant, cylindrical, straight, and 12–24 μ × 2.5–5.0 μ in size. A similar infection of *Colletotrichum gloeosporioides* on capsules resulting in the formation of much large lesions often extending up to three-fourth of the area of the capsules occurs in plantations of Karnataka State. This severe form of anthracnose leads to decay and loss of infected capsules. Sprayed three times at 0.3% concentration, fungicides such as Cuman-L, Foltaf, and Bavistin were found to control the disease.

2.6.2.18 *Capsule Tip Rot*

A characteristic type of rotting of the capsule tip commonly occurs in the state of Karnataka. The disease makes its appearance as small water-soaked lesions at the distal end of the capsule, later spreading downward. The infected capsule decays often up to the middle of the tip. In advanced stages, rotting extends along the entire length of the capsule. *Rhizoctonia solani* causes the capsule tip to rot. A 0.2% spray of Bavistin, copper oxychloride, or Foltaf checks the spread of the disease.

2.6.2.19 Fusarium *Capsule Disease*

Wilson et al. (1979a) reported a type of capsule disease caused by *Fusarium moniliforme* Sheld. Infection appears as small lesions on the capsule rind, which later decays and the periphery of the lesions turns reddish brown. During the monsoon, if the infection is severe, entire capsules decay. The symptoms described by Wilson et al. (1979) closely resemble that of anthracnose, but *Fusarium* infection often leads to capsule decay.

2.6.2.20 *Capsule Canker*

A type of capsule spot suspected to be caused by *Xanthomonas* sp. has been reported by Agnihothrudu (1974). The condition is locally known as Vythiri spot and was initially observed in the district of Wayanad. Subsequently, it turned up in several cardamom plantations. Symptoms develop as raised shining blisters or eruptions on the capsule rind. The blisters are pale to silvery white and sometimes cover almost half the capsule area. The causal pathogen has not been definitively established, since no pathogenic fungi or bacterium was found associated with the spots. The disease occurs only in minor proportions and is not alarming, since no crop loss has been observed due to infection. However, infected capsules fetch lesser price in cardamom auctions, as the blisters are clearly visible on cured capsules.

2.6.2.21 *Capsule Ring Spot*

A rare infection of capsules is noticed in certain plantations in the state of Karnataka. The symptoms are characteristic reddish-brown concentric rings or zonations that develop on the capsule rind. These areas turn necrotic and subsequently dry off. The cause of the infection is suspected to be *Marasmius* sp., although no definitive proof has been established.

2.6.2.22 Erwinia *Rot*

Tomlinson and Cox (1987) reported a serious rot disease of cardamom in PNG, the symptoms of which are seen on the foliage as a yellowing of the leaves of mature plants. Often accompanying this yellowing is a rotting and collapse of leafy stems at the ground level. A pale-brown color develops on rhizomes, later leading to their decay. Roots become blackened and necrotic in advanced stages of infection. The infection is observed in the variety Malabar and often leads to collapse of the entire plant.

2.6.2.23 Causal Pathogen

The disease is reported to be caused by a bacterium that has been identified as a strain of *Erwinia chrysanthemi* Burkholder. Tomlinson and Cox (1987) isolated this bacterium from infected cardamom rhizomes and roots and found it to be pathogenic. The bacterium has been biochemically characterized and identified as a pectolytic bacterium that readily grows on crystal violet peptone agar (cvp agar). Colonies are slightly raised and look like a fried egg with a distinct orange-colored center. Pathogenic isolates are KOH-soluble Gram-negative rod-shaped bacteria. The bacterium survives in infected rhizomes and roots and in the rhizosphere.

2.6.2.24 Diseases of Cardamom Found in Nurseries

The main route of cardamom propagation is through seeds. Seedlings that are raised in nurseries for 10–18 months become plantable. Normally, nurseries are designed to be in two stages: the primary nursery and the secondary nursery. Following are the major diseases occurring in seedling nurseries (Table 2.29).

Damping Off

Wilson et al. (1979b) observed the incidence of damping off in young seedlings of age 1–6 months. Affected seedlings become pale green and wilt suddenly en masse, as their collar portion rots. Overcrowding and excess soil moisture are the factors predisposing seedlings to this disease. The causal organisms of damping off were identified as *Rhizoctonia solani* (Wilson et al., 1979) and *Pythium vexans* (Nambiar et al., 1975).

Seedling Rot, or Clump Rot

This disease is similar to rhizome rot. Normally, it is observed in nurseries, where the seedlings are 6–12 months in age. Often, seedling rot is seen in overcrowded

Table 2.29 Diseases Found in Cardamom Nurseries

Name of Disease	Affected Plant Part	Causal Pathogen
Primary nursery		
Damping off	Young seedlings	*Rhizoctonia solani, Pythium vexans*
Seed or seedling rot	Seeds, young seedlings	*Fusarium oxysporum*
	Leaf or leaf sheath	*Sclerotium rolfsii*
Seedling rot	Pseudostem	
Leaf spot	Young leaves	*Phyllosticta elettariae*
Secondary nursery		
Seedling rot (clump rot)	Rhizomes, tillers, roots, and leaves	*Pythium vexans, Rhizoctonia solani*
Leaf spot	Leaves	*Colletotrichum gloeosporioides*

nurseries during the monsoon. Symptoms are characterized by wilting or drooping of leaves. Leaves turn pale yellow, followed by rotting of the collars of seedlings. As the infection advances, the young tillers fall off and the entire seedling collapses. *Rhizoctonia solani* and *Pythium vexans* are the causal pathogens. Root rot alone affects some nurseries. In such cases, only *Fusarium* sp. was found to be infective. Ali and Venugopal (1993) have reported the association of root knot nematode, *Meloidogyne incognita*, along with *Rhizoctonia solani* and *Pythium vexans*.

Siddaramaiah (1988a) reported the occurrence of seed rot disease resulting in the wilting of seedlings. The disease is caused by *Fusarium oxysporum*. Another seedling disease, caused by *Sclerotium rolfsii*, was reported by the same author (Siddaramaiah, 1988b). This disease results in the rotting of leaves, the leaf sheath, and the leafy stem of the seedlings.

2.6.2.25 Disease Management

Pattanshetty et al. (1973) reported that presowing nursery beds with 2% formaldehyde improved seed germination and reduced the incidence of damping-off disease. Thomas et al. (1988) reported fungicidal control of both seedling rot and damping off by soil drenching with Emisan 0.2%, Mancozeb 0.4%, or Brassicol 0.2%. Seed dressing with *Trichoderma harzianum* followed by one or two rounds of *Trichoderma harzianum* in nursery beds at 30 days intervals has been found to reduce the incidence of seedling rot disease.

2.6.2.26 Nursery Leaf Spot

Leaf spot disease is a serious problem in nurseries, resulting in severe loss of seedlings. Subba Rao (1939) reported the disease, and subsequently Mayne (1942) identified the causal organism as *Phyllosticta* sp. The pathogen was isolated and studied in detail by Chowdhary (1948), who identified it as *Phyllosticta elettariare* Chowdhary. The disease occurs mainly in the primary nursery on tender leaves and appears as minute water-soaked lesions that are almost circular in shape with a light-colored periphery and a depressed necrotic center. This central portion then dries off and becomes papery white. In later stages, shot holes are formed at the center of the lesion. As the disease advances, numerous such lesions of different sizes develop and the entire leaf dries off. Several minute dark pinheadlike pycnidia of the fungus can be seen in the lesions. Older leaves of the seedlings are less susceptible to the disease. As seedlings grow old, they develop resistance to infection and, consequently, to the disease. Leaf spot disease can easily be controlled by fungicide sprays, such as 0.2% Difolatan or 0.1% Bordeaux mixture or 0.2% Dithane, when sprayed at fortnightly intervals (Rao and Naidu, 1974).

2.6.2.27 Leaf Spot in Secondary Nursery

A different type of leaf spot disease is observed in secondary nurseries (6–12 months old). The disease is characterized by the development of many rectangular water-soaked lesions on the foliage. These lesions enlarge longitudinally and are parallely arranged along the sides of the veins. As they mature, they exhibit a muddy red color and become necrotic. The leaves dry off as too many lesions occur side by side. The disease is caused by *Colletotricum gloeosporioides*. A 0.25% Mancozeb spray is effective in controlling the spread of the disease.

2.6.2.28 Conclusions

The fungal diseases of cardamom are far easier to control than the devastating viral diseases. However, the use of fungicides and insecticides is being discouraged owing to the strong antipathy of consumers to phytochemicals. The cultivation of organically grown cardamom is gaining importance. In view of this shift, it is essential to evolve biocontrol strategies against the more serious fungal diseases. A protocol for the production of organic cardamom needs to be developed and popularized in order to cater to the demand in international markets. An intensive search for natural resistance to the pathogens needs to be initiated. Because the Western Ghats are the center for biodiversity for cardamom, the possibility of locating resistance lines there is fairly high. Once such resistant gene-carrying lines are located, the trait can be transferred to locally grown varieties and cultivars through traditional breeding techniques. Where resistant genes are absent, biotechnological approaches may have to be resorted in order to develop resistant genotypes.

2.6.2.29 Viral Diseases of Cardamom

The major production constraint in cardamom production in India is the occurrence of the mosaic virus (car-MV-katte) disease. The cardamom necrosis virus (car-NV-Nilgiri necrosis virus) and cardamom vein-clearing disease (car-VCV-Kokke kandu) are also matters of concern to the cardamom industry in some endemic zones. Diseases of cardamom were reviewed by Chattopadhyay (1967), Agnihothrudu (1987), Naidu and Thomas (1994), and Venugopal (1995). Based on the severity of occurrence and crop losses, four serious viral diseases of cardamom and their integrated management are discussed next.

2.6.2.30 Mosaic or Katte Disease (car-MV)

Locally, mosaic disease is known as "katte," which, in the Kannada language of Karnataka State, means "disorder." It is known as "marble disease" in Anamalai of Tamil Nadu (Varma and Capoor, 1953).

2.6.2.31 Distribution

The earliest reference to katte dates back to 1900, by Mollison. In South India, the disease is widely distributed in all cardamom-growing areas and the incidence ranges from 0.01 to 99% (Venugopal and Naidu, 1981). Until the 1970s, cardamom was free of this viral disease. In 1975, a disease with viruslike symptoms was observed in some parts of the world and within 5 years the disease spread to all nearby cardamom plantations in the South Pacific coastal region, which produces 60% of cardamom in Guatemala (Gonsalves et al., 1986). Surveys conducted in cardamom-growing areas of India on the incidence of car-MV and car-VCV have revealed the prevalence of car-MV in most of the cardamom plantations of Karnataka in South India (Govindaraju et al., 1994).

2.6.2.32 The Extent of Crop Loss

The stage of the plant's growth at which time the infection occurs and the duration the plants are subjected to the infection together decide the extent of crop loss. If the plants are infected at the seedling stage or early prebearing stage, the loss will almost be total (CPCRI, 1980; Samraj, 1970). Infection in the bearing stage results in a gradual decline in productivity (CPCRI, 1984). In cardamom–areca nut mixed cropping systems, crop loss due to viral infection has, respectively, been estimated at 10–60%, 26–91%, and 82–92% in the first, second, and third years of production (Varma, 1962a). Similarly, in monocrop situations, infection of bearing plants led to yield reductions of 38%, 62%, and 68.7% in the first, second, and third years of infection, respectively (Venugopal and Naidu, 1987). Total decline of the plants occurs within 3–5 years from the date of infection.

2.6.2.33 Symptomatology

The first visible symptom appears on the youngest leaf of affected tiller as slender chlorotic flecks measuring 2–5 mm in length. Subsequently, these flecks develop into pale-green discontinuous stripes that run parallel to veins from the midrib to the leaf margins. All of the subsequent emerging leaves show characteristic mosaic symptoms, with stripes of green tissue almost evenly distributed over the entire lamina (Uppal et al., 1945). Often, mosaic-type mottling is seen on leaf sheaths and young leaf shoots. Variations in field symptoms are seen in different cardamom-growing tracts of South India and on inocula of different viral isolates on a common host (Venugopal and Naidu, 1981). Plants of all stages of growth are susceptible to infection; the virus is systemic in nature and gradually spreads to all tillers in a clump. In advanced stages, the affected plants produce slender and shorter tillers with only a few short panicles. The plants degenerate gradually.

2.6.2.34 Transmission

Neither seed, soil, root to root, nor manual operations are the channels for the transmission of the virus (Rao, 1977a,b; Thomas, 1938). The only methods of dissemination are through the banana aphid (*Pentalonia nigroner-vosa* Coq.) and infected rhizomes. The first experimental transmission of katte virus in India was obtained with banana aphids (Uppal et al., 1945). Up to now, 13 aphid species (*Aphis craccivora* Koch, *Aphis gossypii* Glover, *Aphis nerii* B. de F, *Aphis rumicis* L, *Brachycaudus helichrysi* L, *Greenidia artocarpi* W, *Macrosiphum pisi* Kalt, *Macrosiphum rosaeformis* Das, *Macrosiphum sonchi* L, *Schiazaphis cyperi* van der Groot, *Schiazaphis graminum* Rondm, *Pentalonia nigronervosa* f. *typical*, and *Pentalonia nigronervosa* f. *caladii* van de Groot) have been reported to transmit the virus (Rao and Naidu, 1974).

2.6.2.35 Spread of the Disease

Sources of Infection

Both infected and healthy clones, seedlings raised in the vicinity of infected plantations, volunteers working in infected plantations, and a few infected zingiberaceous hosts (*Amomum* sp.) are the sources of infection. In a contiguous area, infected plantations are the reservoirs of virus sources (Naidu and Venugopal, 1987, 1989; Varma, 1962a).

Primary Spread

In plantations, the primary spread occurs at random through the activity of viruliferous alate forms of the vector. Under field conditions, in plantations located 400–600 m from concentrated sources of the virus, the percentage of primary infection varied from 0.07 to 5.19% (Venugopal et al., 1997a). The frequency of random spread depends directly upon access to virus sources.

Secondary Spread

Following the primary spread, secondary spread is mainly internal and the rate of spread is low (Deshpande et al., 1972; Naidu and Venugopal, 1989). A centrifugal influx found around primary foci was due to spread by the activity of apterate adults. In plantations, the disease is concentrated within a 40-m radius, with occasional random spreads up to a 90-m distance. The gradient of infection is steep within 40 m from the initial foci, and it flattens thereafter. In Guatemala, the rate at which the disease spreads is fast, and natural infection may reach 83% within 6 months of planting. A similar situation exists in the case of cardamom–areca nut mixed cropping systems.

Incubation Period

In the field, the incubation period of katte disease varies from 20 to 114 days during different months, and the expression of the disease is directly influenced by the growth of the plants. Normally, young seedlings at the 3–4-leaf stage express the symptoms within 15–20 days of inoculation, whereas mature plants take anywhere from 30 to 40 days for the expression of symptoms during the active growing period and 90–120 days during the winter months (Venugopal and Naidu, 1987). Senile leaf sheaths, which are natural breeding sites of the vector, are poor inoculum sources compared with young, actively growing shoots (Venugopal and Naidu, 1989).

Virus–Vector Relationship

Earlier, it was thought that the aphids found on banana and cardamom were the same, but subsequently it was found that *Pentalonia nigronervosa* f. typically breeds on *Musa* and related genera while *Pentalonia nigronervosa* f. *caladii* breeds on cardamom, *Colacasia*, and *Caladium* (Siddappaji and Reddy, 1972b). In cardamom plantations, the aphid population is seen throughout the year, with one or two peak periods during November–May, and the population is drastically reduced during the monsoon season. All four nymphal instars and the adult are capable of transmitting the disease (Rajan, 1981). Bimodal transmission was examined by using two distinct viral strains with respect to acquisition, latent period, and persistence. Naidu et al. (1985) established the nonpersistent nature of the katte virus.

Host Range of the Virus

Several plants that belong to the family Zingiberaceae, viz., Amomum cannecarpum, Amomum involucratum, Amomum subulatum, Alpinia neutans, Alpinia mutica, Curcuma neilgherrensis, and Zingiber cernuum, and a member of the family Marantaceae (Maranta arundinacea, West Indian arrowroot) were found infected in laboratory inoculation tests (Rao and Naidu, 1973; Siddaramaih et al., 1986; Viswanath and Siddaramaiah, 1974; Yaraguntaiah, 1979).

2.6.2.36 Etiology

The initial evidence pointing to the fact that katte was a viral infection was in 1945, when the virus was successfully transmitted through the banana aphid *Pentalonia negronervosa* Coq. (Uppal et al., 1945). Investigations in Guatemala and India have shown the association of flexuous rod-shaped virus particles measuring 650μ in length and 10–12μ in diameter in dip and purified preparations (Gonsalves et al., 1986; Naidu et al., 1985; Usha and Thomas, unpublished). Purified preparations of six strains also revealed homogenous flexuous particles. The presence of inclusion bodies was reported from leaf tissues of car-MV infected plants. On the basis of the morphology of virus particles and the presence of characteristic pinwheel-shaped inclusion bodies, it was suggested that car-MV be included in the potyvirus group (Naidu et al., 1985). In Guatemala, mosaic-affected cardamom leaves revealed pinwheel-type inclusion bodies, a common feature in other potyviruses. Leaf dip extracts showed particles of 660μ length, and those of purified preparations showed 700- to 720-μ-long particles.

A serological affinity of car-MV of Guatemala with some potyviruses was demonstrated through indirect ELISA testing. Four viruses, namely, Zucchini yellow mosaic, papaya ring spot types w and p, cow pea aphid-borne mosaic virus, and a severe strain of bean common mosaic virus, consistently gave a positive reaction in indirect ELISA testing. The presence of inclusion bodies, particle morphology, and the serological affinity of car-MV has confirmed that it is a member of the potyvirus group (Dimitman et al., 1984; Gonsalves et al., 1986). Sequence analysis of the coding regions for coat protein and the 3-untranslated region of the Yeslur isolate (from Saklespur, Karnataka State) identified the katte virus as a new member of the genus *Madura* of the family Potyviridae (Jacob and Usha, 2001). Considerable genetic diversity was noted among various isolates (Jacob et al., 2002). Some consider katte as a complex disease caused by more than one component or virus (Rao, 1977a). So far, the studies conducted in India and Guatemala do not support a complex nature for katte disease.

2.6.2.37 Strains of car-MV

The presence of three natural strains of car-MV was first reported on the basis of symptomatology on the main host and cross-protection studies (Rao, 1977a).

Further, the occurrence of different natural strains was reported from studies using 68 representative isolates from all cardamom-growing zones of India. Three important biological criteria, namely, symptoms on the main host, transmission through *Pentalonia nigronervosa* f. *caladii*, and reaction on the set of zingiberaceous differentials consisting of *E. cardamomum* Maton var. Malabar, *Alpinia mutica, Amomum microstephanum*, and *Amomum cannaecarpum*, were used to identify the strains (Naidu et al., 1985).

2.6.2.38 Cardamom Vein-Clearing Disease, or "Kokke Kandu" (car-VCV)

Cardamom vein-clearing disease is a new threat to the plant in some endemic pockets in all the main cardamom-growing areas of Karnataka State. Surveys conducted between 1991 and 1993 indicated the prevalence of car-VCV ranging from 0.1 to 80% in plantations and nurseries (Govindaraju et al., 1994; Venugopal and Govindaraju, 1993). The disease is locally (in the Kannada language of Karnataka State) referred to as "Kokke Kandu," meaning "hooklike tiller."

2.6.2.39 Importance of the Disease

In all five cardamom-growing districts of Karnataka State, namely, Coorg, Hassan, Chickmagalur, Shimoga, and North Canara, the disease is widespread. In 381 plantations surveyed, car-MV, car-VCV, and mixed infections were seen in 375 plantations, for a range of incidences of 0.1–82%. A survey in 39 nurseries in hot spots revealed the existence of car-VCV.

2.6.2.40 Extent of Crop Loss

Plants affected by the disease decline rapidly, and the reduction in yield is 62–84% in the first year of peak crop yield (NRCS, 1994). In mixed crop situations with areca nut as the main crop, yield losses vary from 68 to 94% in plants with different stages of infection (IISR, 1995). The affected plants become stunted and perish within 1–2 years after infection sets in. Thousands of hectares of cardamom plantations in the Hongadahalla zone in Hassan district and areca nut-based mixed cropping systems in North Canara district (both in the state of Karnataka) have become uneconomical because of infection with mosaic and Kokke Kandu diseases.

2.6.2.41 Symptoms of the Disease

Symptomatic leaves first reveal characteristic continuous or discontinuous intraveinal clearing, stunting, resetting, loosening of leaf sheath, and shredding of leaves. Leafy stems exhibit clear mottling in all seasons. Clear light-green patches with three shallow grooves are seen on immature capsules. Cracking of fruits and partial sterility of seeds are other associated symptoms. In summer, the newly infected plants reveal only faint discontinuous vein-clearing symptoms. Plants of all stages, from seedling to bearing, show these symptoms. New leaves get entangled in the older leaves and form hooklike tiller—hence the name Kokke Kandu.

2.6.2.42 Transmission and Etiology of the Disease

Car-VCV is not transmitted through seed, soil, leaves, roots, or mechanical contact or through the use of farm implements. Mechanical transmission on a set of differentials through a combination of buffers, antioxidants, additives, and abrasives also was not successful (Anand et al., 1998; Venugopal, unpublished). The disease is transmitted through the cardamom aphid *Pentalonia nigronervosa* f. *caladii* in a semipersistent (IISR, 1996) or persistent (Anand et al., 1998) manner. The incubation period ranges from 22 to 128 days, and a single viruliferous aphid can transmit the virus to plants in all stages of growth. The exact etiology of associated virus has not yet been established. In enzyme-linked immunosorbent assay (ELISA) testing, antigen from infected host parts reacted positively with antibodies raised against potyviruses such as peanut mottle virus, sugarcane stripe virus, and Indian and Guatemalan car-MV isolates (Venugopal et al., 1997b). All these results indicate that car-VCV may be a member of the potyvirus group.

2.6.2.43 Spread of the Disease

As with car-MV, Kokke Kandu is initially spread randomly in distant blocks through the activity of incoming alate viruliferous vectors. Random spread was reported in new plantations located up to 2000 m away from infected plantations. The frequency of primary spread is directly dependent on the distance from the foci of infection (IISR, 1995, 1996; NRCS, 1994). Secondary spread within the infected plantations is both centrifugal and random. Alate forms of the aphid are responsible for random spread and apterate forms for centrifugal spread. In infected plantations, the rate of spread varied from 1.3 to 8.5% per year. The spread of the disease depends on the distance from, and incidence in, the foci of infection. The gradient is steep, concentrated near the sources of virus inoculum (about 100 m); the gradient is shallow in the next 100 m (IISR, 1996).

2.6.2.44 Cardamom Necrosis Disease (Nilgiri Necrosis Disease)

Cardamom necrosis disease was first noticed in a severe form in the Nilgiri district of Tamil Nadu—hence the name Nilgiri Necrosis Virus (NNV). Surveys revealed new pockets of infection in the states of Kerala and Tamil Nadu, and isolated spots in Karnataka State. In Tamil Nadu, the pockets of infection are located in the Nilgiris district, Anamalai, the Cardamom hills, and the Bilgiri Rangan hills.

2.6.2.45 Importance of the Disease

Random surveys in South India revealed a low incidence of 0.1–1% (CPCRI, 1985). Only in an isolated case in Valparai, Tamil Nadu, was an incidence of up to 13% noted. Later surveys in South India indicated that the disease is prevalent in some of the cardamom-growing regions of Tamil Nadu as well, with an incidence ranging from 7.7 to 80% (Sridhar et al., 1990). In Lower Pulneys, Tamil Nadu, out of 24 plantations surveyed, one plantation in Thadiankudisai exhibited a 76% incidence. In Valparai, again in Tamil Nadu, a 7.7–15.07% incidence was recorded. The highest incidence, 80%, was recorded in Coonor of the Nilgiris district in Tamil

Nadu. Some cardamom estates in Munnar and Thondimalai, in Idukki district of Kerala State, recorded incidences of 4.6% and 1.46%, respectively. Unlike plants infected with katte disease, those infected with Kokke Kandu decline rapidly and become stunted and unproductive.

2.6.2.46 Symptomatology and Crop Loss

Symptoms are seen on young leaves as whitish-yellowish continuous or broken streaks proceeding from midrib to leaf margins. In advanced stages of infection, these streaks turn reddish brown. Often, leaf shredding is noticed along the streaks. Leaves are reduced in size and have distorted margins. Plants that are infected early produce only a few panicles and capsules, and in advanced stages of infection, tillers are highly stunted and fail to bear panicles. All the types of cardamom cultivars are susceptible to the disease (Sridhar, 1988).

Plants that are in the early stages of infection recorded less yield reduction compared with those in advanced stages of infection (Sridhar et al., 1991). A 1-year investigation carried out on a diseased plantation indicated a 55% reduction in yield in early infected plants and a total loss of yield when infection took place late in the crop growth.

2.6.2.47 Transmission of the Disease

Seed, soil, sap, and mechanical injury from the use of implements do not transmit the disease. Instead, it is transmitted through planting infected rhizomes. Aphids (*Pentalonia nigronervosa* f. *caladi*), thrips (*Sciothrips cardamomi*), and white flies (*Dialeurodes cardamomi*) were tested for their ability to transmit the disease. No insect transmission of the disease from infected to healthy plants was recorded (Sridhar, 1988).

2.6.2.48 Etiology and Epidemiology

Flexuous particles 570–700 μ long and 10–12 cm broad were seen in dip preparations of NNV-infected leaf tissue. The virus belongs to the genus *Carlavirus* (Naidu and Thomas, 1994). Infected rhizomes or seedlings raised from diseased nurseries are the primary sources of inoculum. Monitoring of new infections at regular intervals in a diseased plantation revealed that the spread of the disease is mainly internal and new infections occur in a centrifugal fashion from the source of inoculum. Most of the infections occurred within a 10- to 15-m radius from the source of inoculum, and the number of new infections decreased as the intensity of the disease increased (Sridhar, 1988). The pattern of the disease spread is similar to that of katte disease. The rate of spread of the disease is a rather low 3.3% for the period of 1 year. The occurrence of a few outbreaks around the infection foci is an indication that the disease can be successfully managed by periodical rouging of infected plants.

2.6.2.49 Infectious Variegation

Infectious variegation was first noticed in Vandiperiyar in Kerala State in a severe form. Subsequently, it was also noticed in the Coorg district, Hassan, and North

Canara of Karnataka State. An incidence of 15% was observed in Vandiperiyar. Infected plants show typical variegated symptoms on leaves, with characteristic slender-to-broad radiating stripes of light and dark green on the lamina. Distortion of leaves and tillers, as well as stunting, are other common symptoms. Within 1 year of infection, the plants become unproductive. Only 2% transmission was obtained through the aphid *Pentalonia nigronervosa* f. *caladii*. Rouging resulted in near total elimination of the disease in all three test plantations.

2.6.3 Integrated Management of Viral Diseases in Cardamom

2.6.3.1 Production and Use of Virus-Free Planting Material

A number of constraints, such as infrastructure, the availability of a suitable nursery, the availability of water, accessibility, the availability of labor, and security, cause seedlings to be raised repeatedly in the same nursery site year after year or, more commonly, in an area adjacent to the plantation. Invariably, the seedlings become infected. Seedlings require 10–18 months to attain plantable age. This prolonged exposure to virus access through viruliferous aphids in the vicinity of concentrated virus sources results in infection at the nursery stage. Further, secondary spread in nurseries through aphids results in the spread of viruses to many plants. As high as 28% of car-MV (Venugopal et al., 1997a) and 73.33% of car-VCV (Govindaraju et al., 1994) infections occur in the nursery stage. None of the three viral diseases (car-MV, car-VCV, and car-NV) are transmitted by seeds, and they all lack long-distance spread beyond 2000 m. Hence, raising nurseries in isolated places is necessary to produce healthy seedlings. For car-MV, isolation of 200 m from virus sources is adequate, and for car-VCV, isolation of more than 200 m is necessary.

Apparently, healthy high-yielding plants are normally subcloned and planted for filling gaps and raising plantations (Varma, 1962). Because the infected plants take 23–168 days to express car-MV symptoms (Venugopal and Naidu, 1987) and 22–128 days to express car-VCV symptoms (Venugopal et al., unpublished), it is not advisable to use clones from infected gardens. Like nurseries, clonal nurseries have to be raised in isolated sites. In micropropagation, starting material has to be checked to make sure that it is free of viruses.

2.6.3.2 Avoidance of Volunteers

Volunteers that sprout from remnants of infected materials are the potential primary sources of spread (Naidu and Venugopal, 1987). Self-sown seedlings in the infected plantations are exposed to virus access for 2–8 months. As high as 28% infection was recorded in the nurseries raised from volunteers. The removal of infected volunteers in replanted areas and the total avoidance of volunteers for nursery activity in hot spots are the most important prerequisites for producing virus-free planting material.

2.6.3.3 Movement of Planting Material

In India, surveys conducted in 1981 (Venugopal and Naidu, 1981, 1987) and 1994 (Govidaraju et al., 1994) have shown that there are many disease-free pockets within an infected zone or plantation. Further, car-VCV is confined to certain endemic pockets. In Guatemala also, large areas are free from mosaic infection (Dimitman, 1981). Creating awareness and preventing the movement of planting material go a long way toward checking the introduction or reintroduction of viruses.

2.6.3.4 Vector Management

a. *Chemical control*: The nonpersistent nature of car-MV and semipersistent nature of car-VCV render chemical control measures less effective in checking or reducing the spread of secondary disease. Nor do insecticides, applied at recommended concentrations, kill aphids rapidly enough to prevent probing. Further, the persistence of aphid vectors throughout the year makes vector control measures almost impracticable. Thirty-four insecticides were evaluated to determine their effect on the transmission and acquisition of the katte virus under laboratory conditions. Transmission results showed that none of the insecticides tested were effective in checking the acquisition and transmission of the virus, even on the day they were applied. On account of the nonpersistent nature of car-MV (Naidu et al., 1985; Rao, 1977a,b), the disease can be transmitted within short periods of probing and feeding. Mere probing is sufficient for transmission of the virus. This could be the reason insecticides are ineffective in checking secondary spread (Rajan et al., 1989). The cardamom aphid, *Pentalonia nigronervosa* f. *caladii*, is photophobic and is found in colonies of 30–50, comprising nymphs as well as alate and apterate adults. Colonies are found in between the leafy stems and loose-leaf sheaths especially of old, partly dried, or damaged parts. Occasionally, colonies are found on the leaf spindles, young suckers, and panicles. Because of their concealed placement in the older parts, there is a lesser likelihood that the aphids will directly contact insecticides or indirectly contact systemic insecticides. As a result of insecticide treatment, the colony might have been disturbed, and their hyperactivity, probing, and intermittent migration in search of suitable hosts may be responsible for the ineffectiveness of some treatments, such as Phorate (granules), Carbofuran (granules), and Phosphamidon.
b. *Removal of breeding sites*: The photophobic vector breeds in senile, concealed parts of the host (Rajan, 1981). Thus, periodic removal of the old parts of the rhizomatous crop is effective in reducing the aphid population and the spread of car-MV (Rajan et al., 1989). Other natural hosts, such as *Colacasia* sp. and *Caladium* sp. (Rajan, 1981; Siddappaji and Reddy, 1972a), which are common weeds in the swampy areas of cardamom plantations, have to be removed periodically to check multiplication of the aphid. In addition, vector control measures must be adopted.
c. *Use of biopesticides*: Extracts of many botanicals were found to be effective in reducing the breeding potential of aphid vectors. Neem products significantly reduced the population of aphids on cardamom leaves, even at 0.1% concentration, and were lethal to aphids at higher concentrations (Mathew et al., 1977, 1999a,b). Aqueous extracts of *Acorus calamus* L. (dried rhizome), *Annona squamosa* L. (seeds), and *Lawsonia inermis* L. (leaves) reduced the settling percentage of aphids on leaves. Vapors of *Acorus calamus* are highly toxic to aphids and lead to their total mortality. Essential oil of turmeric (*Curcuma longa* L.) was also found to be repellant against the aphid (Saju et al., 1998).
   Entomogenous fungi like *Beauvaria bassiana* (Bals-Criv) Vuill, *Verticillium chlamydosporium* Goddard, and *Paecilomyces lilacinus* (Thom.) Samson were promising in suppressing aphid population without causing hyperactivity (Mathew et al., 1998).
d. *Resistant sources*: All of the 168 germplasm accessions constituting Mysore, Malabar, and Vazhukka types at the Research Center, Appangala, Karnataka State, under the administrative control of the Indian Institute of Spices Research in Calicut, Kerala State, which

is under the overall administrative control of the Indian Council of Agricultural Research, New Delhi, India, are susceptible to car-MV. Twenty-one elite accessions—distinct morphotypes such as compound panicle types—are also susceptible (Subba Rao and Naidu, 1981). Seventy natural disease escapes showed field resistance to car-MV in sick plots. These escapes are also high yielding in character and better than the local cultivar (IISR, 1996, 1997). Screening trials consisting of 24 elite accessions of cardamom against car-VCV in sick plots showed that accession 893 (of the Cardamom Research Center under the administrative control of the Indian Institute of Spices Research) is less susceptible compared with all the other test accessions.

e. *Removal of virus sources*: In the management of plant viruses, phytosanitation involves the detection and elimination of virus sources that are present within and outside of the cardamom plantations and the efficiency of phytosanitation in the management of viral diseases is centered on this operation. Attempts to control the spread of katte disease began as soon as researchers recognized the role of the virus involved, and control has been based mainly on sanitation or removal of the virus source. Rouging is reported to be effective in minimizing the spread of the disease and in enhancing the economic life of plantations (Capoor, 1967, 1969; Deshpande et al., 1972; George, 1967, 1971; Naidu and Venugopal, 1982; Varma, 1962b; Varma and Capoor, 1958). However, the intensity of disease and the distribution of disease within a plantation are the prime factors influencing the efficacy of rouging. It is more appropriate to adopt rouging as an effective means of containing the spread of the disease if the intensity is less than 10% (Naidu and Venugopal, 1982; Naidu et al., 1985). In general, in plantations, the disease is seen to be concentrated in patches, with random spread in certain spots. In such concentrated spots, surveying the area to detect fresh infection and rouging may be undertaken at shorter intervals to minimize the chances of secondary spread (Naidu and Venugopal, 1982). This practice may be continued for 3–4 months, until new outbreaks of the disease are reduced; thereafter, survey intervals can be increased to a few more months. Through sustained timely efforts, new infections can be reduced to 2–3% per annum, although it is impossible to eradicate the disease completely in a plantation because of the predominance of small holdings and multiple chances of reintroduction. In contiguous cardamom holdings, isolated attempts are not adequate to contain the disease economically. In such areas, a community approach through the total removal of all the plants, followed by replanting and proper surveillance, has been shown to be more feasible (Varma and Capoor, 1964). In varied field situations, such as new plantations in isolated areas, new plantations in hot spots, plantations with a unidirectional virus source, plantations with a multidirectional virus source, plantations located between two infected plantations, and plantations in a continuous belt, different approaches involving total removal and replanting, selective rouging and gap filling, phased replanting, and so on, were shown to be effective in reducing the secondary spread of the disease.

In car-MV- and car-VCV-infected areas, there are independent as well as mixed infections of both viruses. In such areas, comprehensive efforts involving the use of healthy seedlings, periodical surveys through a trained disease surveillance team, and the prompt removal and destruction of infected plants were reported to be effective in containing both viral diseases (Saju et al., 1997).

f. *Early detection of the disease*: Inoculants take 23–120 days to express the visible symptoms of the disease. Early detection plays an important role in checking the further spread of the disease through virus sources. Attempts were made to test the polyclonal antiserum produced against car-MV before the expression of symptoms, and ELISA testing was performed on the viruliferous aphid vector *Pentalonia nigrovervosa* f. *caladii* for quick detection (Saigopal et al., 1992). Various host plant parts and young and mature seeds of infected and healthy plants were examined. A positive reaction to the presence of the viral antigen was observed before the expression of symptoms in all the host plant parts, but not in the mature seeds of infected plants. Viruses were concentrated more in the roots than in the other plant parts. Testing of viruliferous and nonviruliferous aphid vectors showed that the viral antigen could be detected in viruliferous aphids. The usefulness of quick detection through ELISA testing was further confirmed by indexing the primary cultures after *in vitro* multiplication. ELISA tests can be used for rapid field diagnosis of mosaic infection (Roberto, 1982), and in Guatemala, ELISA testing is being used extensively in the country's virus control program (Gonsalves et al., 1986).

g. The integration of several methods, such as strategies to produce healthy seedlings in isolated places, efforts to reduce the vector population, the use of virus-resistant lines, and the removal of foci of infection, is required to manage the spread of virus diseases in the field. The establishment of plant disease clinics in potential cardamom-growing areas also helps to create an immediate awareness of the dangers of viral diseases and to impart training to the cardamom-growing community to manage those diseases. In India, an attempt was made to contain car-MV through the establishment of "katte clinics" (Nair and Venugopal, 1982). Cardamom planters responded favorably to the initiative, and in about 8 months, 60 plantations covering an area of 393 ha distributed in 30 villages of the Coorg district in Karnataka State were achieved. In India, during British rule, the government of Bombay attempted to eradicate katte disease in North Canara district (now in Karnataka State after the partition of the country into several states) by providing technical assistance to cardamom planters at government expense (Varma, 1962b). Along the same lines today, the Cardamom Board, one of the Agricultural Commodity Boards under the administrative control of the Ministry of Commerce attached to the Government of India, New Delhi, has taken up the eradication of katte disease in contiguous blocks in Karnataka State by providing both financial and technical help to cardamom planters, primarily for rouging diseased plants. These programs have raised awareness of identification and management techniques to contain the spread of the viral diseases.

2.6.3.5 Development of Katte-Resistant Cardamom

An intensive survey of hot-spot areas with a high incidence of katte disease was carried out by the Cardamom Research Center at Appangala in Karnataka State. The survey led to the collection of 138 specimens that had escaped the disease, or "disease escapes." The collection included 4 escapes from var. Mysore, 29 from var. Vazhukka, and 105 from var. Malabar. A clonal nursery was established from these disease escapes, and they were subjected to screening in the greenhouse. Using virulent virus isolates with the aid of the vector showed that 67 collections became infected. These were then discarded, and the escapes that passed the screening tests were planted in sick plots. Screening continued for 6 years. Most tester lines included in the trail became infected in 2 years time, but 23 of them remained totally free from symptoms. Four lines, namely, NKE-11, NKE-16, NKE-22, and NKE-71, expressed faint granular symptoms during the active growth period, but these symptoms vanished in the next period (Venugopal, 1999).

The NKE lines that passed the screening method were planted in hot-spot areas. Nineteen lines remained free from the disease, of which 17 had satisfactory agronomic traits. Venugopal et al. (1999) investigated the breeding potential of the vector *Pentalonia nigronervosa* f. *caladii* on these 17 lines and compared the results with those of a local susceptible check. The aphids colonized and multiplied on all the accessions, thereby indicating that the resistance of the 17 accessions is due, not to deterrence of the vector, but to some other factors associated with the host.

These 17 lines were tested further after interplanting them with known susceptible checks, and they remained free of the disease. Repeated inoculation did not produce any symptoms in these lines, thereby confirming their virus-resistant nature (Venugopal, 1999).

At present, nothing is known about the mechanism of resistance in the aforementioned resistant lines. Different strains of car-MV have been reported from different virus-infected zones. Some zingiberaceous plants, such as *Alpinia mutica*, which was found resistant to Kodagu (Coorg), Hassan, and Chickmagalur isolates (all from the state of Karnataka) and to Wayanad isolate (from the state of Kerala), showed higher susceptibility to Nelliampathy (from the state of Kerala) isolates. Hence, further investigation is important to establish the performance of the 17 virus-resistant clones against other distinct virulent strains from different cardamom-growing areas.

2.6.3.6 Conclusions

In cardamom, viral diseases are responsible for a rapid degeneration of production potential during the early stages of the establishment of the plantation and are a constant threat to sustainable cardamom production. Unfortunately, scant information is available on the characterization of the disease-causing viruses. Systematic efforts are required to characterize the viruses and to identify the virus reservoirs within and outside of the crop in order to reduce the risk of infection in new plantations. Rouging has been reported to be an available and economical strategy to contain virus infestations. However, the persistence of infection and its recurrence from sources outside and within the crop are a matter of great concern to cardamom growers and to commodity development and promotion agencies. Upgrading disease management strategies depends on early diagnosis of viral infection in plants at the incubation stage and on identifying virus carriers. Although attempts have been made in Guatemala, India, and so on, the application of biochemical and immunological techniques for early detection of the onset of disease is not generally being practiced. Successful establishment of the cardamom plantation and its sustained productivity are highly dependent on the production of healthy planting material. Sensitive techniques are yet to be employed in the diagnosis of viral infection and in mass multiplication programs of location-specific, high-yielding lines. Similarly, indexing of rapidly depleting diverse genetic resources is vitally needed to conserve them appropriately.

A number of lines identified from disease escapes have shown field resistance to car-MV. In the long run, a clearer understanding of the genetic mechanisms that impart resistance is the only way to utilize them in cardamom improvement programs through both conventional and biotechnological approaches.

2.7 Cardamom Entomology

Many insect pests infect the cardamom plant, right from the seedling stage up to the time the produce is cured and stored. Pests limit productivity (Anon., 1985c) and are highly destructive. Nearly 60 insect species infest the cardamom plant (Kumaresan and Varadarasan, 1987) at various of its growth stages. On the basis of severity of infestation, the pests are classified as major and minor (Kumaresan et al., 1988, 1989b; Premkumar et al., 1994). Major pests include thrips, shoot borers, root grubs, whitefly and hairy caterpillars; minor ones include capsule borers, root borers, rhizome weevils, shoot flies, lacewings, cutworms, aphids, scale insects, leaf folders, spotted grasshoppers, leaf grubs, midrib caterpillars, red spider mites, and storage pests. Kumaresan et al. (1989b) classified the pests as foliar, subterranean, and pests on reproductive plant parts, on the basis of the plant parts they infest. The pest complex varies with the stage of plant growth.

2.7.1 *Major Pests*

2.7.1.1 *Cardamom Thrips [*Sciothrips cardamomi *(Ramk.)]*

The cardamom thrips is the most destructive of all the pests. Ayyar (1935) first reported its infestation potential in Anamalai Hills of Tamil Nadu, describing it as *Taeniothrips cardamomi*. Subsequently, the insect was renamed *Sciothrips cardamomi*. The nature of the damage and biology of the insect were described by Cheriyan and Kylasam (1941), Kumaresan et al. (1988), and Krishnamurthy et al. (1989a).

The larvae and adults lacerate tissues from leaf sheaths, unopened leaf spindle, panicles, flowers, and tender capsules, and suck the exuding sap, resulting in qualitative and quantitative crop loss. Infestation on panicles results in stunted growth, while infestation on flowers leads to shedding of flowers. Laceration of tissues from tender capsules leads to the formation of small scabs, which develop as prominent ugly growths when the capsules mature. The scabs generally appear as longitudinal lines over the ridges of the capsules or as patches over them. Such capsules appear malformed, shriveled, and with slits on the outer skin. The pest-infested capsules do not have the normal aroma and fetch a poor market price. Unprotected crops may suffer up to 80% damage, causing about 45–48% crop loss.

Several field trials have been conducted to evaluate the efficacy of different insecticides in controlling thrips. Early on in these investigations, certain organochlorine insecticides were tested; later they were abandoned, and low doses of organophosphorus and carbamate insecticides were subsequently used. Among the insecticides tested, Quinalphos and Dimethoate, each at 0.1% concentration, were found to be effective against thrips (Pillai and Abraham, 1971). Results of a field trial by Kumaresan (1982) revealed that Methidathion and Carbosulfan, each at 0.05% concentration, and Bendiocarb at 0.16% concentration were effective in controlling thrips. Kumaresan (1983) tested six insecticides, including two synthetic pyrethroids, and reported that Permethrin and Fenvalerate, each at 0.01% concentration, and Quinalphos at 0.05 concentration effectively controlled thrips infestation when sprayed eight times in 1 year.

Krishnamurthy et al. (1989b) found that an insecticide spray with Monocrotophos/Quinalphos at panicle initiation stage, followed by another spray with Phosalone 30–40 days later, was effective against thrips at Mudigere district in Karnataka State. More recently, organic compounds have been tried and detailed investigation on the efficacy of neem (*Azadirachta indica*) formulations, such as neem oil and other commercial neem formulations, was not effective in controlling cardamom thrips (Gopakumar and Singh, 1994).

To combat thrips, the insecticides sprayed to the tiller only to a height not more than one third from its base, this treatment gives adequate coverage on panicles. Approximately 350–450 ml of fluid sprayed with a high-volume sprayer may be required for a spraying of 50–60 tillers per bush. In times of acute water scarcity, Quinalphos at 1.5% concentration, methylparathion at 2% concentration, or Phosalone at 1.5 concentration may be applied at 25 kg ha^{-1} by means of dusters. Caution should be taken to harvest the produce either before spraying or at least 2 weeks after spraying. Insecticide sprays recommended for the control of thrips are listed in Table 2.30.

2.7.1.2 *Panicle/Capsule Shoot Borer (*Conogethes punctiferalis *Guen.)*

The panicle, capsule, and shoot borer is a serious pest of cardamom in nurseries, as well as in the main field in the southern Indian states of Kerala, Karnataka, and Tamil Nadu. However, infestation is severe in Tamil Nadu and Karnataka states.

Larvae of this pest bore into the panicles, capsules, or shoots and grow by feeding on the internal tissues. Infested tillers and panicles dry off and capsules turn empty. The adult is a medium-sized moth with orange-yellow wings exhibiting several black dots. Eclosion generally happens toward the close of the photophase, and the emerging moths rest on the undersurfaces of cardamom leaves (Varadarasan et al., 1989). The moths feed on nectar and cause no direct damage to cardamom. A female moth lays about 20–35 eggs singly or in groups of two or three on leaf margins or dry leaf sheath or along leaf veins. Full-grown larvae are pale pink, 30–35 mm long, and

Table 2.30 Recommended Insecticides for the Control of Thrips

Name of Insecticide	Strength at Which To Be Used (%)
Quinalphos	0.025
Fenthion	0.05
Phosalone	0.07
Chlorpyriphos	0.05
Dimethoate	0.05
Acephate	0.075
Triazophos	0.04
Monocrotophos	0.025
Methylparathion	0.05

crawl to a place near the borehole within the shoot. During the ensuing prepupal period of a day or two, they remain quiescent within a self-made cocoon and soon become pupae. After 10–12 days of the pupal period, the moth emerges through the borehole. The pest population attains peaks during December–January, March–April, May–June, and September–October in the conditions prevailing in Lower Pulney of Tamil Nadu, whereas it is highest during January–February, May, and September–October in the conditions prevailing in Kerala State (Varadarasan et al., 1989).

The pest is polyphagous, and a number of alternative hosts, such as mango, guava, mulberry, sorghum, pea, and cocoa, exist. In natural conditions, *Conogethes punctiferalis* is host for a number of parasites. *Angitia trochanterata* (family Ichneumonidae), *Threonia inareolata*, *Bracon brevicornis*, and *Apanteles* sp. parasitize its larvae, while *Brachymeria emploeae* parasitizes pupae (David et al., 1964). Patel and Gangrade (1971) noticed *Microbracon hebetor* as the larval parasite of the pest. Joseph et al. (1973) reported two hymenopterans—*Brachymeria nosatoi* and *Brachymeria lasus*—parasitizing *Conogethes punctiferalis*. Jacob (1981) reported *Myosoma* sp., *Xanthopimpla australis*, and a nematode as parasites of *Conogethes punctiferalis*. Varadarasan et al. (1990) reported *Temelucha* sp., *Agrypon* sp., and Friona sp. as parasites. Natural parasitization by *Agrypon* sp. on larvae of *Conogethes punctiferalis* was maximal (19.8%) during the month of November in the Udumbanchola region of Tamil Nadu (Balu et al., 1991).

Several insecticides were evaluated against this pest. Fenthion, BHC, DDT, endrin, malathion, trichlorfon, methyldemeton, Imidan, and Carbaryl were recommended by David et al. (1964). However, in a field in Lower Pulneys in Tamil Nadu, 0.1% Monocrotophos was found to be most effective (Kumaresan et al., 1978). Carbofuran at the rate of 2 kg ha^{-1} was found quite effective (Reghupathy, 1979). In addition to chemical control, physical measures such as the collection and destruction of affected plant parts and the removal of affected tillers in September–October if the infestation is less than 10%, followed by an insecticidal spray of Lebayacid (375 ml in 3000 liter water) or Quinalphos at 0.03% when the infestation is more than 10%, were recommended (Krishnamurthy et al., 1989b). It was found that a higher dosage of insecticides was ineffective when late-stage infestation takes place, with larvae burrowed inside the shoots.

Root grubs (*Basilepta fulvicorne* Jacoby) are a serious subterranean pest of cardamom. The insect damages roots and thereby obstructs the uptake of nutrients, leading to yellowing of leaves and gradual death of plants when the infestation is severe (Gopakumar et al., 1987). The pest affects seedlings both in nurseries and in the main plantation and is commonly found in the states of Kerala, Karnataka, and Tamil Nadu in southern India (Varadarasan et al., 1988). Root grubs were observed to be a very serious pest in Karnataka State (Gopakumar et al., 1987), and Thyagaraj et al. (1991) studied the biology and field management of the pest. The adult form is a small beetle 4–6 mm in length with a shiny metallic blue, green, or greenish-brown color. Females are bigger than males. The beetles are polyphagous. The recorded alternative hosts of the beetle are the jackfruit (*Artocarpus heterophylla*), Indian almond (*Terminalia catapa*), mango (*Mangifera indica*), guava (*Psidium guajava*), fig (*Ficus indica, Ficus bengalensis*), cacao (*Theobroma cacao*), and dadaps tree

(*Erythrina lithosperma*), among others (Anon., 1993). In plantations, infestations of this pest are observed twice a year, in March–April and August–October, with peaks in April and September. Beetles fly about short distances and alight on leaves of shade trees and on cardamom plants. Copulation takes place in the daytime, and after a preoviposition period of 4–6 days, the mated females extrude eggs in groups to a transparent fluid secreted on dry leaf sheaths or leaves, to which the eggs remain glued. Females lay 124–393 eggs in batches of 12–63 during an oviposition period of 8–71 days. Freshly laid eggs are transparent and gradually turn yellow during the incubation period of 8–10 days (at a temperature of 28–31 °C) or 13–19 days (at a temperature of 19–24 °C). Then the eggs hatch, generally during morning hours, liberating small creamy white grubs that fall on the ground, penetrate the soil, reach the root zone of the plant, and start feeding on the roots. Infestations of the grub are greater on cardamom plants that grow under thin shade than on those growing under thick shade. Like beetles, grubs have two periods of occurrence, the first during April–July and the second during August–September and December–January, with peaks during May–June and November–December.

A judicious combination of both mechanical and chemical methods can control the pest problem (Gopakumar et al., 1987). Collection and destruction of the beetles during the periods of their massive emergence and subsequent insecticidal control of grubs are the two methods incorporated into the strategy.

During periods of adult emergence (March–April and August–October), beetles alighting on cardamom plants can easily be collected with an insect net and destroyed. During peak emergence, a laborer could collect 2500–3000 beetles in a day. Such massive destruction of beetles drastically reduces the grub population in the soil to a low level, which would otherwise have been enormous and caused heavy root damage. In endemic areas, suitable measures to control the beetle have to be adopted, as it becomes virtually impossible to destroy the huge masses of the beetle population. Examining chemical control measures, Varadarasan et al. (1990a) reported the application of 20–40 g of Phorate or 0.06% Chlorpyrifos to be effective in controlling the grubs. Subsequently, lowering the dose of Chlorpyrifos to an even lower dose of 0.04% was found effective (Varadarasan et al., 1991b).

The pest has been found to be susceptible to infection of entomopathogenic fungi, at both the grub and adult stages. *Beauveria bassiana* and *Metarhizium anisopliae* were isolated from naturally infected beetles and grubs, respectively (Varadarasan, 1995). The grubs were also infected by the nematode *Heterorhabditis* sp. (Varadarasan, 1995). Laboratory studies, as well as preliminary field trials, with these bioagents have shown their efficacy convincingly, and it is expected that a suitable biocontrol strategy will be developed for the management of root grubs.

2.7.1.3 *Whitefly [*Kanakarajiella cardamomi *(David and Subr.) David and Sundararaj]*

Until the 1980s, infestation of cardamom with whitefly was a rare phenomenon and its sporadic occurrence was limited to the Nelliampathy and Vandiperiyar areas of Kerala State (Anon., 1980). Now infestations are seen in many places in the Udumpanchola and Peermedu taluks of Idukki district of Kerala State and in the Lower Pulneys of Tamil Nadu. The species of whitefly reported to infest cardamom plants are *Dialeurodes cardamomi* David and Subr. [known as *Kanakarajiella cardamomi* (David and Subr.); David and Sundarraj (1993)], *Aleuroclava cardamomi* (David and Subr.), *Aleurocanthus* sp., *Bemesia tabaci* (Genn.), and *Cockerella diascoreae* Sundararajan and David (Selvakumaran and Kumaresan, 1993). However, only *Kanakarajiella cardamomi* is destructive to cardamom plants.

Gopakumar et al. (1988b) and Selvakumaran and Kumaresan (1993) investigated the biology of the pest. Adults produce by parthenogenetic and sexual methods and live for 7–8 days. They are very much attracted toward yellow color, and this behavior is well exploited for trapping them on a yellow surface coated with sticky material. Spraying 0.5% neem oil + triton or 0.5% sandovit on the undersurfaces of leaves twice or three times at fortnightly intervals during periods of pest infestation is effective against nymphs (Gopakumar and Kumaresan, 1991). Acephate at 0.075%, Ethion at 0.1%, and Triazophos at 0.4% were found to be equally effective in controlling the increase in the number of nymphs (Gopakumar et al., 1988; Selvakumaran and Kumaresan, 1993). Under natural conditions, the pest can be controlled by a number of natural predators, including *Mallada boninensis*; unidentified Neuroptera, Diptera, Coleoptera, mites; parasitoids such as *Encarsia septentrionalis* and *Encarsia dialeurodes*; and the pathogenic fungus *Aschersonia placenta* (Selvakumaran and Kumaresan, 1993). The potential of these natural predators to control the spread of the pest has not been explored in any systematic manner.

2.7.1.4 Hairy Caterpillars

The hairy caterpillars are a group of defoliators that appear sporadically and cause severe damage to crops. The incidence of these pests was reported by Puttarudriah (1955). Nine species of hairy caterpillars are known to infest the cardamom plant: *Eupterote canaraica* Moore, *Eupterote cardamomi* Renga, *Eupterote fabia* Cram, *Eupterote testaceae* Walk, *Eupterote undata, Linodera vittata* Walk, *Euproctis lutifacia* Hamp, *Alphaea biguttata* Walk, and *Pericallia ricini* Fabr. The biology of different species of these hairy caterpillars was described by Nair (1975) and Kumaresan et al. (1988). The various *Eupterote* sp. have striking similarities; they are all polyphagous larvae that feed voraciously on leaves of shade trees at early stages and, later on, on cardamom leaves.

Adults of both *Eupterote cardamomi* and *Eupterote undata* emerge from their pupae in June following the onset of the southwest monsoon. *Eupterote canarica* emerge later. Infestations of *Eupterote fabia* normally occur quite late in the monsoon period, by August–October. *Eupterote testaceae* cause only mild damage to cardamom. Its moths have yellowish wings with faint wavy black lines on them. Adults generally emerge in June–July.

The adults of *Lueder vittata* are thickest, with underdeveloped wings. Adults generally emerge in June. The parasite *Carcelia kokiana* is seen to parasitize the larvae (Nair and Kumaresan, 1988). *Euproctis lutifacia* infests the tender foliage of cardamom. The adult is a vinous-brown moth. Adults generally emerge during December, after a pupal period of 16–18 days (Kumaresan, 1988; Nair, 1978). *Alphaea biguttata* is an arctiid black hairy caterpillar that infests cardamom. Its moths are comparatively small. Adults emerge after about 22–23 days of pupation.

Pericalia ricini is medium sized; its dark-brown larvae emerge after an incubation period of 4–5 days (Kumaresan, 1988; Nair, 1978).

Beeson (1941) reported Bombax malabaricum, Careya arborea, Cedrella toona, Dalbergia volubilis, Erythrina indica, Shorea robusta, Tectona grandifolia, Tectona grandis, Terminalia sp., and Vitex negundo as alternative hosts of hairy caterpillars. These caterpillars congregate on tree trunks or cardamom plants in the daytime and can be collected in large numbers and destroyed. Sekhar (1959) recommended fish oil rosin soap sprays against the pest. Nambiar et al. (1975) found 0.2% BHC, 0.1% Malathion, or 0.1% Carbaryl effective in controlling the pest. Collecting and destroying moths with light traps and spraying 0.1% methylparathion are also recommended for controlling the pest (Anon., 1985a). Among the reported natural enemies of hairy caterpillars are Apanteles tabrobanae Cram., Sturmia sericariae, Aphamites eupterotes, and Beauveria sp. (Nair, 1975).

2.7.2 Minor Pests

2.7.2.1 Capsule Borers

Jamides alecto: Damage due to this pest, a lycaenid borer, has been severe in Karnataka State at times (Krishnamurthy et al., 1989; Kumaresan et al., 1988; Siddappaji and Reddy, 1972). The caterpillars bore and feed on inflorescence, flower buds, and capsules. Affected capsules turn empty, decay, and drop in the rainy season. Singh et al. (1993) investigated the biology of the pest. The adult is a swift-flying butterfly having a metallic blue color bordered with a white line and black shading on the dorsal surface of the wings. The ventral surface is colored the same. The larvae feed on immature capsules, rejecting the seeds of ripened capsules after sensing them. Quinalphos spray at 0.05% or methylparathion spray at 0.05% during the early blooming period has been found effective against the pest (Kumaresan et al., 1988). Application of fish oil rosin soap at the rate of 1 kg in 45 liters of water was also found effective against the borer (Kumaresan, 1988).

Thammurgides cardamomi: Adults and larvae of *Thammurgides cardamomi* bore and feed on flowers and immature capsules. In Karnataka, pest infestation is noticeable during July–August, especially on plants that grow under thick shade. The adult is a dark-brown beetle covered over with short, thick hairs. Regulating shade properly and spraying with Quinalphos or methylparathion at 0.05% are recommended for the control of the pest (Anon., 1985a).

Onthophagus coorgensis: This pest bores and feeds on flowers and young capsules during monsoon months. The insect is small and dark brown with short, thick erect hairs over the body. It lays clusters of barrel-shaped eggs, normally 6–12 in a capsule. The pest can be controlled by regulating shade properly and spraying with Quinalphos or methylparathion at 0.05% (Krishnamurthy et al., 1989b; Kumaresan et al., 1988).

Root borer *(Hilarographa caminodes* Meyer.): Caterpillars of this insect pest bore into the roots and feed on them, resulting in yellowing of leaves and gradual dying of plants. Moths emerge during April–May and lay eggs on exposed parts of roots. Emerging caterpillars tunnel into the roots.

Rhizome weevil *(Prodioctes haematicus* Chev. F.): Grubs of this insect tunnel into rhizomes and, rarely, into aerial stems. Infestation is noticed more on seedlings than on mature plants. The adult is a brown weevil 12 mm long with three black lines on the pronotum, one middorsally and other two on either side. On each elytron are three black dots, two anteriorly and one posteriorly. Adults emerge immediately after the summer rains in April. Adults live for 7–8 months. Destruction of infested plants will help reduce the intensity of the pest. Drenching with 0.2% BHC or applying Phorate 10G to the soil at the rate of 20–40 g per clump is also recommended against the pest (Anon., 1985a; Kumaresan, 1988).

Shoot fly *(Formosina flavipes* Mall.): This pest is prominent in the summer months, infesting seedlings and young tillers growing under thin shade. Mature flies lay cigar-shaped white eggs singly in rows of four to six between the terminal leaf sheaths. However, only one maggot is seen to penetrate down the pseudostem, feeding on core tissues, after which the terminal unfurled leaf dries out (dead heart symptom). The pest can be controlled by destroying infected plants and applying Carbofuran at the rate of 8–10 kg acre^{-1}, or by spraying Dimethoate, Quinalphos, or methylparathion at 0.05% (Kumaresan et al., 1988). Sufficient shade should be provided to the plants to guard against the pest.

Lacewing bug *(Stephanitis typicus* Dist): This is a bug with shiny, transparent lace-like reticulate wings. Females lay about 30 eggs singly and then inserted them into the adaxial surfaces of leaves. The eggs hatch in 12 days, and the emerging nymphs congregate on a suitable feeding site on leaves and suck plant sap, resulting in the development of necrotic spots on the leaves. Banana, *Colacasia,* coconut, and turmeric are the alternative hosts of this pest (Kumaresan, 1988; Nair, 1978).

Cutworm (*Acrilasisa plagiata* M.): Cutworms are commonly seen to feed on tender cardamom leaves in nurseries. Infestations are noticed in January–March. The fully grown caterpillar is dark brown with an orange-red head and a humplike projection on the eighth dorsal segment. The caterpillar pupates in soil for a period of 17–18 days (Nair, 1975).

Cardamom aphid (*Pentalonia nigronervosa* f. *caladii* van der Groot): The cardamom aphid is of concern not as a pest of the crop, but as a vector of the virus that causes the devastating katte disease. Siddappaji and Reddy (1972a) confirmed the form of the vector occurring on cardamom as *Pentalonia nigronervosa* f. *caladii.* The virus–vector relationship was studied by Uppal et al. (1945), Varma (1962a), and Naidu et al. (1982). The biology of the insect was investigated and reported by Rajan (1981).

Of the winged and wingless forms of the aphid, the former is longer and slimmer than the latter. Adults are dark brown in color. They reproduce by both viviparous and parthenogenetic means. The population of the insect is high in January–February. The insect is also found on *Colocasia* sp., *Alocasia* sp., and *Caladium* sp. The insect population gets drastically reduced during the monsoon season through infection by *Verticillium intertextum* (Deshpande et al., 1972). *Peragum indica, Coccinella transversalis,* and *Ischiodon scutellaris* were found to prey on the aphids. A spray of 0.05% Phosphamidon or Dimethoate in April and November is recommended to control the aphid (Anon., 1985a).

Scale insect (*Aulacaspis* sp.): Infestations of this pest are noticed during summer. Capsules, panicles, and pseudostems are the usual sites of infestation. An infestation results in shriveling of the capsules.

Leaf folder (*Homona* sp.): Caterpillars of this pest fold the tips of cardamom leaves and feed on the leaves by remaining inside the fold. The fully grown caterpillar is almost 3 cm long and is pale green with a black head.

Spotted grasshopper (*Aularches* sp.): Adults and nymphs of this polyphagous pest scrape and feed on leaves of cardamom voraciously. Infestations are usually noticed in March and continue until November. Adult hoppers have pretty green wings with yellow spots on them. Adults congregate on shade tree tops, presumably to bask in the sun; they descend on cardamom in swarms and defoliate the plants. Exposure of the eggs to desiccation by sunlight and the application of contact insecticides against nymphs are recommended to control the pest.

Leaf grub (*Lema fulvimana* Jacoby): Beetles and grubs of *Lema fulvimana* feed on the tender foliage of cardamom seedlings. The biology of the pest was reported by Singh (1994). Beetles emerge in May. Fully grown grubs have a dull white body with brown streaks, a black head, and a disproportionately swollen abdomen. Usually, they carry faucal matter on their backs. They pupate in soil inside a papery cocoon. *Zingiber cernuum* and *Curcuma neilgherrensis* are alternative hosts of the pest. Removing the alternative hosts and spraying 0.025% Quinalphos or 0.025% Monocrotophos is effective in controlling the pest (Singh, 1994).

Metapodistis polychrysa Meyrick: Caterpillars of this insect feed on the unopened leaves of cardamom (Gopakumar et al., 1989). The caterpillar is pale green and 1 cm long when fully grown. The insect completes its life cycle in 30–35 days.

2.7.3 Storage Pests

Tribolium castaneum and *Lasioderma serricorne* are the major storage pests of cardamom. *Lasioderma serricorne* completes its life cycle in about 115 days on stored cardamom (Balu, 1991). Fumigating with methyl bromide or storing the capsules in alkathene-lined jute bags sprayed with 0.5% Malathion is effective in controlling the pest (Abraham, 1975).

Sporadic infestation by red spider mites on cardamom leaves has been noticed during the summer (February–May) in India. The mites remain on the undersurfaces of leaves within a self-made web. They suck plant sap from the leaves, resulting in the formation of white blotches. The undersurfaces of the infested leaves, with the eggs, excreta of the pest, and nymphs embedded in a web of fine, delicate silken threads, appear ashy white. Affected leaves gradually dry up. Infestations on panicles and tillers are rarely noticed. The mite completes its life cycle in about 20 days. Dicofol at the rate of 2 ml liter^{-1} of water, Sulfur 80 wettable powder (WP) at 2.5 g liter^{-1} of water, Ethion at 2 ml liter^{-1} of water, Dimethoate at 1.67 ml liter^{-1} of water, and Phosalone at 2 ml liter^{-1} of water are effective chemical control measures.

2.7.4 Conclusions

Much information on the biology and control of most of the major and minor pests of cardamom is documented. Most of the control measures are chemical; rarely are they mechanical. Some parasitoids and predators of certain pests have been identified. The immediate need is to eliminate the majority of the insecticides in control measures and substitute plant derivatives and biological control agents. Suitable biocontrol measures must be targeted against the thrips, borers, and root grubs. Excessive dependence on chemicals will bring down the value of the produce in the international market, where there is a growing need for organically produced cardamom. There is also an urgent need to investigate the potential of bioagents that have been already identified as effective control measures, and the impact of agroclimatic conditions prevailing in the cardamom ecosystem on these bioagents needs further investigation. An integrated pest management system focusing on noninsecticidal means of pest control has to be developed. Complementing this endeavor, a keen awareness of the necessity for such a system has to be instilled in cardamom planters in order to enable them to appreciate the importance of sustainable cardamom production.

2.8 Harvesting and Processing of Cardamom

2.8.1 Harvesting

The cardamom plants start bearing 2–3 years after planting seedlings or suckers. Panicles appear from the bases of plants from January onward, and flowering continues from April to August or even later. Generally, flowering is highest during May–June. Fruits mature in about 120 days after flowering. Fruits are small trilocular capsules containing 15–20 seeds. On maturity, seeds turn dark brown to black. On average, a healthy cardamom plant produces about 2000 fruits annually, weighing about 900 g. On drying and curing, this yield gives about 200 g of marketable produce (Photo 2.7).

2.8.1.1 Time and Stage of Harvesting

Flowering in cardamom is a continuous process over several months, resulting in the capsules ripening successively over an extended period. This situation necessitates several pickings. Normally, harvesting commences in August–September and extends into February–March of the next year. Generally, in the peak season, harvesting is done at fortnightly intervals and is completed in 8–10 rounds (Korikanthimath, 1983). In Kerala State and Tamil Nadu, harvesting commences from August–September and continues until February–March of the next year, while in Karnataka State harvesting starts in August and continues until December–January of the next year. Experienced plantation workers do the picking. The physiologically ripened fruits are referred to as "Karikai" in the local language. The maximum dry weight of 285 $g\,kg^{-1}$ of wet (green) capsules is the recovery rate in the Malabar variety, followed by 240 and 140 g recovery in the physiologically mature and physiologically immature stages, respectively

Photo 2.7 Cardamom harvesting by women labourers.

(Korikanthimath and Naidu, 1986). This shows that there is nearly 100% in weight gain from immature to mature stage. Splitting of capsules occurs when harvesting is done at an immature stage of the plant. Also, the essential oil content was found to be more when harvesting was done at a mature, rather than an immature, stage.

About 2860 ripe capsules weigh 1 kg, whereas it takes about 5000 physiologically immature capsules to weigh 1 kg, clearly showing that it is always advantageous to harvest at the right stage of maturation. When the capsules are well matured, the seeds inside will be black.

Two types of picking are practiced: light picking and hard picking. In the first case, only matured capsules are picked; in the second, semimature capsules also are picked. Although hard picking would reduce the curing percentage, it would increase the picking average, secure green-colored capsules, and reduce fruit drop and splitting of capsules in the field, thereby ensuring minimal loss of produce. However, the choice of picking type depends on the availability of manual labor, as no mechanical harvesting is practiced in India. Mostly, women laborers are employed for harvesting.

Postharvest Handling of Cardamom

The color of processed produce is an important factor inasmuch as the consumer is concerned. Most markets, especially in Middle East countries, prefer green-colored cardamom. The cardamom with the highest quality is "Alleppey Green," still regarded as the best. This brand comes from the high ranges of the Idukki district in Kerala State, where the variety that is grown predominantly is Mysore. Investigations have been carried out to understand the mechanism of color retention during the processing of harvested cardamom capsules.

Table 2.31 The Chlorophyll Profile of Different Capsules

Capsule Color	**Chlorophyll Content (ppm)**			
	a	**b**	**Total**	**a/b Ratio**
Light green	509 (186)	561 (167)	1070 (352)	0.9 (1.1)
Medium green	727 (384)	700 (349)	1424 (731)	1.0 (1.1)
Dark green	1677 (446)	1890 (382)	3567 (828)	1.12 (1.2)

Synthesis and degradation investigations indicate that the total chlorophyll content declines about 100 days after flowering. Comparative evaluations of chlorophyll content in dark-green, medium, and light-green capsules show that the depth of green color is directly proportional to the concentration of chlorophyll in the capsules (Table 2.31). Both fresh and cured capsules have more chlorophyll a than chlorophyll b. In the husk, 60% of the total chlorophyll is present in the surface layer. In the three clones tested, namely, Thachangal, Mudigiri, and PV1, total chlorophyll content was most in 100-day-old capsules of Thachangal (2186 ppm), followed by those of Mudigiri (1756 ppm), and least in PV1 (1488 ppm) (Anon., 1991). The investigation suggested that dry matter continued to increase until the capsules reached maturity, whereas chlorophyll content started declining 100 days after flowering (Anon., 1991). The decline in chlorophyll content during the postripening period is greater in Mudigiri (variety Malabar) than in varieties Vazhukka and Mysore, indicating that delay in picking this clone could affect the final greenness of capsules (Anon., 1991).

To retain the green color of the harvested capsules, various chemical treatments have been tried because green-colored capsules fetch a high premium in the market. Among such treatments, soaking the green (wet) capsules immediately after harvest in a 2% sodium carbonate solution for 10 min fixes the green color during subsequent drying and storage (Natarajan et al., 1968). The green color was of greatest intensity in immature capsules. Meisheri (1993) has developed a dehydration unit that can retain the green color and dry the produce rapidly at ambient temperatures (27–40 °C).

Predrying Operations

After harvest, the capsules are washed thoroughly in water to remove any soil that has adhered to them. Then they are taken to drying kilns. In various trials, it was found that presoaking (quickly dipping) capsules in hot water at 40 °C and then dipping them in 2% sodium carbonate for 10 min helped retain the green color of cured capsules. Dipping capsules in hot or warm water at lower temperatures, namely, 30 °C and 35 °C, also was tried. Additives such as 2% sodium carbonate in warm or hot water (in particular, 35 °C) also helped to increase the green color of the capsules. Dipping in hot water may arrest the activity of certain enzymes. Volatiles extracted from capsules presoaked in a solution of hot water and sodium carbonate were subjected to gas–liquid chromatography analysis. Results indicated that there were no significant changes in the oil profile due to hot water or sodium carbonate treatments (Anon., 1991).

Presoaking of capsules in copper formulations and chemicals such as naphthalene acetic acid, ascorbic acid, IAA and gibberellic acid, and magnesium sulfate helped retain more chlorophyll compared with other treatments. However, when the presoaking time was extended to 60 min, a significant depletion of chlorophyll was observed in all except ascorbic acid treatment. Other treatments, namely, urea, 2,4-D, and Cycocel at 100 ppm each, kinetin at 10 ppm, glycerol at 5%, and polyethylene glycol at 5%, recorded either no effect or a marginal negative effect on the stability of chlorophyll (Anon., 1991).

Chlorophyll degradation takes place on exposure to sunlight, which bleaches the green color. Postharvest delay prior to curing is known to cause chlorophyll breakdown, and a better storage system could help to minimize such chlorophyll loss.

Various trials conducted to study the impact of precuring storage indicated the following:

1. Capsules cured immediately after picking retained more of the green color.
2. Loss of the green color was more significant if capsules were stored for more than 12 h from the time they were picked.
3. Bagging capsules helped to minimize the rate of loss of green color.

To store fresh capsules, jute bags were found to be better than polypropylene woven bags. Storing fresh capsules at low temperatures was found to reduce postharvest precuring loss of the green color. Capsules stored in low-energy or zero-energy cooling chambers were found to be distinctly greener than capsules stored in the open. In large cardamom plantations, two reinforced-concrete cement tanks constructed adjacent to each other are used, one for initial washing of the capsules to remove dirt and soil particles that adhere to them and the other for washing with washing soda. Following washing, the capsules are spread in a single layer in portable drying trays to drain the water from them. Subsequently, the trays are arranged in kilns to dry them completely.

2.8.2 Curing

Depending on their degree of maturity, cardamom capsules carry moisture levels of 70–80% at harvest. For proper storage, the initial moisture level has to be brought down to 8–10% (on the basis of wet weight) by curing. Curing also plays an important role in preserving the green color of the capsules, since as much as 60–80% of the initial color is lost during processing. The most widely adopted system is a slow, or passive, process stretching from 18 to 30 h, with an initial temperature around 50 °C. The entire curing time can be divided into four stages:

Stage	Time Lag
I	0–3 h
II	3–6 h
III	6–9 h
IV	9 to final curing

Both the degree of maturity and the curing temperature influence the percentage of splits in cured capsules. However, temperature has a greater influence (Anon., 1991). During curing, if the temperature exceeds the threshold levels or the inflow of air is insufficient, capsules develop brownish streaks as a result of heat injury. In the case of a fairly high temperature, oil from seeds oozes out. Maintaining the temperature at 40 °C in all four stages of the curing process helps produce a greater retention of the green color. The percentage of split and discolored capsules increases with a rise in temperature. Curing at 55–60 °C significantly increases the percentage of yellow capsules.

The husk of raw capsules contains about 80% water, which has to be removed completely in the process of drying. Maximum loss of chlorophyll occurs in the initial 6 h of curing. Higher airflow rates increase the loss of chlorophyll. Less energy is required for extraction of moisture in the initial stages, when the evaporation is from the surface layers of capsules. More energy is required to remove the same amount of moisture in the later stages of curing, when the moisture content of the capsules falls. Cardamom oil extracted from samples dried at 45 °C and 60 °C did not show much difference in the GLC profile (Anon., 1991). Cardamom capsules are moderately hygroscopic and absorb and desorb moisture, depending on changes in the ambient relative humidity and moisture. Two types of drying are generally adopted: natural sun drying and artificial drying using firewood, fuel, or electricity. Drying demands a heavy input of energy. The amount of energy required to dry 1 kg of green cardamom at 100% efficiency can be used to light 250 bulbs of 100 W each for 1 h.

2.8.2.1 Sun Drying

Sun drying is generally undesirable for cardamom. When cardamom capsules are sun dried, there is a bleaching effect due to the action of ultraviolet light present in sunlight. Sun drying requires 5–6 days or more, depending on the availability of sunlight. Because the capsules are frequently turned during sun drying, they split. A cloudy atmosphere and frequent rains hinder proper sun drying. Sun drying is prevalent among small-plantation owners in the Sirsi district and surrounding places in the state of Karnataka.

2.8.2.2 Artificial Drying

Electrical Dryer

A dryer having dimensions of 90 cm × 84 cm is most common. Uniform heat distribution is ensured by means of fans. In this way, 50 kg of fresh capsules can be dried in 10–12 h and medium green-colored cardamom can be obtained by drying at 45–50 °C.

Pipe Curing (Kiln Drying)

This is one of the best methods of drying to obtain high-quality green cardamom. The kiln usually consists of walls made of bricks or stones and a tiled roof with a ceiling. A furnace is situated on one side of the chamber, and heat is generated by burning farm waste. The fire in the furnace maintains the temperature between 45 °C and 50 °C, and high-quality green cardamom can be prepared in 18–22 h by this method. A drying chamber of dimensions 4.5 m in length and 4.5 m in breadth is sufficient for a plantation producing 1800–2000 kg of raw cardamom. Some of the kilns make use of brick-constructed heat conveyer lines (Kachru and Gupta, 1993).

2.8.2.3 Bin Dryer

The bin dryer was designed by the University of Agricultural Sciences, Bangalore, Karnataka State, India. The drying unit consists mainly of a blower with a motor, an electrical heating unit, and a drying chamber. The dryer is made of mild steel, asbestos sheet, and wood. Aluminum or steel trays 0.4 m $\times$ 0.6 m size can be arranged one over the other. Cardamom capsules are spread uniformly on these trays. Hot air passing through pipes increases the temperature from 30 to 80 °C. Good-quality cardamom can be produced by drying capsules at 55 °C by maintaining the volume of air at $3.7\,m^3\,s^{-1}$. The cost of drying by this method comes to less than 1 cent per kilogram, compared with about 2 cents per kilogram in the conventional method (Gurumurthy et al., 1985).

2.8.2.4 Melccard Dryer

The Melccard dryer is a firewood-operated dryer that is commonly used in some parts of Tamil Nadu. It consists of a fully insulated (firebricks with a mud coating) oven kept 3 m below the dryer. The hot flue gas from the oven is passed to an iron tank through insulated pipes. All the trays carrying cardamom capsules move smoothly on rails fixed inside the dryer. Trapdoors that are attached can be opened periodically to clean off the soot formed in the interior of the flue pipes (Palaniappan, 1986). Dried capsules are rubbed by hand or with coir mat or wire mesh and winnowed to remove other plant residues and foreign matter. They are then sorted out according to size and color.

2.8.2.5 Cross-Flow Electric Dryer

The cross-flow dryer is a tray type of dryer having capacities ranging from 25 to 400 kg. The air is heated by 15-kW electric heaters and circulated over the material by a 0.5-hp electric fan. The drying-time requirement at the full loading condition is about 18–20 h (Kachru and Gupta, 1993).

2.8.2.6 Solar Cardamom Dryer

A direct-type solar dryer developed for copra (a dried coconut ball without the outer shell) drying by Central Plantation Crops Research Institute at Kasaragod in Kerala State can also be used for cardamom. The dryer has a drying surface area of $1\,m^2$ made of black painted wire mesh tray over black painted corrugated GI sheet inclined at 12.5°. Aluminum foil reflectors $1.5\,m^2$ are provided from three sides of the dryer. The material load density can be three times that used in an open drying system. Complete drying of cardamom could be achieved within 3 days with this dryer, in comparison to 5 days in the open sun. Bleaching of cardamom capsules due to the action of ultraviolet rays in sunlight is a disadvantage of this dryer.

2.8.2.7 Mechanical Cardamom Dryer

Developed by the Regional Research Laboratory in Trivandrum, Kerala State, the mechanical cardamom dryer consists of a centrifugal blower, an electrical furnace, a conducting arrangement for a uniform flow of hot air, and a drying chamber. It can be used for cardamom drying at a load of 120 kg fresh cardamom per batch. It takes about 22 h for complete drying at a temperature of 50 °C. The final product is claimed to possess superior green color, flavor, and appearance (Kachru and Gupta, 1993).

2.8.2.8 Through-Flow Dryer

The through-flow dryer is fabricated by the Central Food Technology Research Institute, Mysore, Karnataka State. The dryer consists of a centrifugal blower, electrical furnace ducting with arrangements to distribute the flow of hot air uniformly, and a drying chamber where 120 kg of fresh cardamom capsules can be loaded to a bed thickness of 20 cm. The air velocity is 60 cm s^{-1} and the drying temperature is thermostatically controlled. The hot air carrying the humidity is not allowed to recycle. This type of dryer was found to take about 22 h to dry 120 kg of fresh capsules at a temperature of 50 °C. Because cardamom plantations are generally located in forest areas where electricity is not available, flue pipe dryers are more dependable and suitable to ensure continuous working. There is still good scope for developing a dryer that is ideal for producing green capsules without any loss of volatile oil and with minimum expenditure for drying.

2.8.2.9 Bleached Cardamom

Bleached cardamom is creamy white or golden yellow in color. Bleaching can be done with either dried cardamom capsules or freshly harvested capsules as starting material.

Bleaching of Freshly Harvested Capsules

Fresh capsules soaked for 1 h in a 20% potassium metabisulfite solution containing a 1% hydrogen peroxide solution degrade the chlorophyll. Drying of these capsules yields a golden yellow color.

Bleaching of Dry Capsules

Sulfur bleaching: This techniques involves sulfur fumigation with alternate periods of soaking and drying. Capsules are soaked in 2% bleaching powder (20 g $liter^{-1}$ of water) for 1 h and spread on wooden trays, which are arranged inside airtight chambers. Sulfur dioxide is produced by burning sulfur (15 g kg^{-1} of capsules) and made to pass over the trays. The process of soaking and drying is repeated three to four times, depending on the intensity of white color required.

Potassium metabisulfite bleaching: In this method, capsules are treated with 2% potassium metabisulfite containing 1% HCl for 30 min. Then they are transferred to a 4% hydrogen peroxide solution for 6 h.

Hydrogen peroxide bleaching: Hydrogen peroxide at low concentration (4–6%, pH 4) can bleach capsules in 6–8 h of soaking. The capsules are then dried to 10–12% moisture content. The bleached capsules contain sulfur, which protects cardamom from pests. However, it was found that bleaching led to a loss of volatile oil.

Conventional bleaching: In Karnataka State, bleaching of cardamom is carried out by steeping the dried capsules in soap nut water. The fruits of soap nut (*Sapindus saponaria*) are mixed with water in a large vessel and stirred vigorously to produce plenty of lather. Dried cardamom capsules are then steeped in this water, with occasional stirring. After 1 h or so, the fruits are collected in wicker baskets, the water is allowed to be completely drained off, and the fruits are then spread out in mats for drying. Clean water is occasionally sprinkled over the cardamom capsules. The process of sprinkling water and drying is continued for a couple of days, until a good-quality bleached product is obtained. In general, bleaching of dried capsules leads to a loss of volatile oil, probably because the bleaching process makes the husk brittle. However, bleached cardamom has a white appearance and is resistant to weevil infestation due to its sulfur dioxide content (Govindarajan et al., 1982; Krishnamoorthy and Natarajan, 1976).

2.8.3 Moisture Content

The moisture content of commercial samples from the market ranges from 7 to 20%, depending on the regions and mode of curing (Varkey et al., 1980). It has been found that 10% moisture is ideal for the retention of green color, which also depends on the type of drying. Well-dried capsules produce a typical tinkling sound on shaking.

2.8.4 Grading

The quality requirement of a produce varies with the primary raw-material producer, the intermediary collector, and the trader, exporter, importer, processor, distributor, and final consumer. The moisture level, cleanliness, content of a substandard product, extraneous matter, appearance, and color determine the final quality of the product. The processor determines the values of the extractives, volatile oil, and specific ingredients. Specifications are restricted to attributes that can be simply and rapidly analyzed. Many of them are related to physical parameters, such as color, size, weight per specified volume, and freedom from contamination by microbes, insects, and filth (Govindarajan et al., 1982b). Specifications for Indian varieties of cardamom are given in Table 2.32.

Alleppey Green cardamom is the kiln-dried capsule of *E. cardamomum* grown in South India. The capsule has a reasonably uniform shade of green color, is three cornered, and has a ribbed appearance. Coorg clipped cardamom is the dried capsules of *E. cardamomum* variety Malabar grown in Coorg in Karnataka State. Its color ranges from pale yellow to brown, and it is global in shape and skin ribbed or smooth. The pedides are separated. Bleachable white cardamom is the fully developed dried

Table 2.32 Specifications (Physical Characteristics) for Indian Cardamom Varieties

Grade	Description	Size (mm)	Minimum Weight (g liter^{-1})	Color	General Characteristics
AG, Alleppey Green					
AGB	Extra bold	7	435	Green	Kiln dried, three cornered, with ribbed appearance
AGS	Superior	5	385	Green	Same as previous
AGS 1	Shipment	4	320–350	Light	
AGL	Light	3.5	260	Green	
CG, Coorg Green					
CGEB	Extra bold	8	450	Golden to light green	
CGB	Bold	7.5	435		
CG 1	Superior	6.5	415		Round, ribbed or smooth skin
CG 2	Mota Green	6	385		
CG 3	Shipment	5	350		
CG 4	Light	3.5	280		
Bleached or Half Bleached					
BL 1		8.5	340	Pale	Fully developed
BL 2		7	340	Creamy	Round, three cornered, ribbed or smooth skin
BL 3		5	300	Dull white	

Source: Indian Standard Specification for Cardamom. IS: 1907–1966. Indian Standards Institution, New Delhi.

capsule of *E. cardamomum* grown in the state of Karnataka to a reasonably uniform shade of white, light green, or light gray.

In India, "Agmark" grades are commonly used to specify the quality of products. The Agmark grades of Coorg clipped cardamom and bleachable white cardamom are given in Tables 2.33 and 2.34.

2.8.5 Bleached and Half-Bleached Cardamom

Cardamom should be fully developed dried capsules, bleached or half-bleached by sulfuring; the color should range from pale cream to white, and the shape should be globose or three cornered, with skin ribbed or smooth.

Table 2.33 The Agmark Specifications of Coorg Clipped Cardamom

Grade Designation	Trade Name	Empty and Malformed Capsules by Count (max. %)	Unclipped Capsules by Count (max. %)	Immature and Shriveled Capsules by Weight (%)	Size (mm)	Weight Minimum (g liter^{-1})
CCS 1	Bold	5.0	0.0	0.0	8.5	435
CCS 2	Coorg green or Mota green	5.0	3.0	4.0	6.0	385
CC 3	Shipment	3.0	5.0	7.0	4.0	350
CC 4	Light			3.5	260	

Table 2.34 The Agmark Specifications of Bleachable White Cardamom

Grade Designation	Trade Name	Empty and Malformed by Count (max.)	Immature and Shriveled by Weight (%)	Size (mm)	Weight Minimum (g liter^{-1})
BW 1	Mysore/Mangalore bleachable cardamom, clipped	1.0	0.0	7.0	460
BW 2	Mysore/Mangalore bleachable cardamom, unclipped	1.0	0.0	7.0	460
BW 3	Bleachable bulk cardamom, clipped	2.0	0.0	4.3	435
BW 4	Bleachable bulk cardamom, unclipped	2.0	0.0	4.3	435

Allepey cardamom seeds are the decorticated and dry seeds of *E. cardamomum* grown in Coorg and adjoining districts of Karnataka State. The specifications are given in Table 2.35.

Mangalore cardamom seeds are the decorticated and dry seeds of *E. cardamomum* grown in Mangalore, Karnakata State. The specifications are given in Table 2.36.

Following are some of the general specifications for cardamom:

1. The capsules should be well formed and packed with sound seeds inside. The cardamom may be graded on the basis of place of origin, color, size, mass per liter, bleach level, proportion of lower grades, and extraneous matter.

Table 2.35 Agmark Specifications of Alleppey Cardamom Seeds

Grade Designation	Trade Name	Extraneous Matter by Weight (%)	Light Seeds by Weight (%)	Weight Minimum (g liter^{-1})
MS 1	Prime	1.0	3.5	675
MS 2	Shipment	2.0	5.0	460
MS 3	Brokens	5.0		

Note: Extraneous matter includes calyx pieces, stalk bits, and other foreign matter. Light seeds include seeds that are brown or red in color and broken, immature, and shriveled seeds.

Table 2.36 Agmark Specifications of Mangalore Cardamom Seeds

Grade Designation	Trade Name	Extraneous Matter by Weight (%)	Light Seeds by Weight (%)	Weight Minimum (g liter^{-1})
AS 1	Prime	1.0	3.5	675
AS 2	Shipment	2.0	5.0	460
AS 3	Brokens	5.0		

2. The aroma and taste of cardamom in capsules and seeds should be characteristic, fresh, and free from any unwelcome aromas or tastes, including rancidity and mustiness.
3. Cardamom capsules and seeds should be free from living insects and molds and should be practically free from dead insects, their fragments, and rodent contamination.
4. The mass of cardamom capsules or seeds contained in 1 liter should be as specified for the different grades.
5. Cardamom should be free from visible dirt or dust. Extraneous matter, such as bits of calyx, stalks, and other kinds of matter, shall not be more than 5% by weight in cardamom capsules and 0.5–2% by weight in different grades of cardamom seeds.
6. The proportion of empty or malformed capsules, determined by opening and examining 100 capsules taken from the sample, should not be more than 1–7% by count, varying with grade specifications.
7. The proportion of immature and shriveled capsules, separated according to specified methods, should not be more than 2–7% (m m^{-1}).
8. Capsules having a black color and those that are split open at corners for more than half the length should not be found in bold grades and should be found in not more than 10% and 15% by count in the "shipment" and "light" grades.
9. The proportion of cardamom seeds that are light brown, broken, or immature (shriveled) should not be more than 3–5% (m m^{-1}).

The chemical and physical specifications for whole and ground cardamom are given in Tables 2.37 and 2.38.

Table 2.37 The Physical and Chemical Specifications of Whole Cardamom

Specification	Suggested Limit
ASTA cleanliness specifications	
Whole dead insects by count	4
Mammalian excreta (mg lb^{-1})	3
Other excreta (mg lb^{-1})	1.0
Mold (% by weight)	1.0
Insect defiled, infested (% by weight)	1.0
Extraneous matter (% by weight)	0.5
FDA DALs	None
Volatile oil	3% minimum
Moisture	12% maximum
Ash	10% maximum
Acid-insoluble ash	2% maximum
Average bulk index (mg 100 g^{-1})	
Bleached	320
Green	250

Table 2.38 The Physical and Chemical Specifications of Ground Cardamom

Specification	Suggested Limit
FDA DALs	None
Volatile oil	3% minimum
Moisture	12% maximum
Total ash	10% maximum
Acid-insoluble ash	2% maximum
Military Specifications	
(EE-S-631J, 1981)-Decorticated Cardamom	
Volatile oil (ml 100 g^{-1})	3% minimum
Moisture	12% maximum
Total ash	7% maximum
Acid-insoluble ash	3% maximum
Granulation	95% minimum through US$40
Bulk index (2 ml 100 g^{-1})	190

2.8.6 Commercial Cardamom Grades in Sri Lanka

Most of the cardamom produced in Sri Lanka is exported. The traders use various designations for cardamom, such as the following ones (Guenther, 1952):

1. Green cardamom
   Kandy type: Relatively large, dark greenish in color
   Copernicus type: Slightly smaller than the Kandy type, generally green in color
   General faq. Type: Small cardamom, grayish green in color
2. Bleached cardamom: Malabar half-bleached: Fair and of average quality, small capsule
   Curtius: Fair in size, rather long capsules
   Cleophas: Fair in size, roundish capsules
3. Seeds
   Crispus type: Freshly removed seeds obtained by the dehusking of either green or bleached capsules

In general, the trade distinguishes among decorticated cardamom, green cardamom, and bleached cardamom.

2.8.7 Grading and Packing

Cardamom, which is sun dried or dried in a dryer, has to be protected from absorption of moisture, contamination with extraneous matter that might impart an unpleasant odor, microorganisms, and insects. The specific requirement pertaining to the packing of cardamom is that the product be protected from sunlight in order to maintain the husk color and the green or golden color of the bleached cardamom. Cardamom is a high-value crop, the second most important spice crop of the world, and every care has to be taken to process and grade it efficiently. Sieves of different mesh sizes, namely, 6, 7, 7.5, and 8 mm, are available for sieving, which is done manually. After sieving the capsules and grouping them into different grades, it is essential to sort out the "splits" and the thrips and borer-infested capsules separately. Currently, sorting is done by skilled women laborers. Because the harvest alone demands about 60% of the labor force, there is an urgent need to fabricate mechanical sorting machines in order to obtain the produce of different sizes, with "splits" and capsules infested with insects separated out.

Cardamom needs to be stored for a specified length of time after grading. The graded produce is normally stored in double-lined polyethylene bags. The quality of the stored material could be impaired by storage pests. Hence, there is also an urgent need to devise storage systems to minimize storage pest infestation.

Equilibrium relative humidity studies have shown that cardamom dried and maintained at or below 10% moisture retains the original color and precludes infestation by molds (Govidarajan, 1982). If black polyethylene is used to store the produce, the adverse effect of light is further minimized and safe storage is possible for 4 months, a period required for port storage and transshipment. It is advisable to make use of the dried cardamom capsules within 12–15 months of harvest, failing which the pleasant flavor and aroma are likely to be adversely affected. The stored samples must be frequently tested for storage pests.

2.8.8 Conclusions

Close monitoring of the various operations, right from harvesting to drying and final grading, is required to obtain high-quality end produce. Any initial moisture that the capsules release while drying should be removed immediately by providing cross ventilation and exhaust fans. Any breakdown in the heat energy supply affects the appearance and quality of the produce. Excessive heat adversely affects the quality of the cardamom capsules. Even in the pipe dryers, where heat is generated by firewood, a mechanism should be developed to regulate the temperature suitably. Small and marginal farmers, who constitute nearly 70% of the cardamom growers, face difficulties in processing their own drying kilns due to economic constraints. Since cardamom estates are located in deep interior forests, the transport of wet capsules to distant places causes much practical difficulties. Hence, there is an urgent need to design and fabricate an efficient cardamom dryer that costs less than those currently in operation and that can provide efficient drying.

Cardamom capsules need to be dried within 24 h of harvest, and any delay will result in deterioration of the green color and appearance. It is uneconomical to operate cardamom dryers with smaller quantities of the harvested produce. Hence, research efforts need to be made to store wet capsules for 2–3 days without impairment of quality. The use of nonconventional energy sources for drying cardamom merits consideration in the fast-changing agroecological conditions of cardamom tracts in India. The exploration of locally available farm wastes and their biorecycling to generate heat energy merits immediate attention.

2.9 Industrial Processing of Cardamom and Cardamom Products

Cardamom, the dried fruit (capsule) of the cardamom plant, is processed into various products, such as cardamom seeds, cardamom powder, cardamom oil, cardamom oleoresin, and encapsulated cardamom flavor. Proper maturity with a good characteristic aroma and a high volatile oil content are the prime considerations in processing cardamom into various products. Although cardamom is sold mainly as the dried capsule, both in the national and international markets, there is some demand for cardamom seeds in the American and Scandinavian markets. The shelf life of cardamom seeds is poor because the aromatic volatile principle is present in a single layer just below the epidermis. The flavor of cardamom is due entirely to its volatile oil content, and the strength of the flavor is directly related to the quantity of oil that is present. Hence, a suitable packaging material has to be employed to store the seeds. By contrast, cardamom capsules with the husk intact can be stored for a year without losing any volatiles (Gerhardt, 1972; Guenther, 1952). Whole cardamom does not deteriorate in storage because of the natural protection of the outer cover. Cardamom is stored in gunny bags (jute bags lined with 300-gauge polyethylene and sometimes packed in wooden chests lined with moistureproof kraft paper or polyethylene).

Table 2.39 Quality Specifications for Cardamom Seeds

Component (%)	Requirement	Country
Moisture (maximum)	9.0	USA, UK
Volatile oil (minimum)	4.0	SriLanka, UK, India IS: 1797–1961
Total ash (maximum)	5–6	USA, UK
Acid-insoluble ash (maximum)	3–3.5	USA, UK

Source: Govindarajan et al. (1982).

Cardamom seeds, powder, volatile oil, oleoresin, and encapsulated flavors obtained from cardamom are the important products in the trade. Industrial processing and related technological aspects of these products are discussed here.

2.9.1 Cardamom Seeds

Cardamom seeds are obtained by decorticating the dried capsules. Decortication is done by using a flour mill or plate mill, also known as a disc mill. The distance between the discs plays a crucial role in the decortication process. The gap is adjusted in such a way that only husk is detached, without damaging the seeds. With proper disc adjustment, there should be only a minimal loss of the material during the dehusking operation. Good-quality seeds will be black to brown in color. The seeds of variety Malabar are sweet to the taste because of the presence of a sweet mucilaginous matter (Purseglove et al., 1981). The quality specifications for cardamom seeds are given in Table 2.39.

2.9.2 Packaging and Storage of Cardamom Seeds

Much greater attention is required in the storage of cardamom seeds than in the storage of dried capsules. This is because, unlike the situation with capsules in which the husk provides a natural protection to the seed, there is no husk in the case of seed. Bulk packaging of seeds is done in wooden chests lined with aluminum foil laminate. Loss of oil from seeds is reported to be as high as 30% in 8 months under ambient conditions, whereas the loss from dried capsules is negligible (Guenther, 1952). Clevenger (1934) observed that there was a 30% loss of volatiles from the seeds in 8 months of storage. The importance of distilling seeds immediately is well understood, and Table 2.40 summarizes the details of the volatile oil that is distilled. Freshly harvested and processed capsules gave a yield of 9.8%, and capsules yielded only 2.9% volatile oil after exposure to air for 1 month, indicating a loss of 70% of the oil. The loss of oil from ground seed is rapid when the seed is not properly protected, and in 13 weeks time only traces of volatile oil were found in the seed.

Griebel and Hess (1940), Gerhardt (1972), and Koller (1976) have reported the effect of different storage conditions on the rate of oil loss from seeds and ground

Table 2.40 Effect of Storage Periods on the Yield of Cardamom Oil

Type of Material	Yield of Volatile Oil (%)
Freshly gathered whole fruit	9.8
Seeds exposed to air for 1 month	2.9
Seeds exposed to air for 6 months	2.4
Seeds exposed to air for 14 months	2.0
Seeds exposed to air for 1 week	2.4
Seeds exposed to air for 6 weeks	2.4
Seeds exposed to air for 13 weeks	Traces
Seeds freshly removed from capsules	4.8
Ground seeds exposed to air for 1 week	2.4
Ground seeds exposed to air for 13 weeks	Traces

Source: Mahindru (1978); Wijesekera and Nethsingha (1975).

cardamom. Koller in particular states that the storage temperature has a greater influence on the rate of oil loss than does the type of container or the period of storage.

2.9.3 Cardamom Powder

In its powder form, cardamom gives the maximum flavor to food products. But the disadvantage with powder is that it loses its aroma through the rapid loss of volatiles. Hence, the powder needs more protection than the whole capsules or seeds. The industrial and institutional requirements of cardamom are met by grinding seeds just before use (ITC/SEPC, 1978; ITC, 1977).

2.9.4 Grinding

Grinding is an important step in the process of converting a spice into powder, and one has to be very cautious with a spice such as cardamom because it has a very delicate aroma. The aroma principles of cardamom seed are present near the surface; hence, more attention is needed during grinding because of the heat produced in attrition. The temperature during grinding can go up to as high as 95 °C in mass production (Pruthi, 1980; Wistreich and Schafer, 1962). For grinding, conventional mills such as plate mill, hammer mill, or pin mill are employed. The particle size of the ground spice may vary from 250 to 700 μ, whereas, as an additive flavorant in food products, the preferred size will be 250–300 μ. A finer particle size promotes the release of the spice's aroma and better mixing with food products.

Investigations into the grinding of cardamom by using plate mill at ambient conditions and centrifugal mill at low temperatures were carried out by Gopalakrishnan et al. (1990). In ambient conditions, using a 0.25-mm sieve led to loss of volatiles to the extent of 52.8%, while using 0.50-mm sieve resulted in a smaller loss, 34%. However, with a coarse powder obtained by using a 0.75-mm sieve, the loss of volatiles was lowered to 26.2%, but when a 1-mm sieve was used with coarser powder, the trend in the loss of volatiles was reversed. The higher loss or poor recovery in the latter case was attributed to the incomplete release of oil from the very coarse powder. Using a 0.25-mm sieve to grind frozen cardamom seeds or to grind seeds with liquid nitrogen resulted in 35.4% and 37.8% loss of volatiles, respectively. However, cryogrinding seeds with dry ice gave the best results, and the loss of volatiles was only 8.74%, but moisture absorption by the material was noticed during the grinding. Other studies have also shown that the loss of volatiles is considerably minimized by prechilling the spice and grinding at low temperature (Anon., 1975, 1977).

Cryogrinding, or freeze grinding, of spices is a novel approach to get spice powder of better quality along with enhanced retention of volatiles (Wistreich and Schafer, 1962). Advantages of cryogrinding are minimal oxidative losses of volatiles, increased output of the powder (the end product), and the prevention of screens or discs from gumming up during milling (Russo, 1976). The product so obtained has good dispersibility in food preparations. It is also reported that low temperature reduces the microbial load on spices. The cost of the cryoprocess gets reduced when milling operations are carried out on a bigger scale and with efficient recycling of the refrigerant. The maximum yield of oil has been obtained when the cardamom seeds are precooled by using liquid nitrogen within a temperature range of −180 °C to −190 °C and grinding the seeds to a size of 250 μ, which is, indeed, a fine powder size.

2.9.5 Storage Powder

Ground cardamom loses its aroma rapidly through the loss of volatiles; hence, proper care should be taken during storage. Gerhardt (1972) found that lacquered cans, polyvinylidene chloride (PVDC), and high-density polyethylene (HDPE) were suitable for the storage of powder. Koller (1976) found that vacuum-packaged ground cardamom stored at 5 °C retained flavor for longer periods. Polyester/aluminum foil/polyethylene laminate, with its outstanding moisture, oxygen, and odor barrier properties, can offer a long shelf life of over 180 days under normal conditions for cardamom powder. For a shorter storage life of 90 days and below, metalized polyester/polyethylene laminate can be considered.

2.9.6 Cardamom Oil

Cardamom oil is obtained by the distillation of powdered seeds of cardamom. Steam distillation is the most common method employed for the production of cardamom oil. Use of the cohabitation technique for distillation has been discontinued because the process hydrolyzes esters. The quality of the oil that is obtained depends on the variety, rate, and time of distillation. The important trade varieties are Alleppey Green, Coorg Green, and Saklespur bleached. Yields of volatile oil from the seeds of these three varieties were 10.8%, 9.0%, and 8.0%, respectively (Lewis et al., 1967).

Table 2.41 The Flavor Profile (Main Components) of Cardamom Oil

Component	Content (%)	Trace Components	
α-Pinene	1.5	Hydrocarbons	Alcohols and phenols
β-Pinene	0.2	α-Thujene	3-Methyl butanol
Sabinene	2.8	Camphen	*p*-Menth-3-en-l-ol
Myrcene	1.6	α-terpinene	Perillyl alcohol
α-Phellandrene	0.2	*cis*-Ocimene	Cuminyl alcohol
Limonene	11.6	*trans*-Ocimene	*p*-Cresol
1,8-Cineole	36.3	Toluene	Carvacerol
y-terpinene	0.7	*p*-Dimethylstyrene	Thymol
p-Cymene	0.1	Cyclosativene	Carbonyls
Terpinolene	0.5	α-Copaene	3-Methyl butanal
Linalool	3.0	α-Ylangene	2-Methyl-butanal
Linalyl acetate	2.5	γ -Cadinene	Pentanal
Terpinen-4-ol	0.9	γ -Cadinene	Furfural
α-Terpineol	2.6		8-Acetoxycarvotanacetone
α-Terpinyl acetate	31.3	Acids	Cuminaldehyde
Citronellol	0.3	Acetic	Carvone
Nerol	0.5	Propionic	
Geraniol	0.5	Butyric	Others
Methyl eugenol	0.2	2-Methyl butyric	Pinole
Trans-nerolidol	2.7	3-Methyl butyric	Terpinyl-4-yl acetate
			α-Terpinene propionate
			Dihydro-alfa-terpinyl acetate

External appearance, size, and bleached color are not the parameters to be considered in selecting cardamom for distillation. High-grade cardamom is not economical for distillation, since it fetches a better price as whole cardamom in the trade. Lower grades, which do not fetch higher value because of their defective appearance, but are still good from the point of view of flavor, are ideally suited for distillation. The husk is almost devoid of any volatile oil (Anon., 1985). The flavor of cardamom is due mainly to 1,8-cineole, terpinyl acetate, linalyl acetate, and linalool. (Table 2.41, which also gives the total flavor profile.)

The United Kingdom was earlier distilling oil from the cardamom obtained from India, Sri Lanka, and Tanzania (British Pharmacopoeia, 1980, 1993). The oil used was termed “English distilled cardamom oil” and priced higher compared to the oils produced from these cardamom-growing countries. With the advent of better

Table 2.42 The Composition of Different Varieties of Cardamom

Variety	Husk (%)	Seeds (%)	Volatile Oil in Seeds (% v/w)
Kerala State			
Alleppey Green	26.0–38.0	62.0–72.3	7.5–11.3
Karnataka State			
Coorg	25.2–28.0	69.6–73.3	7.5–9.1
Tamil Nadu			
Yercaud	24.0–33.0	73.0–76.0	6.5–9.6

Source: Data compiled from Nambudiri et al. (1968) and Shankaracharya and Natarajan (1971).
Note: Moisture in the aforementioned raw materials Husk and seeds ranged from 8 to 12%.

technology for the distillation of cardamom oil, the production of oil in the United Kingdom has been considerably reduced, and the oil is being imported now.

2.9.7 Industrial Production of Cardamom Oil

Cardamom capsules of proper maturity and that have a moisture content of 10–12% are selected for oil distillation. The capsules are cleaned with a destoner (which removes small stones mixed with the capsules) and an air classifier to remove undesirable extraneous matter. The cleaned capsules are dehusked in a disc (plate) mill. The gap between the discs is critical in order to avoid damage to seeds. Seeds and broken husks are separated in a vibratory sieve. The average composition of capsules of different varieties of cardamom is given in Table 2.42.

Cardamom seeds that are free of husk are passed through the plate mill, wherein the gap between the discs is brought closer to get coarse powder to pass through a 2-mm sieve. The oil glands exist just below the epidermal layer; hence, great care should be exercised while powdering. Fine milling should be avoided to prevent the loss of volatiles. The powdered material is subjected to distillation as quickly as possible. If, for any reason, there is a delay in distillation, the ground powder is packed into airtight containers until it is used. Distillation of a 500-kg batch powder usually takes 5–6 h and may even go up to 10 h. The rate of distillation and the condensate temperature are carefully regulated, and it has been observed that keeping the condensate warm helps to separate oil from water (Nambudiri et al., 1968). After commencement of the distillation, about 60–70% of the oil is collected in the first 1 h. It has been observed that early fractions are rich in low-boiling terpenes and 1,8-cineole, and subsequent fractions are rich in esters such as terpinyl acetate (Krishnan and Guha, 1950). Between varieties Malabar and Mysore, the former contains much larger amounts of 1,8-cineole, and this makes it more harsh and camphoraceous, while the oil from the latter has a sweet and fruity floral odor due to the lower amount of cineole and higher amounts of terpinyl acetate, linalool, and linalyl acetate (Lewis, 1973). Variety Mysore is the largest selling Indian cardamom, named Alleppey Green.

Table 2.43 The Specification of Cardamom Oil

Definition and source	Volatile oil distilled from the seeds of *Elettaria cardamomum* (Linn) Maton; family Zingiberaceae; cardamom grown in South India, Sri Lanka, Thailand, Guatemala, South China, and Indonesia
Physical and chemical constants	Appearance: colorless to very pale yellow liquid. Odor and taste: aromatic, penetrating, somewhat camphoraceous odor of cardamom; persistently pungent; strongly aromatic taste. Specific gravity: 0.917–0.947 at 25 °C (temperature correction factor 0.00079 $°C^{-1}$).Optical rotation: +22° to +44°.Refractive index: 1.463–1.466 at 20 °C
Descriptive characteristics	Solubility: 70% alcohol: in 5 volumes; occasional opalescence: benzyl alcohol: in all proportions diethyl phthalate: in all proportions fixed oil: in all proportions glycerin: insoluble mineral oil: soluble with opalescence propylene glycol: insoluble stability: unstable in presence of strong alkali and strong acids; relatively stable to weak organic acids; affected by light
Containers and storage	Glass, aluminum, or suitably lined containers, filled full; tightly closed and stored in a cool place protected from light

Source: Adopted from EOA, 1976.

2.9.8 Improvement in Flavor Quality of Cardamom Oil

The quality of the flavor of cardamom oil containing high amounts of 1,8-cineole has been improved by fractional distillation (Narayanan and Natarajan, 1977). In their experiment, the authors subjected 200 g of cardamom powder to distillation and, in the first 2.5 min, collected 6.5 ml oil, of which 78.86% was 1,8-cineole and traces of α-terpinyl acetate. In the next period of distillation, in the time range of 2.5 min to 2 h, of the 10.5 ml collected, 47.5% was 1,8-cineole and 36.8% α-terpinyl acetate. Hence, it is possible to get good-quality cardamom oil by using inferior-grade cardamom by suitably collecting the oil fractions at different intervals of time. Careful blending of the fractions is carried out by keeping the aroma profile and specifications in view. The oil yield will be less by about 25% by this method, but will be economical, since the subsequent fraction fetches a higher price. The specification of cardamom oil is given in Table 2.43.

Raghavan et al. (1991a) have standardized a method for the separation of 1, 8-cineole from cardamom oil by adduct formation using orthophosphoric acid. In this method, 100 ml of cardamom oil is treated first with 30 ml of orthophosphoric acid and then with 50 ml petroleum ether with constant stirring. The adduct (precipitate) formed is then filtered. The precipitate is air dried and extracted with 500 ml of hot water. The cineole fraction is released as a separate layer and recovered. The aqueous layer is extracted with 200 ml of petroleum ether and desolventized to get a terpinyl acetate-rich fraction. Gas chromatography analysis of these fractions showed that the cineole fraction (28 ml) contained 80% cineole and 18% terpinyl acetate while the terpinyl acetate fraction (58 ml) contained 76% terpinyl acetate and 16% cineole.

Oil yield from husk is reported to vary from 0.2 to 1%, and the oil possessed properties similar to those of seed oil (Rao et al., 1925; Rosengarten, 1969). Nambudiri et al. (1968) found that husk does not give more than 0.1% volatile oil; the higher values reported may be due to the admixture of seeds along with the husk during sieving. The chemical quality of the oil obtained from seeds and from husk was evaluated by Verghese (1985), using GLC and infrared methods. Although there was excellent correlation and the spectra were superimposable, the organoleptic profiles differed. The author concluded that distillation of oil from seeds along with husk is detrimental, as it is likely to impair the flavor spectrum of the oil. Purseglove et al. (1981) mention that oils obtained from green and bleached cardamom, respectively, will be similar in composition.

Hydrodistillation of cardamom is not practiced commercially because the distillation time is long and the release of oil is slow because of the gelatinization of starch. Besides, hydrodistillation hydrolyzes the esters that are present in the oil (Wijesekera and Nethsingha, 1975). Another disadvantage is that the resulting mass is not easily amenable for oleoresin extraction with solvents.

2.9.9 Storage of Cardamom Oil

Before storage, cardamom oil should be free of trace amounts of moisture. This is accomplished by the addition of any hydrous sodium sulfate. The oil is stored in aluminum containers. Polyethylene terepthalate bottles that possess very good odor barrier properties can also be considered. Food-grade, high-molecular-weight, high-density polyethylene containers are also used. The oil is filled to the capacity of the container, stored at 8–10 °C, and protected from light.

2.9.10 Cardamom Oleoresin

Oleoresin is made of two components: volatile oil and resin. The former represents the aroma, while the resin is made up of nonvolatile matter such as color, fat, pungent constituents, and waxes. The total flavor effect of a spice is obtained only after blending the oil with the resin. Volatile oil is obtained by steam or hydrodistillation, whereas resin is obtained by solvent extraction. Of late, supercritical fluid extraction is also being adopted.

Because cardamom oleoresin not like the oleoresin obtained from black pepper, demand for the former is rather limited. Cardamom oil itself represents almost all the aroma and flavor of the capsules. The consumption of cardamom oleoresin is slowly picking up, probably because of its mellower and less harsh flavor characteristics (Sankarikutty et al., 1982). Although the cardamom oil represents the flavor of cardamom, it lacks the "richness" that is attributed to the absence of nonvolatile components (Lewis et al., 1974). Sensory differences have also been noticed between oils and oleoresins of cardamom (Govindarajan et al., 1982b).

For oleoresin extraction, either freshly ground cardamom or essential oil-free cardamom powder (cardamom powder from which oil has been distilled off) is employed. The main considerations involved in the oleoresin preparations are the selection of suitable raw material, grinding to the optimum particle size for extraction, the choice of solvent, the type of extraction, miscella distillation, and blending.

2.9.11 Solvent Extraction

Cardamom seed is ground to a coarse powder of particle size 500–700 μ, which helps in the rupture of flavor cells and is amenable to ready extraction by solvents. Fine grinding should be avoided because it not only results in the loss of volatiles, but also creates problems during extraction, such as slow percolation of the solvent, and channeling and engagement of the extractor for longer periods. The powdered spice is loaded into the extractor, which is also called a percolator, and extracted with a suitable solvent. The solvent can be acetone, alcohol, methanol, ethyl acetate, ethyl methyl ketone, and so on, or a mixture of these solvents. The selection of the solvent for extraction is a crucial step, and it should be standardized on a small scale at the laboratory level before venturing on to commercial production.

The selected solvent is allowed to percolate through the bed of material by keeping the bottom drain valve open for the escape of air. When the entire material is soaked in solvent, the bottom drain is closed and sufficient contact time is given for the solutes to leach into the solvent. After the contact time, the extract, called "miscella," is drained and collected.

For oleroresin production, either Soxhlet extraction (Goldman, 1949) or batch countercurrent extraction is industrially practiced (Nambudiri et al., 1970). The concentrated miscella obtained from each extractor is carefully collected and distilled to obtain the finished product. Most of the solvent (about 90–95%) present in the miscella is recovered by normal atmospheric distillation, while the remaining solvent is taken off by distillation under reduced pressure. Great care should be exercised during distillation to minimize heat damage to the product. After the completion of solvent stripping, the product, while hot, is discharged from the bottom still and stored in suitable containers. It has been observed that in a 100-kg batch extraction, the retention of solvent in the spent material is on the order of 60–70 kg and about 95% of this quantity is recovered during the desolventization process. The spent meal after the recovery of solvent is discharged from the bottom side vent of the extractor and dried. Spent meal contains starch, fiber, carbohydrate, protein, and so on, and finds application in animal feed composition. It can also be used as a broiler feed and as a source of manure for crops. Cardamom spent meal has been used in the manufacture of scented sticks, used in most Indian (Hindu) homes and temples for worship and known locally as "Agarbathi" (Suresh, 1987). The quality and yield of cardamom oleoresin depend on the variety of raw material, the solvent used, and the method of extraction. By using hydrocarbon solvents, oleoresin having 10–20% fixed oil has been obtained, whereas with a polar solvent such as alcohol, a fat-free product is obtained (Naves, 1974). Oleoresins containing 54–67% volatile oil have been obtained by Salzer (1975), wherein the fixed oil content varied according to the extracting solvent. The color of the product varies from brown to greenish brown. Kasturi and Iyer (1955) used carbon tetrachloride as solvent to extract cardamom seeds from which volatile oil was already distilled and got a 4% yield of fixed oil. On analysis, the fixed oil was found to contain 62.6% oleic acid, 18.3% stearic acid, 8.4% palmitic acid, 10.5% linoleic acid, and 0.3% caprylic and caproic acids. Miyazawa and Kameoka (1975) and Marsh et al.(1977) found palmitic (28–38%), oleic (43–44%), and linoleic (2–16%) acids as the major fatty acids present in the fatty oil. The Central Food Technological Research Institute in Mysore, Karnataka State, has developed analytical processes for the production of cardamom oil, spice oleoresins, and encapsulated spice flavors. These processes have been commercially exploited by companies involved in the spice trade.

A company in the United Kingdom claims to have produced good-quality cardamom oil by extracting seeds with a hydrofluoro solvent having a boiling point of about –26 °C. During extraction, damages due to heat, oxygen, or high pH are eliminated (Anon., 1996).

2.9.11.1 Supercritical Carbon Dioxide Extraction of Cardamom

The use of liquid and supercritical carbon dioxide as a solvent for flavor extraction from plant materials has been a subject of intense study (Schultz and Randall, 1970). Four or so decades back, Shultz et al. (1967) used carbon dioxide to extract spices such as cardamom, clove, nutmeg, coriander, and celery. The use of carbon dioxide for flavor extraction has several advantages over the traditional methods that use other solvents.

Carbon dioxide is cheap, abundantly available, not flammable, nontoxic, and noncorrosive as a solvent. It has been widely accepted as a safe solvent for flavor extraction and does not leave any residue. It behaves either as a polar or nonpolar solvent, depending on the pressure and temperature employed. It is liquid below its critical point (31.2 °C, 7.38 mPa pressure) and is safe above its critical point. Under normal conditions, the density of carbon dioxide is less than 100 g liter^{-1}, while under supercritical conditions its density varies between 200 and 900 g liter^{-1}. Naik and Maheswari (1988) used liquid carbon dioxide (20 °C, 55–58 bar pressure) and a modified high-pressure Soxhlet apparatus to extract cardamom and obtained a 9.4% yield in a 2.5-h extraction period, whereas with steam distillation in a 5-h extraction period, only a 9% yield was obtained. GLC and TLC analysis of the extracts showed that liquid carbon dioxide extract contained slightly higher amounts of cineole, terpinyl acetate, geraniol, and α-terpineol (35.72%, 24.87%, 4.53%, and 11.06%, respectively), compared with steam-distilled oil, for which the corresponding values were 30.25%, 22.05%, 4.22%, and 7.88%. The extraction of cardamom under different conditions of pressure, temperature, contact time, and moisture content did not have much influence on the yield and quality of the product. However, the extraction of nonvolatiles and chlorophyll content increased with increases in pressure and time. Although oil freshly extracted with carbon dioxide is of excellent quality, it lost its fine aroma during the 90 days of storage. There was less of a deterioration in quality of the commercial steam-distilled oil under similar conditions (Gopalakrishnan, 1994). Table 2.44 details the quality of cardamom extracts by different methods of extraction.

Table 2.44 Quality of Cardamom Extracts by Different Methods of Extraction

Parameter	Supercritical Fluid Extract	Hexane Extract	Clevenger Distilled Oil
Yield (%)	7.7	6.2	8.3
Nonvolatile matter (%)	4.6	22.5	–
Color	Pale green	Pale green	Colorless
Aroma	Superior, close to that of fresh cardamom	Residual solvent note, fresh aroma absent	Varied because of artifact formation,
Major components 1,8-cineole	29.7	16.6	31.2
a-Terpenyl acetate	37.0	57.3	35.5
a-Terpeneol	4.6	5.0	2.4
Linalool	2.6	2.3	3.8
Sabinene	4.1	2.2	3.4
b-Pinene	2.8	1.7	2.8
D-Limonene	2.4	2.2	3.3
Linalyl acetate	1.6	1.6	2.1

2.9.11.2 Encapsulated Cardamom Flavor

Encapsulation is a technique in which the flavorant is covered by a suitable material, thereby protecting the flavor from exposure to the environment. In this method, the liquid aroma concentrate is converted to a solid stable powder that will have a long shelf life. Bakan (1973) reported that some volatile liquid flavors are retained in microcapsules for periods upto 2 years. The flavor-protecting material is called "wall material" or "encapsulating material" and is generally either gum, acacia, or a starch derivative such as maltodextrin. The actual flavorant that is to be encapsulated is called "core" material. The selected wall material should be food grade, and an effective film former should stabilize the emulsified flavor in the process of encapsulation. The encapsulated product is spherical and miniature in size, ranging from submicrons to several millimeters (Bakan, 1978). When the particle size of the capsules is less than 500 μ, they are called microcapsules (Anandaraman and Reineccius, 1980).

Of the different methods of encapsulation, spray drying is the most widely used with cardamom. Encapsulation is also done for black pepper (Raghavan et al., 1990; Sankarikutty et al., 1988). In addition to spray drying, other methods are phase separation, adsorption, molten extrusion, spray cooling or chilling, inclusion complex formation, and so on. Spray drying is the most popular method. The basic steps involved are (1) preparation of the emulsion, (2) homogenization of the emulsion with the flavorant, and (3) atomization of the mass into the drying chamber. Raghavan et al. (1990) used a small spray dryer (Bower Engineering, NJ, USA), as well as a pilot spray dryer (Anhydro), to spray dry cardamom oil. Up to a 7.2-kg yield of the dried product having 4% moisture and 8.5% volatile oil was obtained. After several trial batches, the optimum oil-to-encapsulant ratio was found to be 1.4, which very much agrees with findings from the investigations of Sankarikutty et al. (1988). These authors used hot water (50–60 °C) in their experiments to aid dispersion of the gum. After the cardamom oil was mixed and emulsification took place, the globule size was 2 μ. The material was spray dried at an inlet air temperature of 155 ± 5 °C and exit air of 100 ± 5 °C. [For detailed information on the technique, see Sankarikutty et al. (1988)]. Under ideal conditions, the encapsulated powder should not have any flavor. However, in practice, it has been found that the product has a mild odor due to some amount of flavor left unencapsulated and due to the rupture of a few capsules. Attempts have been made to remove surface flavor by washing the particles with hexane (Omanakutty and Mathew, 1985). About 1–2% oil was found on the spray-dried flavors made with different encapsulations.

Gas chromatographic examination of the cardamom oils obtained from the emulsion and their spray-dried product showed a similar pattern when compared with that of the original oil. However, the oil derived from the spray-dried encapsulated product showed a slight decrease in cineole content (44%) and increase in terpinyl acetate content (38.9%), while the corresponding values in the oil obtained from emulsion were 47.4% and 34.8% (Raghavan et al., 1990; Sankarikutty et al., 1988). Cineole is more volatile, and has a lower molecular weight, than terpinyl acetate. The stability of the encapsulated cardamom oil product was found to be satisfactory when the product was stored in airtight glass containers at room temperature. The moisture pickup was negligible, and there was only a 5% loss of volatile oil during the 2-year storage period (Raghavan et al., 1990).

The following are the advantages of spray-dried encapsulated cardamom flavors: (1) They are nonvolatile, dry, and free flowing; (2) they can be readily incorporated into food mixes to obtain a uniform flavor effect; (3) their flavor stability is good over longer storage periods, even at higher temperatures; and (4) in aqueous systems, the capsules break and the flavor is released.

2.9.12 Large Cardamom (Nepal Cardamom)

Another type of cardamom of commercial importance is the Sikkim, or Nepal, large cardamom, which is almost equal to the small cardamom produced in the southern states of India. Thailand, Indonesia, and Laos also produce large cardamom to a limited extent. Large cardamom is used as a flavoring agent in curry powders, sweet dishes, and cakes, and for masticatory and medicinal purposes. Physicochemical studies on five cultivars of large cardamom, namely, *Ramsey, Golsey, Sawney, Ramla,* and *Madhusey*, have been carried out (Pura Naik, 1996). The studies revealed that the percentage of husk varied between 27% and 31.5% and that of seeds between 68.2% and 72.0%. The volatile oil content in seeds ranged from 2.7 to 3.6%. Large cardamom contains less volatile oil than small cardamom and is more camphoraceous and harsh in aroma, with a flat cineole color. The oil is rich in 1,8-cineole and devoid of α-terpinyl acetate (Govindarajan et al., 1982b). The chemical composition of large cardamom oil is well documented (Lawrence, 1970). Products such as volatile oil, oleroresin, and encapsulated flavor can be produced from large cardamom with the processing methods described for cardamom.

2.9.13 Other Products

A number of products having cardamom as the major flavorant can be prepared. Some of these products with commercial value are as follows (Raghavan et al., 1991b):

1. *Sugar cardamom mix:* This blend of sugar powder with encapsulated cardamom flavor, along with sunset yellow colorant, may contain tricalcium phosphate as an anticaking agent. The product finds application in culinary sweets and in the flavoring of milk and milk products. Incorporation of the mix with malted *ragi* (finger millet, *Eleusine coracana,* a popular food in the state of Karnataka) flour makes a good *ragi* beverage that is highly nutritious.
2. *Cardamom-flavored cola beverage:* This is an amber-colored sparkling carbonated beverage containing sugar, caramel, acid, and flavors. A market survey on the product was very encouraging, and the product was quite acceptable to consumers.
3. *Cardamom flavored flan:* This product is made from milk, sugar, and starch, with added color, flavor, and gelling agents. It tastes like a custard dairy dessert. The formula for its preparation has been standardized by the Central Food Technological Research Institute in Mysore, Karnataka State.
4. *Cardamom chocolate:* Cardamom-flavored milk chocolate is prepared by using cocoa mass with butter, sugar powder, milk powder, encapsulated cardamom flavor, and emulsifiers. The resultant product has good consumer acceptability.
5. *Cardamom Plus*: In West German markets, "Pepper Plus" has found favor with consumers. This product is prepared by fortifying black pepper powder with encapsulated pepper flavor. The advantage of such a blend is its rich naturalness with high-flavor strength. Along similar lines, "Cardamom Plus" can be prepared by mixing freshly ground cardamom with encapsulated cardamom flavor. This cardamom residue (subsequent to oil distillation), which contains fixed flavors, and resinous mass, which can be removed only by solvent extraction, can be used in place of cardamom powder. Spent residue is a valuable source for making a "Cardamom Plus" preparation.
6. *Cardamom tincture:* This product is prepared by the extraction of crushed cardamom seeds, along with other spices such as caraway and cinnamon, and cochineal using 60% alcohol as solvent. Five percent glycerin is added to the extract, which is used as a carminative mixture (British Pharmacopoeia, 1980).
7. *Cardamom coffee and tea:* The cardamom flavor is quite compatible with coffee and tea. The major use of cardamom in Middle East countries is as an additive to coffee. This cardamom is called "Gahwa" and it is a traditional drink with Arabs of the Gulf region (Survey of India's Export Potential of Spices, 1968). Encapsulated cardamom flavor is very handy and useful in making Gahwa coffee. The flavor is incorporated into roast and ground coffee, which, on brewing, imparts a predominantly cardamom-flavored extract. Cardamom coffee can also be consumed black, without milk, as in Europe and North America. Similarly, cardamom tea in which cardamom powder is mixed with tea powder, is available on the market. As with cardamom coffee, encapsulated cardamom flavor can conveniently be used to make cardamom tea, which has a long shelf life.

2.9.14 Conclusions

Cardamom flavor finds application in different items, such as processed foods, beverages, confectionery, health foods, medicines, perfumery, and cosmetics. Cardamom products, such as essential oil, oleoresin, and encapsulated flavor, may find good potential in food and nonfood industries because of advantages such as a standardized flavor strength, a good shelf life, hygienic quality, ease of handling, and stabilized flavor. However, the increase in consumer demand for processed cardamom products is related to the global increase in population and the spread of new markets. Both research and marketing are crucial in creating new forms of value-added products of cardamom and in the exploration of new markets. It is also important for the cardamom-producing countries to create and sustain internal demand for cardamom-based by-products.

2.10 The Economy of Cardamom Production

Cardamom has been an important spice commodity of international significance since Greek and Roman times. Until 1979–1980, India was the largest producer of cardamom, dominated international trade in the spice and earning valuable hard currency for the nation. More than 90% of international trade, both in small and large cardamom, originated in India. However, during the past two decades, Indian cardamom has faced serious competition in the international market from Guatemala, which has steadily eroded India's share of world trade in cardamom. With an average annual production of more than 13,000 metric tons, Guatemala has replaced India as the top producer and exporter in the world and India has dropped to the second position. The cost of cardamom production in India is relatively higher compared with that in Guatemala, mainly on account of poor yields and low productivity. India's highest productivity level in the years of a good crop is 300% less than that of Guatemala. Senility and poorly selected varieties, prolonged drought and overdependence on the monsoon, the predominance of small holdings, problems of land tenure, inefficient and inadequate attention to management in both production and protection practices, and faulty postharvest practices are some of the important reasons for the poor yield (Anon., 1996). The situation warrants a critical evaluation of the limiting factors mentioned earlier. To make India's cardamom production competitive, the first step required is an in-depth analysis of the economics of production, marketing, and other aspects affecting the cardamom economy of the country. This chapter attempts to do so in a holistic manner.

Table 2.45 depicts the production of cardamom during the past quarter century.

India and Guatemala are the major producers and players in the world economy of cardamom. Other, smaller producers are Tanzania, Sri Lanka, PNG, Honduras, Costa Rica, El Salvador, Thailand, and Vietnam. India accounted for nearly 65% of world production in the early 1970s, but by 1997–1998 production had plummeted to 28%. Guatemala, by contrast, stepped up its production beginning in the middle of the 1960s; by the 1970s it contributed 21.5% of world production, and today its share

Table 2.45 Cardamom Production in the Major Cardamom-Producing Countries of the World

Period	Percent Share in Total World Production (mt)			World Production (mt)
	India	Guatemala	Others	
1970/1971–1974/1975	65.4	21.5	13.1	4678
1975/1976–1979/1980	53.7	34.5	11.8	6628
1980–1981	42.9	48.8	8.3	10,250
1984–1985	31.9	60.3	7.8	12,220
1985/1986–1989/1990	26.5	67.5	6.0	14,392
1990/1991–1994/1995	28.4	65.6	6.0	19,470
1995/1996–1997/1998	29.8	64.2	6.0	24,953

Source: *Cardamom Statistics, 1984– 1985*, Government of India, Cardamom Board, Cochin, Kerala State, India.
Spices Statistics, 1991, Government of India, Spices Board, Cochin, Kerala State, India.
Spices Statistics, 1997, Government of India, Spices Board, Cochin, Kerala State, India.
All India final estimate of cardamom— 1997/ 1998, Government of India, Ministry of Agriculture, New Delhi.
Note: Estimated figures (actual figures are not available).

is 65%. Unlike India, Guatemala has negligible local demand and all the cardamom produced is exported. Operating in accordance with the "law of comparative advantage," Guatemala has increased its share of cardamom production in the world market from 30% to more than 90% in the recent past and has already captured the traditional markets for the spice. In India, the weak infrastructure support, insufficient credit, and substandard marketing facilities in the cardamom belt have adversely affected the prospects for cardamom. A vicious cycle of "low price–low production, high price–high production" came into operation. In 1982–1983, there was a steep fall in production, resulting in a high price in the domestic market. Then, at this stage of recession in production and export, the cardamom industry was badly hit by severe drought in 1983–1984, destroying a substantial portion of the productive area, which was gradually made up by replanting and gap filling in the years that followed. An analysis of area under crop, production, productivity, export, export earnings, prices, and so on in recent years the cardamom industry would be relevant in planning future programs for the revival of the crop in India. Table 2.46 shows the area, production, and productivity in cardamom starting from the 1970s, when Guatemala's push to increase its production of the spice began to felt on the world stage.

2.10.1 *Emerging Trends in Cardamom Production*

2.10.1.1 Area

The data presented in Table 2.46 can be categorized into three periods: (1) 1970–1971 to 1977–1978, a period of no change in area; (2) 1978–1979 to 1988–1989, a period of increasing area; and (3) 1989–1990 to 1997–1998, a period of declining area. Although the area under the crop remained unchanged during the first period of 8 years, there were year-to-year fluctuations in quantity produced, indicating the impact of climate on productivity.

Table 2.46 Area, Production, and Productivity in Cardamom

Year	Area (ha)	Growth Index	Production (mt)	Growth Index	Productivity (kg ha^{-1})	Growth Index
1970–1971	91,480	100.00	3170	100.00	34.65	100.00
1971–1972	91,480	100.00	3785	119.40	41.38	119.42
1972–1973	91,480	100.00	2670	84.23	29.19	84.24
1973–1974	91,480	100.00	2780	87.68	30.39	87.71
1974–1975	91,480	100.00	2900	91.48	31.70	91.49
1975–1976	91,480	100.00	3000	94.64	32.79	94.63
1976–1977	91,480	100.00	2400	75.71	26.24	75.73
1977–1978	91,480	100.00	3900	123.02	42.63	123.03
1978–1979	92,760	101.40	4000	126.18	43.12	124.44
1979–1980	93,950	102.70	4500	141.96	47.90	135.64
1980–1981	93,950	102.70	4400	138.80	46.83	135.15
1981–1982	93,950	102.70	4100	129.33	43.64	125.95
1982–1983	93,950	102.70	2900	91.48	30.87	89.09
1983–1984	93,950	102.70	1600	50.47	17.03	49.15
1984–1985	100,000	109.31	3900	123.03	39.00	112.55
1985–1986	100,000	109.31	4700	148.26	47.00	135.64
1986–1987	105,000	114.78	3800	119.87	38.00	109.67
1987–1988	105,000	114.78	3200	100.95	30.48	87.96
1988–1989	81,113	88.67	4250	134.07	40.48	116.83
1989–1990	81,113	88.67	3100	97.79	38.22	110.30
1990–1991	81,554	89.15	4750	149.84	58.24	168.08
1991–1992	81,845	89.47	5000	157.73	61.09	176.31
1992–1993	82,392	90.06	4250	134.07	51.58	148.86
1993–1994	82,960	90.69	6600	208.20	79.56	229.61
1994–1995	83,651	91.44	7000	220.82	83.68	241.50
1995–1996	83,800	91.60	7900	249.21	94.27	272.06
1996–1997	72,520	79.27	7290	208.99	100.52	290.10
1997–1998	69,820	76.32	7150	225.55	102.40	295.53
1998–1999	72,135	75.85	7170	226.18	135.00	389.61
1999–2000	72,451	79.20	9290	293.06	173.00	499.28

Source: Data from Spices Statistics, various years, Spices Board, Government of India, Cochin, Kerala State, Agricultural Production Statistics, Ministry of Agriculture, Government of India, New Delhi.

2.10.1.2 *Production and Productivity*

A significant feature of cardamom production in India is cyclical fluctuations in yield. After a succession of increases in production and productivity for 2–3 years, a decline set in and continued for a time before production increased again. Climate is the most important factor in growing cardamom, which is a shade-loving plant. There were cyclical fluctuations in world production during 1973–1976, 1978–1981, and 1985–1987. India's production had been showing a consistent increase from 4250t in 1992–1993 to 7900t in 1995–1996, but declined to 7150t in 1997–1998.

Still, the rate of decline was not as rapid as it was in the 1970s and 1980s, likely because of improvements in productivity. The use of improved varieties and better production technology contributed much to the overall rise in productivity. After 1997–1998, production exhibited an increasing trend once more.

Cardamom yields, which were around 34.65 kg ha^{-1} during 1970–1971, did not show much improvement until the end of 1980, except for an increase up to 48 kg ha^{-1} in 1979–1980. It appears that the increase in yield during this period did not contribute to the overall increase in production, which was entirely accounted for by the increase in area planted. However, productivity increased from the 1990 onward and reached a peak of 102.4 kg ha^{-1} in 1997. The productivity in 1997–1998 was a threefold increase over that obtained in 1970–1971. Record productivity of 173 kg ha^{-1} was achieved in 1999–2000 through better varieties of the spice and improved crop practices (Spices Board, 2000).

2.10.1.3 Growth Estimates

In order to get summary measures of long-term trends in area under cultivation, production, and productivity of cardamom in India, semilogarithmic equations are used. A decadal analysis showed that there was a positive growth rate of 0.2% and 0.1% in the area under cardamom cultivation in the 1970s and 1980s, respectively. A negative growth rate of –1.9% prevailed in the 1990s. Production had a positive growth rate in all the three decades, reaching a maximum of 8.3% in the 1990s. The figures for productivity are a matter of major concern, as productivity has a direct bearing on the cost efficiency and profitability of cardamom cultivation. The estimated negative growth rate in area and positive growth rate in production in the 1990s indicate an improvement in productivity, that is, more production is achieved with less area under cultivation.

Table 2.47 presents data on area under cultivation, production, and productivity for three Indian states: Kerala, Karnataka, and Tamil Nadu. The cardamom belt of India is located in the Western Ghats regions of these three states. Kerala State accounts for the major share of area and production of cardamom in India, and this has remained more or less unchanged over the last three decades. Karnataka State stands second, followed by Tamil Nadu. Over the years, productivity per unit area has gone up both in Kerala State and in Karnataka State, but declined in Tamil Nadu.

2.10.1.4 Production Constraints

The major reasons for the low productivity of cardamom in India are as follows:

1. Recurring vagaries of climate, especially drought and the absence of irrigation practices.
2. The absence of regular replantation. In the mixed cropping system, as is practiced with cardamom and coffee, the farmer is content with the additional income from the companion crop and is not disappointed with the lowering of yields in the main cardamom crop due to aging. Replanting with improved varieties would have led to better yields.
3. Deforestation and the resultant changes in ecological conditions prevailing in the growing area—leading to the conversion of land originally growing cardamom to the cultivation of competing crops such as black pepper.

Table 2.47 Area, Production, and Productivity of Cardamom in Three States of India

Period	Variables	Kerala		Karnataka		Tamil Nadu		India
		Actual	%	Actual	%	Actual	%	%
1970–1971	Area (ha)	55,190	60.33	28,220	30.81	8070	8.81	91,480
	Production (t)	21.30	67.19	805	25.39	235	7.41	3170
	Productivity (kg ha^{-1})	38.59		28.53		29.12		34.65
1980–1981	Area (ha)	56,380	60.01	28,220	30.03	9350	9.95	93,950
	Production (t)	3100	70.45	1000	22.73	300	6.82	4400
	Productivity (kg ha^{-1})	54.98		28.22		32.09		46.83
1990–1991	Area (ha)	43,826	53.74	31,605	38.75	6123	7.51	81,554
	Production (t)	3450	72.63	800	16.84	500	10.53	4750
	Productivity (kg ha^{-1})	78.72		25.31		81.66		58.24
1997–1998	Area (ha)	43,050	61.66	21,410	30.66	5360	7.68	69,820
	Production (t)	5430	75.94	1240	17.34	480	6.71	715
	Productivity (kg ha^{-1})	126.13		57.92		89.55		102.41
1999–2000*	Area (ha)	41,522	57.31	25,882	35.72	5047	6.97	72,451
	Production (t)	6550	70.51	1950	20.99	790	8.50	9290
	Productivity (kg ha^{-1})	213		103		205		173

Source: Spices Statistics, various years, Government of India, Spices Board, Cochin, Kerala State.
Note: Yield is arrived at by dividing total production by area; (*) midterm estimate.

4. Disinterest among planters in adopting high-production technology, although better varieties and a practically proven package of practices are available to enhance yields up to $600\,kg\,ha^{-1}$.
5. The problems caused by pests and diseases.
6. The remote locations of plantations.
7. A system of land tenure that does not allow long-term planning for improvement by the actual producer who works on the land (Anon., 1996; Cherian, 1977; George, 1976).

Coffee, black pepper, and areca nut are the other plantation crops raised along with cardamom. A comparative picture is given in Table 2.48.

2.10.1.5 *Cost of Production*

Productivity and cost of production play crucial roles in determining the competitiveness of a product in the global market. The cost of production of cardamom in Guatemala is much lower than it is in India, for two reasons: higher productivity and lower wages for labor. Productivity in India was a mere $47\,kg\,ha^{-1}$ in the 1980s, when it was $91\,kg\,ha^{-1}$ in Guatemala. Guatemala obtains more than $200\,kg\,ha^{-1}$ (dry capsules), whereas in India the figure is only $120\,kg\,ha^{-1}$. This in itself gives Guatemala a cost advantage of more than 255%. Yet Guatemala has an advantage over India in cost of production as well. For instance, production cost was just about US$1 per kg during the 1980s in Guatemala, while in India it was double that (Bossen, 1982).

Consequently, Guatemala has been able to edge out Indian cardamom from the world market. The price of cardamom from Guatemala in recent years has been around US$5–$7\,kg^{-1}$ lower than that in India, and the quality of the produce is quite comparable to that from India. Such advantage helped exporters from Guatemala to penetrate markets held earlier by India. Still, advances in cardamom production technology helped India increase its productivity per unit area. The highest yield achieved was $2475\,kg\,ha^{-1}$, by a farmer in the Idukki district of Kerala State in 1999–2000. Other farmers have achieved yields greater than $1400\,kg\,ha^{-1}$ (Korikanthimath, 1992). However, the labor component in the total cost of production accounts for more than 60% during the establishment stage of the plantation and more than 40% thereafter, making Indian cardamom a lot costlier in the international market. Studies have shown that expenditure on labor is positively correlated with yield per hectare (Mahabala et al., 1991). The current estimated cost of labor in India varies from Rs 150 (approximately US$4) to Rs 200 (approximately US$5) to produce 1 kg, depending on the cropping system followed. Thomas et al. (1990) have concluded that low productivity and high cost of production vis-à-vis stiff competition in the international market rendered Indian cardamom less competitive and subsequently unremunerative for the planters. Because of the nonavailability of skilled labor for harvest and postharvest handling, including on-farm processing, the employment of unskilled laborers resulted in a lower recovery rate of 19%, as against the desired rate of 25%. Thus, an avoidable postharvest loss of around 6.8% is also responsible for India's reduction in productivity.

Table 2.48 Quinquennial Averages of the Indexes of Area, Production, and Productivity of Major Commercial Crops (1960–1961 = 100)

Period	Arecanut			Coffee			Cardamom			Black Pepper		
	A	P	Y	A	P	Y	A	P	Y	A	P	Y
1960/1961–1964/1965	111	131	117	106	101	95	100	113	113	100	93	93
1965/1966–1969/1970	139	258	185	131	157	126	126	113	91	99	85	86
1970/1971–1974/1975	165	448	270	128	209	163	140	120	84	117	94	80
1975/1976–1979/1980	187	569	304	166	266	159	165	160	97	109	91	84
1980/1981–1984/1985	236	647	274	194	293	151	185	273	146	106	97	91
1985/1986–1989/1990	276	930	336	213	388	182	211	320	150	144	152	105
1990/1991–1994/1995							148	173	167	181	182	101

Source: Radhakrishnan (1993).
Note: A, area (ha); P, production (t); Y, productivity (kg ha^{-1}).

2.10.1.6 Domestic Market Structure and Prices

Cardamom trade in India is a regulated trade. In 1987, licensing and marketing rules were introduced to streamline the system of marketing cardamom in general and to bring about control in the form of restricting the entry of persons into the different functional categories, namely, exporters, dealers, and auctioneers. The declared purpose of such regulation is to ensure a fair price for the product and timely payment of the proceeds from sales. Export marketing of cardamom began to be regulated by the Spices Board (Registration of Exporters) in 1989. The Board issues the following certificates or licenses:

1. Cardamom dealer license
2. Cardamom auctioneer license
3. Certificate of registration as exporter of spices
4. Registration-cum-membership certificate to exporter.

Market intelligence officers have been posted in important marketing/auction centers to collect reports on crop purchases, sales, movement, and price trends.

2.10.1.7 Price Analysis

Analysis of the structure and behavior of farm prices is of considerable interest in achieving the aim of finding ways and means to increase production and productivity. Prices often act as a guide that indicates changes in production decisions.

Cardamom is a moderately storable export commodity. Long-term storage is not possible, as it is in the case of black pepper. Thus, markets must be cleared within the crop year, thereby ruling out price speculation. Within these limits, the formation of prices in the domestic market takes place in the following manner: Depending on the length of the summer, severity of drought, premonsoon showers, and amount of rainfall during June–July, well-experienced traders forecast the crop size and its prospects for the forthcoming season. Their forecasts are aided by the fact that many of the dealers and exporters of the crop are themselves cardamom planters. If the expected production is much lower than the normal production, a significantly higher price than that which ruled during the previous season is set at the beginning of the current season. If, however, production is much higher than the normal production, a considerably lower price is set (Nair et al., 1989).

The export value of cardamom usually depends on its color, the spice's major quality. Traders are keen to acquire as much as possible of the output in the peak harvest season, as high-quality harvest (with good color) comes in the middle of the season. This is what makes for the peak prices in the peak harvesting season, which in turn becomes the peak sales period. Broadly speaking, this period is between September and December. An investigation by Joseph (1985) indicates that the export price leads the domestic price with a lag of about 1 month, but according to Nair et al. (1989), although export price trends and auction price trends are synchronized, a month-to-month correspondence does not hold. However, there exists an asymmetry in that a rise in the export price is not always paralleled by a corresponding increase in the domestic price whereas a fall in the export price is transferred entirely to the domestic price.

The estimated growth equation indicated that, while auction prices have registered an average annual growth rate of 6.8%, both wholesale and export prices increased at the rate of 6.4% during the period from 1971 to 1997–1998.

2.10.2 Export Performance of Cardamom

Cardamom began to be cultivated in India as an export crop, with little attention paid to domestic consumption. Until the end of the 1960s, the crop was in the sellers' market in world trade and India was the leader in production and export. Guatemala stepped up its production starting in the mid-1960s and began capturing India's international market in the Middle East, where cardamom is a priced spice. In 1977–1978, while India could export only 2763 million tons, Guatemala exported 3610 million tons, edging out India for the world's number-one position as exporter. In India, export peaked at 3272 million tons in 1985–1986, representing a growth index of 191.91, with 69.62% of the domestic production exported. With a further increase in production in Guatemala, India's export has been seriously affected. The lowest export from India was 180 mt in 1989–1990, just 5.81% of world production. Indian exports were in proportion to the amount produced until 1985–1986; thereafter, exports remained at a low level, with consecutive drops during 1986–1987 and 1987–1988. Although the level of production had gone up, the quantity exported was low. The gap between production and export started widening, and the trend continued. Heavy exports from Guatemala had their effect on India. Even export promotion schemes, such as air freight subsidy to the Middle East, where there was a big market for Indian cardamom, exemption of cess (a levy by the government on export trade), and export assistance of Rs 35 kg^{-1} (about US 85 cents, which, according to the Indian scale of economy, is a substantial figure), could not boost exports from India. Today, almost all cardamom production in India is absorbed by the domestic market, as a result of deliberate efforts by the Spices Board to raise domestic consumption. Consequently, the enlarged domestic market was able to sustain the cardamom industry during the years in which international prices were low. Occasionally, prices went up above that of the international market. Although it is, in general, good for the cardamom industry to depend on the domestic market, doing so does pose one particular problem. Since exports account for only a portion of total production, prices are influenced more by the strong internal demand than by external demand. In the 1990s, high domestic prices paved the way for smuggling cardamom from Guatemala through Nepal into India, and an estimated 2000 t came into the country in this manner (Anon., 1998). The smuggling trade continues and has developed into a substantial menace, with local producers adversely affected. The government has been unable to check this menace effectively because there is always a "hand-in-glove" design in such an enterprise.

2.10.3 Direction of Indian Export Trade

Until the end of the 1980s, more than 80% of Indian cardamom exports were to the Middle East, accounting for over 50% of world imports. Because of the high degree

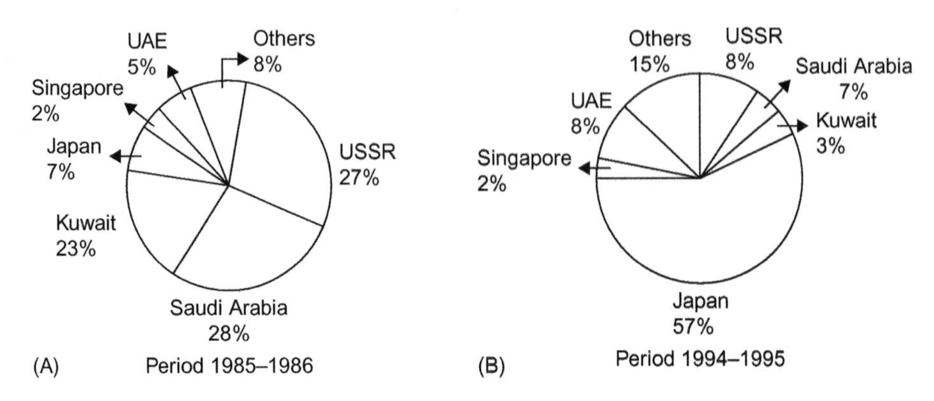

Figure 2.3 Direction of Indian export of cardamom during 1986–1987 and 1994–1995.

of quality differentiation of the product in the Middle East market, higher trade in cardamom produced in India was finding a ready market in the Arab world. Demand in this region increased vastly with the oil boom, which left a lot more cash in the pockets of Arabs. In recent years, especially in the 1990s, the share of the Middle East market in Indian cardamom exports has shrunk as the oil boom has tapered off. The shrinkage of the market produced a change in thinking from quality consciousness to price sensitivity. Thus, cardamom from Guatemala, although of inferior quality compared with that of the Indian varieties and priced about US\$4–5 kg^{-1} less than that of Indian cardamom, got preference. As a result, India lost its traditional Middle East market. India also lost its East European market after the collapse of the Soviet Union. Japan has supplied the only consistent market for Indian cardamom in recent years. In 1995–1996, more than 60% of India's total export of 226 mt went to Japan. Among the suppliers of cardamom to Japan, India tops the list at 66%, followed by Guatemala and Vietnam. Japan buys the second-grade Alleppey Green from India. It is worth noting that Japanese buyers were consistent in their purchases of Indian cardamom, even in years of high price. Saudi Arabia and Kuwait are the other markets. Figure 2.3 gives a pictorial representation of India's export performance.

2.10.4 India's Competitive Position in the International Cardamom Market

It is the relative price of a product that determines a country's competitive position in the international market. An examination of Indian versus Guatemalan cardamom is informative in this regard. Indian cardamom has always been priced higher than Guatemalan cardamom and has never been cheap. By contrast, Guatemalan productivity has been consistently higher than that of India, and in 1985–1986, when India's production peaked at 3272 mt, Guatemala' production was 8845.20 mt, more than 90% of which was exported, compared with India's 70%. The 70% figure was the peak export percentage for India; in Guatemala, exports reached 98% in 1990. Guatemala's higher productivity and price advantage contributed to this disparity.

But Guatemalan cardamom is smaller in size and lower in quality than Indian cardamom. Yet, the price advantage has helped boost the Guatemalan cardamom industry.

2.10.5 Demand and Supply Pattern

Cardamom production is influenced by national and international price movements. The response to price changes occurs every 5–6 years. Although demand is influenced by many economic factors, including the overall economic development of the country, supply is influenced by both economic and noneconomic factors, the latter including agroclimatic, biotic, and abiotic stress factors in the growing region. Hence, it is necessary to consider a multitude of factors in order to forecast the future of cardamom. The type of data available does not permit the development of sophisticated forecast models, which would provide a precise picture of cardamom trade. What data India possesses are historic data on the area under cardamom cultivation, on production, and on prices. Still, suitable predictive models have been identified on the basis of that information, and identifying these models may be seen as only a preliminary step in the development of more precise and dependable subsequent models.

2.10.6 Model Identification

A number of statistical techniques, ranging from simple to highly sophisticated, are available to forecast international trade and price movements. All of them attempt to capture the statistical distributions in the data provided and to present future uncertainty quantitatively. In view of the body of data available and its quality, the following three models have been employed:

1. Simple moving-average models
2. Exponential smoothing models
3. Box–Jenkins models.

To identify the appropriate model, the data have to be first explored.

Exploring the data: Time-series data on area under cultivation, production, and prices for cardamom were plotted in order to identify their specific characteristics and select an appropriate model. The characteristics observed in the time-series data for cardamom can be listed as follows:

1. There is an overall positive trend (i.e., the trend cycle accounted for nearly 80% of the area, production, and prices).
2. The series is nonseasonal in nature that is, it is neither consistently high nor consistently low, and the annual pattern repeats.
3. The series is unsatisfactory in both mean and variance, that is, the mean is too low (area under cultivation and production) or too high (prices) and the variance is too great.

To forecast demand and quantity, an exported Box–Jenkins model has been used with log-transformed data.

Holt's (1957) exponential smoothing model uses a smoothed estimate of the trend, as well as the level, to produce forecasts. The forecasting equation is

$$Y(m) = S_t + mT_t \tag{2.4}$$

The current smoothed level is added to the linearly extended current smoothed trend as the forecast into the indefinite future. We have

$$S_t = \alpha Y_t + (1 - \alpha / \text{ie}(1 - \alpha))(S_{t-1} + T_{t-1}) \tag{2.5}$$

and

$$T_t = \gamma(S_t - S_{t-1}) + (1 - \gamma)T_{t-1} \tag{2.6}$$

where m is the forecast lead time, Y_t denotes the observed value at time t, S_t is the smoothed level at the end of time t, T_t designates the smoothed trend at the end of time t, γ is the smoothing parameter for the trend, and α is the smoothing parameter for the level of the series.

Equation (2.5) shows how the updated value of the smoothed level is computed as the weighted average of new data (the first term) and the best estimate of the new level based on old data (the second term). In much the same way, Eq. (2.6) combines old and new estimates of one period of change of the level, thus defining the current linear (local) trend.

2.10.7 *The Forecast*

Before embarking on the presentation of figures for likely future developments, it needs to be stressed again that the forecast presented is not an accurate exercise based on systematically collected elaborate data. However, the forecast, based on historical data, helps us to understand the overall direction in which the supply (area and production) will move and the price will fluctuate.

Forecasts are produced with upper and lower confidence limits. The upper confidence limit is 97.5% and the lower 2.5%, that is, the actual number calculated should fall inside the confidence band 95% of the time.

2.10.8 *Demand*

The major markets for Indian cardamom are Saudi Arabia, Kuwait, Jordan, United Arab Emirates, Qatar, the erstwhile USSR, and Western Europe. Other important importers include West Germany, Pakistan, the United Kingdom, Japan, and Iran. The highest consumption of cardamom takes place in the Middle East, where it is used in the preparation of that region's traditional drink "Gahwa." According to the United Nations Commission on Trade and Development in Agriculture, the Middle East market accounts for 80% of total world consumption. In Europe and Scandinavian countries, cardamom is used to flavor bread and pastries. Cardamom is imported in raw and ground forms for use in food manufacturing and special blends. Among the producing countries, India consumes the largest quantity of cardamom in the world.

There was a sharp increase in world demand in the late 1970s. Since then, demand has remained more or less stagnant because of declining purchasing power in the Middle East on account of the slump in the oil boom. Although the market has shifted to a preference for lower quality cardamom supplied by Guatemala, the total quantity imported has remained almost unchanged. With the main determinative factors of consumption, namely, population, age, and income, on the increase, demand for cardamom in countries such as Japan has gone up. New uses for cardamom in the food and industrial sectors are triggering accelerated demand in international market, and that demand is growing almost at the rate of global population growth. World demand was estimated at about 9000t in 1985–1986, excluding India's domestic consumption. With this figure as the baseline, demand is projected to grow about 2% per annum, on average, a percentage that is proportional to the growth rate of the world's population. Demand for 2000–2001 was estimated at 12,000t. According to the Spices Board (1990), the total cardamom requirement in India, with an estimated annual per capita consumption of 8.5kg, was estimated at 8700t in 2000 and is projected to escalate to 10,800t by 2010 A.D. As far as India is concerned, the demand for cardamom is increasing more than the population growth. Compared with the 1500t consumed in 1985–1986, domestic consumption went up to more than 6850t in 1997–1998. If the current trend had continued, world demand (including India's) was expected to touch 15,000t by the turn of the century. When actual data are not available, deducting the quantity exported from total production gives approximate consumption in the domestic market. Fitted for the consumption trend in India, the growth equation is

$$\text{In } D_t = 6.7110 + 0.0725T, \quad R_2 = 0.815$$

The estimated growth rate is then calculated to be 7.3% per annum. This growth rate in demand is much more than the growth rates in production. Hence, under the circumstances, it is unlikely that India can regain its earlier position as the world's largest producer and exporter of cardamom, because an increasing percentage of production will go into domestic consumption, leaving very little for export. Although there has been a shift in the consumption pattern in the Middle East countries from high-quality Indian cardamom to cheaper Guatemalan cardamom, there is no reduction in quantity imported. Further, new markets are emerging for cardamom worldwide. While the household consumption sector remains intact, increasing along with increases in population, recent developments in the industrial use of cardamom (in both the food and nonfood sectors) are also expected to push up global demand for the commodity. By the end of 2005, global demand is projected to be 20,000mt.

2.10.9 Projections of Supply

Figures 2.4 and 2.5, respectively, show projections of area under cardamom cultivation and production of cardamom in India, based on a model that was developed for

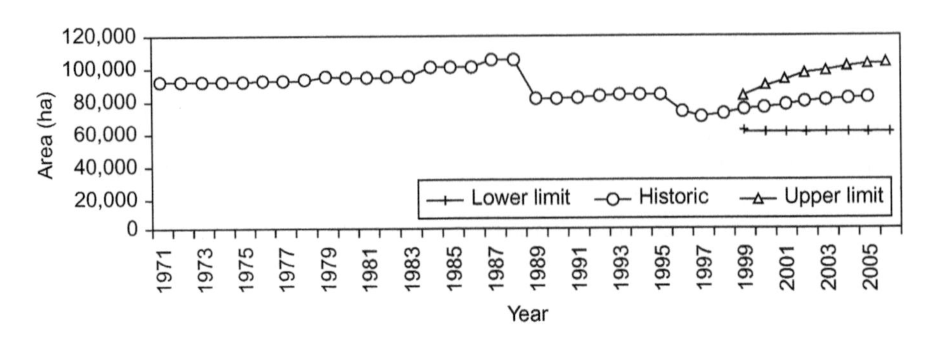

Figure 2.4 Trend in area under cardamom cultivation in India.

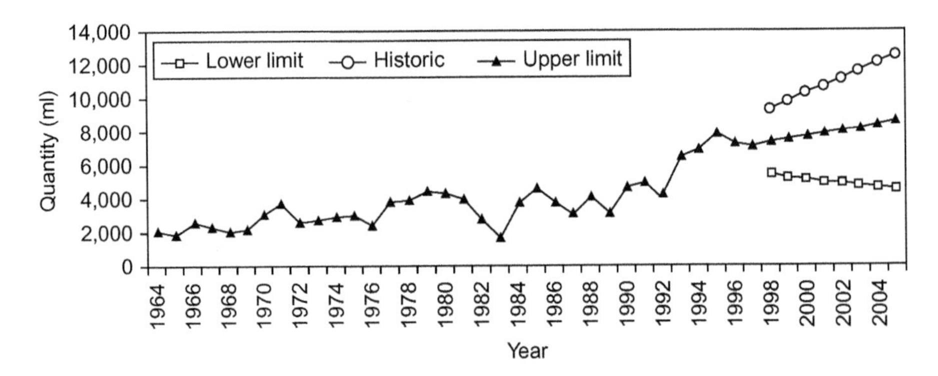

Figure 2.5 Production trend and forecast for cardamom in India.

the purpose. The two figures present both historic and fitted values, along with values forecast beyond 1997. Both area under cultivation and production are expected to grow slowly in the immediate future. The growth in production is expected to be more pronounced than the growth in area, indicating an improvement in yield per unit area. As per the cyclical movements discussed earlier, after the peak achieved in 1995–1996 the 3-year period of decline is already over, and an increasing trend points to the next peak in the cycle. In the early part of the twenty-first century (2000–2001), the expected production level was between 8000 and 10,000 t, and the increase in cultivated area was expected to touch 90,000 ha. The improvement in internal and international prices will spur supply to jump in the usual fashion discussed earlier.

The forecast and actual price movement from 1990 onward is given in Figure 2.6. Since the level of production forecast is insufficient to create enough export surplus, and since reported declining production in Guatemala is already reflected in the form of less supply on the world market by that country during the 1998–1999 crop year, the resulting increase in prices will be favorable to Indian cardamom producers. The prevailing higher market price is expected to continue into the near future, and there is also a possibility that the price will rise enough to cross the Rs 1000 kg^{-1} (US$23

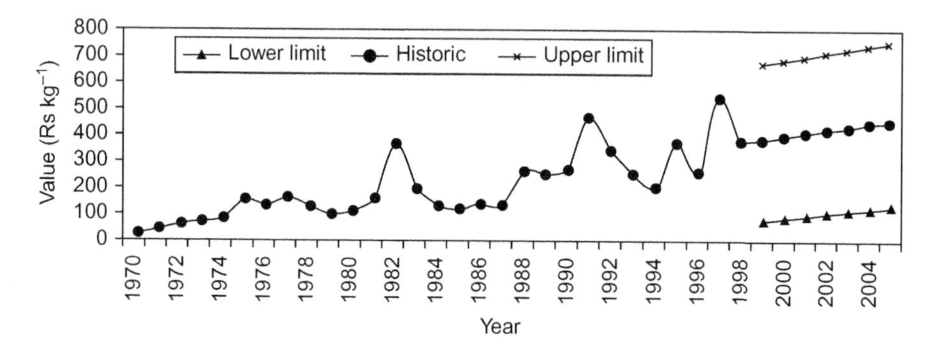

Figure 2.6 Trend in domestic price of cardamom in India.

approximately) mark before declining as per the usual cyclical fluctuations. The decreased availability of exportable surplus will have a direct effect on exports. The standard forecast scenario indicates that the trend prevailing during the last 5 years will continue for the next 5 years, meaning that exports will remain below the 300-t mark. Estimated exports are expected to lie between the upper confidence limit of 97.5% and the actual forecast.

2.10.10 Conclusions

Following are the main findings of this chapter.

1. The supply of cardamom is increasing at a fast rate to meet growing world demand. Much of the increased supply is from Guatemala. While consuming more than half of total world production, India now contributes barely 2% to the world market.
2. Although the increase in production during the 1980s was due mainly to an increase in area under cultivation, the increase in production during the 1990s was due to improvements in productivity. However, the gap that exists between the potential productivity level and the average achieved level of productivity in India indicates the possibility of improving yield levels considerably.
3. On the export front, India has lost most of its traditional markets to Guatemala because of the latter's price advantage. Guatemala derives its price competitiveness mainly from a low production cost and high productivity per unit area.
4. In the changed scenario, Japan is the lone steady and reliable market for Indian cardamom. However, because of the decline in the oil boom and the consequent fall in purchasing power of the Gulf market, Japan's preference has changed from the higher quality afforded by Indian cardamom to the lower prices fetched by Guatemalan cardamom.
5. Cyclical fluctuations in prices have an implicit bearing on the level of supply through farmers' responses. Forecasts indicate that there will be a steady increase in supply (production and yield), and prices are expected either to move up from their present level or to remain as they are now. This future scenario, however, can change dramatically as a result of innovative product development or the diversification of some entirely new application to cardamom or its products. However, imaginative product development programs will have to be initiated if consumption is to be increased. Attractive formulations backed up by catchy advertisements can do wonders in this field.

2.11 Pharmacological Properties of Cardamom

Spices and herbs used to season food often have a mild, broad spectrum of antimicrobial activity. Many crude drugs are used medicinally because of their volatile oil content or other chemical constituents that possess biological activity. Cardamom is very popular as a spice and food additive because of its delicious flavor. The constituents of its volatile oil are responsible for the flavor and fragrance. Cardamom also possesses carminative, stomachic, and antimicrobial actions. These biological activities bring about many advantages to seasoned and prepared foods. Apart from this use, cardamom finds application in the indigenous systems of medicine.

2.11.1 Pharmacological Properties

In the ancient Indian systems of medicine *Ayurveda* and *Siddha,* cardamom finds application as a component of several therapeutic formulations. *Charaka samhita,* the ancient Indian medical text, describes the use of cardamom as an antidote to food poisoning. Cardamom is a constituent of Brahmi rasayana, which is used as a treatment for inflammations. Cardamom is also used as a component of many balms, ointments, and therapeutic oils used to treat cramps, rheumatic pain, inflammations, and so on. In Ayurvedic texts, the properties of cardamom seeds are described as aromatic, acrid, sweet, cooling, stimulant, carminative, diuretic, cardiotonic, and expectorant. Cardamom is used as an ingredient in preparations for the treatment of asthma, bronchitis, hemorrhoids, renal and vesicle calculi, cardiac disorders, anorexia, dyspepsia, gastropathy, debility, and vitiated conditions of *vata* (arthritic pain in the knees and joints). No pharmacological investigations have been carried out to validate the preceding properties. An aqueous extract of cardamom seeds is given to nursing mothers to treat ringworm infection of the children (Aloskar et al., 1992). Roasted cardamom seeds are boiled with betel leaves, and the resulting extract is used to treat indigestion and worm infection. However, such uses of cardamom in the indigenous system of medicine have not been evaluated pharmacologically.

2.11.2 Carminative Action

Both the cardamom seeds and their oil have carminative action. Tincture of cardamom (tinctura cardamomi) is used in many medicinal preparations that have carminative, stomachic, and colic-relieving properties (British Pharmacopoeia, 1993). Tincture of cardamom and compound tincture of cardamom are included as official preparations in the British Pharmaceutical Codex 1963 and the British Pharmacopoeia 1993, as well as in the Chinese, Hungarian, and Japanese Pharmacopoeias. Martindale (1996) also describes preparations of cardamom fruits, cardamom oil, and cineole, the major constituent of cardamom oil, as carminative and flavoring agents. Cineole has been used as a counterirritant in ointments and in dentifrices, and has also been used in nasal preparations. Jain et al. (1994) have shown that cardamom essential oil-containing preparations, such as Brahmi

Table 2.49 Minimum Inhibitory Concentration (%) of Cardamom in Comparison to Other Spice Extracts

Material	pH	BS	Sa	Ec	St	Sm	Pa	Pv	Pm
Cardamom	7	2.0	2.0	<4	<4	<4	<4	<4	<4
	5	0.1	0.5	<4	<4	<4	<4	<4	<4
Cinnamon	7	4.0	2.0	4	4	4	4	2	4
	5	0.5	2.0	4	4	4	4	1	2
Clove	7	1.0	1.0	1.0	1.0	1.0	2.0	1.0	1.0
	5	0.5	2.0	1.0	1.0	1.0	1.0	0.5	0.5
Mace	7	0.2	0.05	<4	<4	<4	<4	<4	<4
	5	0.1	0.5	<4	<4	<4	<4	<4	<4

Source: Hirasa and Takemasa (1998).
Note: BS, *Bacillus subtilis*; Sa, *Staphylococus aureus*; Ec, *Escherichia coli*; St, *Salmonella typhimurium*; Sm, *Salmonella marcescens*; Pa, *Pseudomonas aeruginosa*; Pv, *Proteus vulgaris*; Pm, *Proteus morganii.*

rasayana, suppresses castor oil-induced diarrhea in experimental rats. This activity points to its possible beneficial use in humans as well.

2.11.3 Antimicrobial Activity

The terpenoid constituents in cardamom are responsible for the antifungal and antibacterial effects. Mishra and Dubey (1990) and Mishra et al. (1991) studied the effect of cardamom on *Aspergillus flavus,* the fungus that produces the deadly aflatoxin B1. Mycostatic activity was observed at 400 ppm. Cardamom was found to be as potent as the synthetic antifungals that are commonly used (Hirasa and Takemasa, 1998). The flavor components also showed antibacterial effects against several food-borne microorganisms (Kubo et al, 1991). Another study proved that the growth of *Morgenella morganii* was moderately inhibited by the application of cardamom oil or powder (Shakila et al., 1996). This organism is a potent histamine-producing bacterium growing on stored fish. The minimum inhibitory concentration (MIC percent) of cardamom extracts for bacteria and fungi in comparison with a few other common spices are given in Tables 2.49 and 2.50.

2.11.4 Anticarcinogenic Activity

Banerjee et al. (1994) found that cardamom oil enhances glutathione transferase enzyme and acid-soluble sulfhydryl activities. These compounds mediate the oxidation and detoxification of xenobiotics. Cardamom oil was fed as gavage at $10\,\mu l\,day^{-1}$ for 14 days, and hepatic microsomal enzymes were measured. GST and acid-soluble sulfhydryl were found to be significantly elevated (at a high statistical confidence limit of $p = 0.001$).

Hashim et al. (1994) reported that cardamom oil suppresses DNA adduct formation by aflatoxin B1 in a dose-dependent manner. The effect appeared to be modulated through the action of microsomal enzymes.

Table 2.50 Minimum Inhibitory Concentration (%) for Some Fungi

Material	Sc	Cp	Ck	P sp.	Ao
Cardamom	4.0	4.0	<4	<4	<4
Cinnamon	1.0	1.0	1.0	1.0	1.0
Clove	0.5	0.5	0.5	0.5	0.5
Mace	<4	<4	<4	<4	<4

Source: Hirasa and Takemasa (1998).
Note: Sc, *Saccharomyces cerevisiae*; Cp, *Candida parakrusei*; Ck, *Candida krusei*; P sp., *Penicillium* sp.; Ao, *Aspergillus oryzae*.

2.11.5 Anti-Inflammatory Activity

Yamada (1992) reported that cardamom showed potent complement system-activating properties. Complements represent the humoral arm of the natural immunological host defense mechanism and are essential for survival. Once activated, the complement system kills certain bacteria, protozoa, fungi, and viruses, as well as cells, of a higher organism. Thus, complement activation forms a major part of our natural defense, affording a range of mediators possessing immune inflammatory potency. Jain et al. (1994) found that the drug *Brahmi rasayana* (an Ayurvedic preparation), containing cardamom, cloves, and long pepper, exhibited a dose-dependent anti-inflammatory activity in the case of carrageenan-induced rat paw edema. The drug also inhibited nystatin-induced inflammation in rats. Al-Zuhair et al. (1996) have shown that cardamom oil administered at 175 and 280 $\mu l\,kg^{-1}$ of body weight inhibited the growth of carrageenan-induced paw edema in rats by 69.2% and 86.4%, respectively. The anti-inflammatory activity of cardamom oil is comparable to that of indomethacin (indometacin). El-Tabir et al. (1997) investigated the pharmacological action of cardamom oil on various animal systems, such as the cardiovascular system of rats, nictitating membrane of cats, isolated rabbit jejunum, isolated guinea pig ileum, and frog sciatic nerve. The essential oil (5–20 $\mu l\,kg^{-1}$ IV) decreased the arterial blood pressure and heart rate in rats in a dose-dependent manner. The effects are antagonistic to treatment with cyproheptadine (1 $mg\,kg^{-1}$) for 5 min. Atropine was also antagonistic to cardamom-induced bradycardia. The oil did not have any effect on isolated, perfused rat heart and did not affect electrically induced contractions of the cat nictitating membrane. At concentrations of less than 0.08 $\mu l\,ml^{-1}$, the oil induced contractions of the jejunum, but larger doses relaxed it. A larger dose of the oil was antagonistic to the action of acetylcholine, nicotine, and barium chloride on the rabbit jejunum. At concentrations of 0.01–0.04 $\mu l\,liter^{-1}$, the oil induced contractions of the isolated guinea pig ileum; this effect was suppressed by atropine and cyproheptadine. Exposure of frog sciatic nerve to 0.2–0.4 $\mu l\,liter^{-1}$ cardamom oil suppressed the frog limb withdrawal reflex, exhibiting a local anesthetic effect (El-Tabir et al., 1997). Al-Zuhair et al. (1996) found that cardamom oil exhibited analgesic properties and inhibited spontaneous and acetylcholine-induced movements of rabbit intestine *in vitro* in a dose-dependent manner.

2.11.6 Other Pharmacological Studies

From these pharmacological studies, the beneficial effects of cardamom and its oil were established. Cardamom is not a mere flavoring agent: It has carminative, fungicidal, and bactericidal effects, and it activates the complement system, thereby enhancing the immunobiological defense mechanism of the human body. Two other studies of extracts of cardamom show another aspect of the spice's therapeutic utility. In one of them, Yamahara et al. (1989), using a mouse skin model to examine the dermal penetration of prednisolone, reported that terpineol and acetylterpineol are the active constituents of cardamom extract, facilitating the absorption of medications. In the other study, Huang et al. (1993) used a rabbit skin model and conducted *in vitro* and *in vivo* investigations. They observed that extract of cardamom enhanced penetration into the skin in both situations. Hence, the addition of cardamom extract, terpineol, or terpineol acetate in balms and ointments enhances the absorption of medications through the skin. Terpineol and bornyl acetate together exhibit disinfectant and solvent properties, and hence are used with other volatiles for cough and respiratory disorders. Cineole is an ingredient that is used along with other volatile substances for the treatment of renal and biliary calculi (Martindale, 1996).

Yaw Bin et al. (1999) investigated the effect of cardamom extract on the transdermal delivery of indometacin. After pretreatment with cardamom oil, the permeation of indometacin was significantly enhanced both *in vitro* (rat, rabbit, and human skin) and *in vivo* (rabbit) studies. The indometacin flux decreased as the length of the pretreatment increased. Both natural cardamom oil and a cyclic monoterpene mixture composed of the components of the oil showed similar enhancement of indometacin permeation, indicating that cyclic monoterpenes are the predominant components that alter the barrier property of stratum corneum. This study also showed that the three minor components of cardamom oil (α-pinene 6.5%, β-pinene 4.8%, and α-terpineol 0.4%) had a synergistic effect with 1,8-cineole (eucalyptol) and D-limonene to enhance the permeation of indometacin.

2.11.7 Toxicity

There is no toxicity in the use of cardamom by-products. The main uses of cardamom are as a spice and as a flavorant. When flavor substances are added to food items, no health hazard should arise at the concentrations used, which are small, normally not exceeding 10–20 ppm of the total quantity of the food item. Higher concentrations cannot be used because of the intense odor and taste. Most of the individual components of cardamom oil were studied to assess their toxicological actions on experimental animals. The investigations were conducted under the auspices of an international food safety program. In the series of technical reports published by the joint FAO/WHO Expert Committee on Food Additives, cardamom oil and its chemical constituents were found to have no toxicological effects. In allopathy, cardamom is used only as a carminative in certain medical formulations.

Table 2.51 Antioxidant Activity of Cardamom and a Few Other Selected Spices against Lard (Con. Added 0.02%)

Spice	Ground Spice (*POV, meq kg^{-1})	Petrol Ether, Soluble Fraction (POV meq kg^{-1})	Petrol Ether, Insoluble Fraction (POV meq kg^{-1})
Cardamom	423.8	711.8	458.6
Black pepper	364.5	31.3	486.5
Cinnamon	324.0	36.4	448.9
Clove	22.6	33.8	12.8
Turmeric	399.3	430.6	293.7
Nutmeg	205.6	31.1	66.7
Ginger	40.9	240.5	35.5

Source: Hirasa and Takemasa (1998).
POV, peroxide value, which is negatively correlated with the antioxidant property.

2.11.7.1 Antioxidant Function

Cardamom exerts only a mild antioxidant function and hence is not effective in preventing food spoilage. The antioxidant function of cardamom in comparison with a few selected major spices is shown in Table 2.51.

2.11.7.2 Pharmaceutical Products

Blancow's (1972) Martindale details the following preparations that use cardamom:

1. Aromatic cardamom tincture (BPC, tincture cardamom aromatic, carminative tincture). This is prepared in the proportion of 1 part cardamom seed to about 15 parts of strong ginger tincture, alcohol (90%) and oil caraway, and cinnamon and clove.
2. Compound cardamom tincture (BP, tincture cardamom compound). This is prepared from cardamom, cochineal, and glycerin by percolation with 60% alcohol. Often, the tincture is decolorized by alkaloidal salts, bismuth carbonate, calcium ions, and sodium bromide.
3. Compound cardamom tincture (USNF). This is prepared by macerating 2 g of cardamom seed, 2.5 g of cinnamon, and 1.2 g of caraway with 5 ml of glycerin and is diluted with 100 ml of alcohol.

2.11.8 Other Properties

2.11.8.1 Effect on Stored-Product Insect Pests

Huang et al. (2000) investigated the contact and fumigant toxicities and antifeedant activity of cardamom oil on two stored-product insect pests: *Sitophilus zeamais* and *Trilobium castaneum*. Topical application was employed for contact toxicity studies, and filter paper impregnation was employed to test fumigant action. The adults of both *Sitophilus zeamais* and *Trilobium cataneum* were equally susceptible to the

Table 2.52 Mortality Rate of *Dermatophagoides farinae* Following the Use of Cardamom Essential Oil in Comparison with Other Selected Spice Oils

Essential Oil of	**Mortality Rate (%)**
Cardamom	4.7
Clove	97.3
Mace	0.5
Nutmeg	–
White pepper	0.1
Anise	56.5
Garlic	72.8

contact toxicity of the oil at the lethal dose (LD) 50 values of 56 and 52 $\mu g\,mg^{-1}$, respectively. For fumigant toxicity, the adults of the former were more than twice as susceptible than the adults of the latter at both LD 50 and LD 95. Twelve-day-old larvae of *Trilobium castaneum* were more tolerant than the adults to the contact toxicity of the oil. The susceptibility of the larvae to contact toxicity increased with age. Cardamom oil applied to filter paper at concentrations ranging from 1.04 to 2.34 $mg\,cm^{-2}$ significantly reduced the hatching of *Trilobium castaneum* eggs and the subsequent survival rate of the larvae. Adult emergence was also drastically reduced by cardamom oil. When applied to rice or wheat, cardamom oil totally suppressed F1 progeny production of both insects at a low concentration of 0.0053 ppm. Cardamom oil did not have any growth-inhibiting or feeding-deterrence effects on either adults or larvae of *Tribolium castaneum.* However, the oil significantly reduced all the nutritional indexes of the adults of *Sitophilus zeamais* (Huang et al., 2000).

2.11.8.2 Effect of Cardamom on House Dust Mites

It is generally held that 70% of the allergies caused in humans show positive antigenic reaction for house dust mites *(Dermatophagoides farinae* and *Dermatophagoides pteronyssinus).* Yuri and Izumi (1994) studied the effect of essential oil of spices on *Dermatophagoides farinae* and reported that some of the spice essential oils were effective against this mite (Table 2.52). They used a concentration of 80 $\mu g\,cm^{-2}$ on filter papers and counted the mortality rate after 24 h. The essential oil of cardamom exerted only a low mortality rate.

The antihelminthic activity of spice extracts was studied by Tsuda and Kiuchi (1989), who found that methanol extract of cardamom exhibited an antihelminthic effect on dog roundworm.

2.11.8.3 Cardamom in Traditional Systems of Medicine

In the ancient Indian systems of medicine, namely, *Ayurveda, Siddha*, and *Unani*, cardamom is used as a powerful aromatic stimulant, carminative, stomachic, and diuretic. It also checks nausea and vomiting and is reported to be a cardiac stimulant. Powdered cardamom seed mixed with ground ginger, cloves, and caraway is helpful in combating digestive ailments. Tincture of cardamom is used chiefly in medicines for windiness or as a stomachic. A good nasal application is prepared from extracts of cardamom, neem, and myrobalan, along with animal fat and camphor. Cardamom seeds are chewed to prevent foul breath, indigestion, nausea and vomiting due to morning sickness in pregnancy, excessive watering in the mouth (pyrosis), and so on; gargling with infusion of cardamom and cinnamon cures pharyngitis, sore throat, and hoarseness of voice during infective influenza. The use of cardamom as a daily gargle protects against influenza infection (Pruthy, 1979).

Powdered cardamom seeds boiled in water with tea powder impart a pleasant aroma to tea, and this is a highly popular practice in the Arab world, and the concoction can be used as a medicine for scanty urination, diarrhea, dysentery, heart palpitations, exhaustion due to excessive work, depression, and so on (Singh and Singh, 1996). Eating a cardamom capsule daily with a tablespoon of honey improves the eyesight, strengthens the nervous system, and keeps one healthy. Some people believe that excessive use of cardamom causes impotency.

One of the main properties of cardamom is its effect on dermatological disorders. Medicated cardamom oil and cardamom powder can retard the spread of different hypopigmentation on the face (Nair and Unnikrishnan, 1997). Cardamom powder is a safe emetic that can be used in bronchial asthma patients when an excess of sputum is present in the lungs. Further, cardamom powder is a very good cough suppressant. Cardamom finds a place in the formulation of lozenges for the management of the common cold and associated symptoms (Nair and Unnikrishnan, 1997). In the form of tincture or powder, cardamom is a frequent adjunct to other stimulants, bitters, and purgatives. A decoction of cardamom with its pericarp mixed with jaggery is a popular home remedy that relieves giddiness caused by biliousness. A mixture of cardamom seeds, ginger, clove, and caraway in powder form in equal parts is a good stomachic for atonic dyspepsia. A powder made of equal parts of parched cardamom seed, aniseed, and caraway seed is a good digestive. Cardamom is used in as many as 24 important preparations in *Ayurveda* in the form of decoctions, oils, and powders, as well as in medicated fermented beverages such as *Arishta* and *Aasava* (Sahadevan, 1965a). Along with saffron *(Crocus sativus)*, galengal *(Alpinia galanga)*, and "nealgor of the corryrium" (Ayurvedic preparation), cardamom seeds cure cataract and other eye ailments, such as tumors in eyelids, fleshy growths, and ophthalmia.

Cardamom fruit is an emmenagogue, the only spice that qualifies as having this property. Cardamom, cinnamon, tejpatta *(Cinnamomum tamala)*, and ironwood tree *(Mesua ferrea)* taken together constitute a mixture known as *Chaturjata*. The mixture is used to flavor electuaries to promote their actions (Warrier, 1989). Cardamom is also a component of medicinal preparations used to cure skin diseases, poisons, colds, and inflammation. Preparations such as *Eladigana* (Ela, cardamom) are a common cure for *vata* (arthritis) and *kapha* (congestion) diseases and are used to counter the effects of poison, to enhance one's complexion, and to cure itching. Cardamom is also an ingredient of mixtures that improve digestion, cure vomiting, suppress coughs, and so on. Cardamom stimulates diuresis, particularly in the case of snake-bite. A group of medicines known as *Ariyaru kashayam* (six grains) for the skin diseases of children contains cardamom.

The Burmese (Myanmar) traditional medicine formulation O2 (tmf-O2) consists of four basic plant ingredients, one of which is cardamom. (The others are *Anacyclus pyrethrum, Glycyrrhiza glabra*, and *Syzygium aromaticum.*)

2.11.9 Cardamom as a Spice

Cardamom for culinary use: The major use of cardamom worldwide is in whole or ground form for culinary purposes. In Asian kitchens, cardamom is an important ingredient used to prepare a variety of dishes, such as spiced rice, vegetables, and meat preparations. Cardamom can add a lingering sparkle to many dishes, both traditional and modern. International trade in cardamom is dependent, however, on the demand created by specialized applications that have evolved in two distinct markets: the Arab market of the Middle East and the Scandinavian market.

Cardamom provides a warm, comforting feeling and is responsible for the unique and exotic flavor of Bedouin coffee. In the Middle East, religious ceremonies, social functions such as marriages, and celebrations are incomplete without the use of cardamom and the famous Arab coffee *Gahwa* (cardamom-flavored coffee). It is believed by the Arabs that *Gahwa* cools the body heat in a region where extreme heat is a daily feature of life. It is also believed that *Gahwa* aids digestion and acts as an aphrodisiac. Cardamom is used in indigenous Arab cooking. The Arabs have adopted a number of Indian delicacies, especially meat-based ones, and the Arab *Biryani* (a rice-based dish with different kinds of meat, principally lamb or chicken) is incomplete without the sprinkling of cardamom capsules. In the Islamic Republic of Iran, cardamom is used in making confectionery, bakery, and meat preparations to add flavor and aroma to those products. Invariably, cardamom is found in the spice chest of an Indian kitchen. The Indian housewife uses this unique spice in a variety of vegetable and meat dishes, including those enhanced by the flavoring of sweets, and in rice porridge (known locally as *Payasams* in southern India and *Kheer* in northern India).

In European countries and North America, cardamom is used mainly in ground form by food industries as an ingredient in curry powder, some sausage products, fruit cups, green pea soups, curry-flavored soups, spice dishes, rice Danish pastry, buns, breads, rolls, cookies, desserts, coffee cakes, orange salad, jellies, baked apple coffee, honey pickles, pickled herring, canned fish, and, to a small extent, tobacco and cigarettes, as a flavor enhancer. Cardamom cola, instant *Gahwa,* carbonated *Gahwa,* biscuits, Spanish pastries, toffees, chewing gum, and so on, are other products in which cardamom is an ingredient. Among the new products developed recently are various breakfast foods that use encapsulated cardamom oil. Cardamom is also used in spiced wine and to flavor custard (by steeping crushed cardamom seeds in hot milk). Arabs use cardamom in coffee; Americans use it in baked foods; Russians add it to pastries, cakes, and confectionery to impart unique flavors and aromas; Japanese use it in curry, ham, and sausage; and Germans use it in curry

Table 2.53 Nutritional Composition of Cardamom per 100 g

Ingredients	USDA Handbook 8–21[a]	ASDTA[b]
Water (g)	8.28	8.0
Food energy (Kcal)	311.00	360.0
Protein (g)	10.76	10.0
Fat (g)	6.70	2.9
Carbohydrates (g)	68.47	74.2
Ash (g)	5.78	4.7
Calcium (g)	0.383	0.3
Phosphorus (mg)	178.00	210.0
Sodium (mg)	18.00	10.0
Potassium (mg)	1119.00	1200.0
Iron (mg)	13.97	11.6
Thiamine (mg)	0.198	0.18
Riboflavin (mg)	0.182	0.23
Niacin (mg)	1.102	2.33
Ascorbic acid (mg)	–	ND
Vitamin A activity (RE)	Traces	ND

Source: Stobart (1982).
ND, not detected.
[a]*Composition of Foods: Spices and Herbs.* USDA Agricultural Handbook, 8–21, January 1977.
[b]*The nutritional composition of spices*, ASTA Research Committee, February 1977.

powders, sausages, and processed meat. The list goes on and on: The spice is used in countless food items.

In Scandinavia, cardamom is widely used in bakery products. Ground cardamom is mixed with flour to add flavor to most baked products, and it adds an exotic taste to apple pie (Rosengarten, 1973). In Sweden, cardamom use is most popular with baked foods, the per capita consumption of which is about 60% greater than that in the United States. Ground cardamom is also used to flavor hamburgers and meat loaf.

Indian cardamom is low in fat and high in protein, iron, and vitamins B and C (Pruthy, 1993). Table 2.53 gives the nutritional value of cardamom. In India, it is used as a masticatory and in flavoring culinary preparations. In several cities of India, especially in the North Indian belt, cardamom is used in the preparation of all kinds of puddings, which are mandatory in both social functions, such as marriages, and religious festivities. Cardamom is also used to lace tea, with or without lime; the drink is a popular and refreshing one in North India (Philip, 1989). On many occasions, cardamom seeds are offered to be chewed after sumptuous marriage feasts.

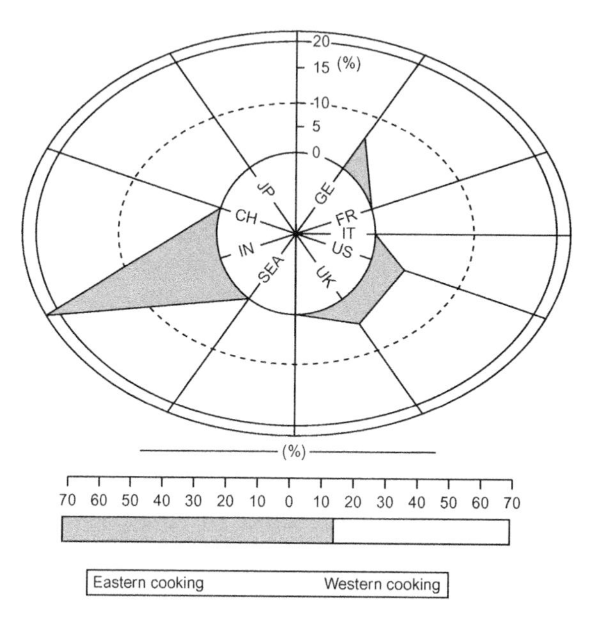

Figure 2.7 Suitability pattern for cardamom.
Source: Hirasa and Takemasa, 1998.

Also, cardamom-flavored hot water is provided in many North Indian hotels. In India, a variety of cardamom-flavored products, ranging from biscuits, to cheese, to milk-based drinks, and so on, are lately being marketed. Finally, cardamom is used to make garlands for special occasions to greet dignitaries both in India and in the Arab world.

2.11.9.1 The Pattern of Suitability for Cardamom

Figure 2.7 depicts the suitability pattern for cardamom. Cardamom is most suitable in Indian cooking and does not exhibit above-average suitability for any other national cuisines. In fact, cardamom is totally absent from the cuisines of most European and American countries. Hirasa and Takemasa (1998) indicated that cardamom is more useful in meat, milk, and fruit preparations. It is also more useful for simmered, baked, fried, deep-fried, and pickled food and less suitable for steamed food. The spice is used in the preparation of food mainly for the following purposes:

1. To flavor the dish directly
2. To mask undesirable flavor and to deodorize
3. To add color to the food
4. To add pungency.

Since a single spice cannot contribute all qualities to a meal, a combination of spices is often used in the preparation of food. Cardamom is used mainly for its direct effect of imparting flavor.

Cardamom is used for any of the above-mentioned purpose in blended spice mixtures, such as curry powder, *Garam masala* (a strong mixture of several spices, including cardamom to impart aroma and pungency to Indian cuisine), mixed pickles, and so on. Masking undesirable flavor odors is important in the use of spices. According to the Weber–Fechner law, the strength of an odor perceived by the sense of smell is proportional to the logarithm of the concentration of the smelled compounds. In other words, the strength of the sensation perceived by the five senses is proportional to the logarithm of the actual strength of the stimuli. Thus, even if 99% of the total compounds smelled are eliminated chemically, the strength of the sensation perceived is reduced by only 66% (Hirasa and Takemasa, 1998). Hence, it is more effective and easier to mask the remaining 1% through aromatic spice. Tokita et al. (1984) investigated the deodorizing efficiency of various spices. The deodorizing rate (measured by the percentage of methyl mercaptan (500 mg) captured by a methanol extract of the spice) of cardamom is low (9%) compared with that of other spices (99% for thyme, 97% for rosemary, 90% for mint, 79% for clove, 30% for black pepper). Ito et al. (1962) calculated the deodorizing points for major spices in masking mutton odor. Cardamom showed a deodorizing point of 30, as against 600 for black pepper, 90 for ginger, 50 for cinnamon, 25 for celery, 23 for garlic, 5 for coriander, 4 for caraway, 3 for clove, 2.5 for thyme, and 0.7 for sage. Thus, cardamom is more effective than many other spices in masking certain odors, although its masking capacity is poor compared with that of some other flavors.

Desrosier (1978) investigated the relative flavor intensities (RFIs) of various spices, and cardamom was found to have an RFI of 125, compared with 200 for turmeric, 260 for curry powder blend, 300 for celery, 400 for cinnamon, 450 for black pepper, 475 for ginger, 600 for clove, 900 for cayenne pepper, and 1000 for fresh red pepper.

2.11.9.2 Spice Blend and Garam Masala

Ready-to-use spice formulations are available in the market under many brand names, either as dry powder or as soluble seasonings (spice extracts on salt or dextrose carriers). Most common spice blends are curry powders, pickling spice mixes, fish or meat *masala* mixes, and *Garam masala* (see shortly), and all are available under a variety of brand names.

Curry powders are used most extensively in Indian cuisine, and there are virtually hundreds of them in the market, for many specific needs. The basic ingredients in most brands are coriander, cumin, turmeric added for color, and red chilies for pungency. Other spices, such as black pepper, cardamom, and cinnamon, are added to enhance the flavor of the curry powder and to suit the Indian palate. In many such formulations, cardamom is used to impart a specific flavor, but only in small quantities. The relative proportions of cardamom in typical curry powder formulations are given in Table 2.54. Formula No. 2 is considered a mild curry, formula No. 3 a sweet curry; formula No. 4 is a hot type, and formula No. 5 a very hot, pungent, Indian-style curry suited for use in the more tropical regions of India. The US federal specification for curry powder is shown in Table 2.55.

Table 2.54 Formulations for Typical Curry Powder Blends

Freshly Ground Spices	US Standard Formula No. 1[a](%)	General-purpose Curry Formulas			
		No. 2 (%)	No. 3 (%)	No. 4 (%)	No. 5 (%)
Coriander	32	37	40	35	25
Turmeric, Madras	38	10	10	25	25
Fenugreek	10	0	0	7	5
Cinnamon	7	2	10	0	0
Cumin	5	2	0	15	25
Cardamom	2	4	5	0	5
Ginger, Cochin	3	2	5	5	5
Pepper, white	3	5	15	5	0
Poppy seed	0	35	0	0	0
Clove	0	2	3	0	0
Cayenne pepper	0	1	1	5	0
Bay leaf	0	0	5	0	0
Chili, hot red pepper	0	0	0	0	5
Allspice	0	0	3	0	0
Mustard seed	0	0	0	3	5
Dried lemon peel	0	0	3	0	0
	100	100	100	100	100

[a] From the US Military Specification Mil-C-35042A, December 30, 1964.
Source: Farrell (1985)

Table 2.55 Federal Specification ES-S-631 J for Curry Powder

Ingredient	Limit (%)
Turmeric	37.0–39.0
Coriander	31.0–33.0
Fenugreek	9.0–11.0
Cinnamon	Not < than 7.0
Cumin	Not < than 5.0
Black pepper	Not < than 3.0
Ginger	Not < than 3.0
Cardamom	Not < than 32.0

Source: Tainter and Grenis (1993).

Garam masala is a blend of spices having an approximate composition of the following ingredients: cumin seeds, 1/4 oz; cardamom, 1/2 oz; black pepper, 1/2 oz; cinnamon, 1/4 oz; and clove, 1/4 oz. Other spices used in the preparation of *Garam masala* add unique tastes for various dishes, but cardamom is the major component.

Kalra et al. (1991) and Premavalli et al. (2000) analyzed the composition of *Garam masala* brands available in the Indian market. Kalra et al. found 11 spices and Premavalli et al. found 9–11. The type of ingredients and the proportions in which they are used is a trade secret. Premavalli et al. found that 27 spices in all were used in the preparation of *Garam masala* of which 5 (coriander, cumin, black pepper, clove, and cardamom) are common to all brands. Cardamom imparts a special flavor to the *Garam masala* mix and its use is mainly as a taste and flavor enhancer.

2.11.9.3 Cardamom Oil, Oleoresin, and Soluble Cardamom

Cardamom oil is colorless or pale yellow with an aromatic pungent odor and taste. It is produced in small quantities in some Western spice-importing countries and in India, Guatemala, and Sri Lanka. The essential oil finds its main application in the flavoring of processed foods, but it is also used in certain liquid products, such as cordials, bitters, and liquors, and occasionally in perfumery. The perfumery and cosmetic industries employ the oils of several spices, including cardamom, in the blending of volatile and fixed oils to make dozens of alluring perfumes, far superior in quality than that of the crude scents of the ancients (Bhandari, 1989).

Oleoresin of cardamom is produced in some spice-producing Western countries and in India, and has applications similar to those of the essential oil in flavoring processed foods, but is less extensively used. Both the essential oil and oleoresin tend to develop "off flavors" when exposed to air for prolonged periods, and their usage is generally confined to meat products and other products with a short shelf life, such as sausages. Cardamom oil and oleroresins are used in the processed foods industry, mostly as a soluble spice mixed with a carrier such as common salt or dextrose. It is easy to use soluble spices, as they are a dry, free-flowing powder compared with liquid essential oil or oleoresin. However, much care is needed when replacing ground spice with oil, oleoresin, or soluble spice, because such products often do not represent the freshly ground spice in its richness of flavor, owing to the loss of some delicate components. Much investigation needs to be done to match the flavor of oleoresin and oil with that of the ground spice. The replacement ratios of cardamom in comparison to other spices are given in Table 2.56.

The Central Food Technological Research Institute in Mysore, Karnataka State, India, has carried out innovative research-and-development efforts to diversify cardamom uses in order to widen the domestic and export markets for cardamom products. Cardamom flavor has been encapsulated with the use of an innovative technology. Encapsulated spices possess unique features, such as a free-flowing nature, uniform flavor strength, and convenience of use. The flavor that is encapsulated is released instantly on contact with water (Pruthy, 1993). Cardamom cola (a fizzy drink with cardamom flavor), a cardamom-flavored *flan* mix (an instant dessert mix with cardamom flavor), cardamom tea, cardamom coffee, cardamom milk, dry cardamom powder for a soft-drink mix, and an instant *Pongal* (a favorite delicacy of Tamil Nadu) mix flavored with cardamom are some other products developed by the Institute (Pruthy, 1993). Innovative product development programs can help diversify cardamom use, leading to the creation of new dishes and food items.

Table 2.56 Replacement Ratios of Cardamom in Comparison to Other Spices

	Replace 1# of Ground Spice With		
Spice	**# Oil**	**# Oleoresin**	**Remarks**
Allspice	0.020	0.035	
Anise	0.020	0.050	
Basil	0.005	0.050	
Cardamom	0.030	0.015	
Caraway	0.010	0.050	
Celery	0.010	0.100	
Cinnamon	0.025	0.025	Oil use based on volatile oil
Clove	0.140	0.050	Stem, leaf, or bud oil can be used
Coriander	0.003	0.070	
Cumin	0.020	0.040	
Dill seed	0.020	0.050	
Fennel	0.010	0.050	
Ginger	0.015	0.035	
Mace	0.140	0.070	Nutmeg oil should be used
Marjoram	0.008	0.050	
Nutmeg	0.060	0.080	Spice and oil must be from the same origin
Oregano	0.015	0.040	
Black pepper	0.015	0.050	Oil does not provide piperine bite
Rosemary	0.008	0.040	Oleoresin is deflavored for antioxidant preparation
Sage	0.010	0.050	
Savory	0.005	0.065	
Tarragon	0.002	–	

Source: Tainter and Grenis (1993).

2.11.10 Conclusions

Cardamom has been used since ancient times as a flavoring agent and also as a component in many indigenous medicines of India. Cardamom-flavored tea and coffee form part of the daily routine of people in the Middle East.

Chewing cardamom after a meal is a habit with a large number of people, especially in northern India. Cardamom also is used to flavor oral formulations of many medicines, bakery products, and milk. Cardamom is unique because of its delicate blend of aromas. The potential uses of this great spice have not been fully exploited.

Table 2.57 Cardamom Production (t) in India and Guatemala

Period	India	Guatemala
1990–1991	4750	11,500
1991–1992	5000	11,120
1992–1993	4250	13,500
1993–1994	6600	13,500
1994–1995	7000	14,200
1995–1996	7900	15,300
1996–1997	6625	17,000
1997–1998	7900	15,000
1998–1999	7170	13,000
1999–2000	9330	10,000
2000–2001	10,480	11,800
2001–2002	11,365	13,500

Source: India, estimates by Spices Board; Guatemala, United Nations Food and Agriculture Organization, Rome, Embassy of India in Mexico.

Cardamom is important only in Indian and South Asian cooking, not in continental, American, or Japanese cuisine, where it has only minimal influence. The use of eye-catching advertisements to promote the diversification of cardamom-based products can create fresh demand for this unique spice. The reported property of cardamom extract in enhancing the percutaneous absorption of medications can be made use of in the preparation of skin ointments and balms, as well as in oral formulations. Innovative technologies are needed to enhance global demand for cardamom. Product development, including the development of novel cardamom-flavored dishes, would be an important component of this drive.

2.12 A Peek into the Future of Cardamom

It was in the early 1930s that cardamom spread to Guatemala from its original home in India. India continued to be the largest producer until the 1980s. Now the situation is reversed, and it is Guatemala that leads in production and export. During the last two decades, while Indian cardamom showed fluctuating trends in production and export, Guatemala forged ahead in production, productivity, and export. For the sake of comparison, the scales of production and export from India and Guatemala are given in Tables 2.57 and 2.58, respectively.

World production of cardamom during 1980–1981 was about 10,250t, and after a decade, in 1990–1991, it rose to 16,000t—a 56% jump. A decade later, in

Table 2.58 Cardamom Export (t) from India and Guatemala

Period	India	Guatemala
1990–1991	400	11,114
1991–1992	544	13,163
1992–1993	190	13,000
1993–1994	387	13,000
1994–1995	257	14,000
1995–1996	527	15,
1996–1997	226	014,50
1997–1998	297	014,50
1998–1999	476	012,00
1999–2000	646	8536
2000–2001	1100	D.N.A

Source: India, DGCI & S, Kolkata/shipping bills/exporters returns. Guatemala: up to 1991–1992, Banco de Guatemala. From 1992 to 1993: estimates based on past trends. D.N.A., data not available.

2000–2001, production reached 22,800 t, an increase of 42%. The next year, in 2001–2002, cardamom production peaked at 25,800, the highest level ever recorded and a further jump by 10%. In other words, during the last two decades of the past century, cardamom production increased by 144%, an average increase of more than 7% per annum. India was the largest exporter until 1980–1981, when its supremacy began to be challenged by Guatemala, especially during the last decade of the past century. Practically with no domestic demand, Guatemala exported the entire quantity it produced. By contrast, cardamom finds a ready domestic market in India. Owing to a well-developed internal market, domestic prices are high—often higher than international prices—so that export from India frequently is not possible because of the comparatively low international price influenced by the steady supply of cardamom from Guatemala. Table 2.58 gives data on exports.

India and Saudi Arabia are the largest cardamom consumers in the world. Both countries, put together, consume more than 50% of total world production. Because Indians enjoy self-sufficiency in cardamom, import is not encouraged by the government. Yet, cardamom from Guatemala comes through the Nepalese border into India and is posing a threat to India's production, as this import, although clandestine, is price competitive, even though the cardamom from Guatemala is of lower quality. The government of India has been unable to check this economic threat. Table 2.59 gives a comparative picture of cardamom consumption by India and Saudi Arabia.

Saudi Arabia imports its entire cardamom requirement from Guatemala. Other important importing countries are Kuwait, Jordan, Qatar, United Arab Emirates, Japan, Singapore, Russia, the United Kingdom, Germany, and the Scandinavian countries of Sweden, Norway, Finland, and Denmark. Saudi Arabia, where cardamom is used

Table 2.59 Cardamom Consumption (t) by India and Saudi Arabia

Period	India	Saudi Arabia
1990	4350	5272
1991	4456	6639
1992	4060	6000
1993	6213	3853
1994	6743	2709
1995	7373	7488
1996	6399	8524
1997	7603	7603
1998	6694	6569
1999	8684	6249
2000	9380	6628

Source: Spices Board of India; UN Statistics: 1990–2000.
Note: Based on import statistics.

extensively in the preparation of the traditional drink *Gahwa*, is the world's highest consumer of the spice. *Gahwa* is also popular in Kuwait, Muscat, and Qatar. It is reported that *Gawha* normally contains 30% cardamom and the balance is made up of coffee powder. Often, however, the ratio of cardamom to coffee powder can be 50:50 or even 60:40. It is mostly the elderly and conservative Arabs in the Middle East who enjoy *Gawha* (Sahadevan, 1965b). Cardamom tea is also popular in the Middle East, as well as in India (Anon., 1952b). In India, cardamom is consumed not only in households, but also in businesses and research institutions. A survey conducted by the Spices Board of India indicates that cardamom finds manifold applications (Anon., 1977a). The main reason for the choice of cardamom is its cool, refreshing aroma with a pleasant, sweet taste. Average household consumption of cardamom in India in both urban and rural areas is about 35 kg year^{-1}. Household consumption is estimated to grow at a rate of 3.7% per annum, to attain a level of 6150 t in the early part of the twenty-first century (George and Johan, 1998). Although the total consumption is less, businesses are the bulk consumers of cardamom in India. Their preparations comprise *Pan masala* (a betel nut-based chewing mixture that is very popular in northern India and Pakistan), other *Masala* products, herbal medicines, tobacco products, biscuits and similar items, and cardamom oil. Demand from the industry was around 2050 t in 2000, at a growth rate of 15% per annum.

Chief among the Indian businesses that consume cardamom are hotels, restaurants, bakeries, sweetmeat shops, and so on. Demand from this sector was increasing at the rate of 10% per annum and reached 1250 t in the year 2000. A number of measures have been taken in India to widen the demand base. The Regional Research Laboratory, Trivandrum, Kerala State, and the Central Food Technological Research

Institute, Mysore, Karnataka State, have initiated studies to develop new cardamom-based products. The Indian Institute of Nutrition, Hyderabad, has conducted an investigation into the nutritional and medicinal values of cardamom. Ayurveda College, Trivandrum, has carried out a study on the use of cardamom to develop *Ayurvedic* (*Ayurveda*, the ancient Indian system of medicine) medicines for different common ailments. The Arya Vaidya Sala, the premier research and medical care institute in Kottakkal, Kerala State, has done pioneering work on the usefulness of cardamom-based soaps for good skin care (George and John, 1998).

The Spices Board of India, which comes under the administrative control of the Ministry of Commerce, Government of India, has been in contact with manufacturers of various food products to promote cardamom use. Some manufacturers have begun to use cardamom flavor in their products, while many have conducted tests to assess the suitability of cardamom oil as a flavoring agent in their manufactured products. As a result of all these efforts, a number of end products that use cardamom flavor, such as cardamom-flavored biscuits, toffee, flan, tea, and coffee powder, as well as cardamom-based concentrates, have been launched in the market (George and John, 1998).

2.12.1 Potential Applications

The future of any commodity depends on both its present and its potential uses. Since cardamom is a weak flavorant, it does not have adequate strength to displace a strong flavorant such as vanilla. But its use as a breath freshener is widespread in India, where it is preferred by many in place of commercial chewing gum. Although cardamom seed has no chewing gum properties, because of its mild, yet exciting, taste and acceptable smell, the habit of chewing cardamom is spreading in many parts of India. Promotional efforts by the Spices Board of India have also tapped the potential of cardamom as a substitute for smoking cigarettes and *Beedi* (the Indian version of the cigarette into which tobacco is stuffed and rolled into ripened and cured leaves from the *Kendu* tree, a perennial tree growing in eastern India) In cola drinks, the aroma of cardamom is highly acceptable, and the testing of a consumer preference for cardamom-flavored cola in many cities in India has been very positive (Anon., 1996).

The medicinal value of cardamom has not been studied fully in any country. Only little work has been done in India. *Ayurveda* mentions the use of cardamom, and the spice has been found effective as a carminative, a body massage oil, and a suppressant of colds and coughs, but more research needs to be carried out to understand the full potential of cardamom (Anon., 1952a, Sahadevan, 1965a).

2.12.2 Future Outlook

2.12.2.1 Research and Development

The most critical of the research-and-development efforts in cardamom production is aimed at widening and enhancing the genetic potential of the existing germplasms. Supportive crop management technology also plays a vital role. Unless a large increase in productivity is attained, Indian cardamom cannot be competitive in world markets (Anon., 1988). Fertility management of cardamom-growing soils is a very important component of this task. As of now, prescriptive soil management principles, especially with regard to cardamom nutrition, revolve around classic textbook knowledge, according to which empirical fertilizer recommendations generated from microplots are extrapolated to large-scale field conditions. These experimental microplots are nothing but artifacts, and many of the recommendations that emanate from studies involving them turn out to be quite off the mark when applied to large-scale plantations, even failing to reproduce quantifiable results. This often shatters the farmers' confidence. A significant departure is being made following my own efforts, which have led to the development of an entirely new approach to soil testing and fertilizer management based on nutrient buffering. The concept is now universally known as "The Nutrient Buffer Power Concept." A detailed discussion of the concept and its relevance to cardamom nutrition, especially with regard to potassium, which is required by cardamom in large quantities for high productivity, is given in Section 2.5 of this chapter. In India, in the state of Kerala, where cardamom is grown to its largest extent in the country, it has been observed that fertilizer input can be significantly reduced by taking into consideration the buffer power of the nutrient under question. Experimental results of Nair et al. (1997) show that, in contrast to the routine soil-testing and fertilizer recommendations, the addition of potassium fertilizer, a crucial input in cardamom production, can be substantially reduced by following the buffer power concept (Nair, 2002). There is an urgent need to extend the concept to other nutrients as well, possibly phosphorus and nitrogen, because these are also crucial in cardamom production. Of the two, phosphorus is more important in view of the fact that the soils of Kerala State, the home of cardamom, are lateritic and much of the applied phosphatic fertilizer, based on routine soil testing, is rendered unavailable to the crop. It is in this context that the buffer power concept holds out much promise in cardamom farming. However, it must be emphasized that the success of a new approach rests to a great extent with the ingenuity of those applying it to suit the demands of a new situation. This principle is no exception to making "The Nutrient Buffer Power Concept" succeed with a perennial crop such as cardamom, as has been the case with other crops, such as black pepper, maize, rye, and white clover (Nair et al., 2002). The fact that cardamom is a perennial crop makes it all the more important because, unlike the situation with an annual crop, to which a midcourse correction can be effected the next season, in a perennial crop such as cardamom the fertilizer regime has to be precise right from the beginning because cardamom grows upward of 25 years. Unlike routine soil testing, the new approach calls for an accurate determination of the buffer power of the nutrient in question at the very start of the fertilizer regime. Once this is accomplished, the buffer power factor can be incorporated into the computations with the routine soil test data, and accurate fertilizer recommendations can be made on the basis of the new information obtained. Of course, this implies that, in addition to obtaining routine soil test data, one needs to know the buffer power. I have obtained very encouraging results with the new concept in cardamom production in the state of Kerala, India, with regard to potassium fertilization. Hopefully, the new concept could successfully be extended to other important plant nutrients as well.

Because cardamom is a shade-loving plant, its production physiology is different from that of other zingiberous plants. What is more, little is known about these aspects. An increase in productivity can best be achieved by genetic upgradation through new gene combinations, heterosis breeding, and the subsequent production of hybrid seeds. The production of genetically homozygous lines for heterosis breeding is thus an urgent need. Hence, an area of great importance is the production of haploids and diploids for hybrid seed production. This step itself can revolutionize cardamom production.

An intensive search is required to locate heat- and drought-tolerant lines. Heat and drought susceptibility are the most serious constraints facing cardamom production in India. Once is the requisite lines are achieved, the incorporation of such resistance into elite genotypes can be achieved by utilizing conventional breeding or the haploid–diploid hybrid system.

Resistance to biotic stress factors, especially from viral diseases, is another, equally important, aspect that merits great attention by cardamom researchers. Surveys of natural disease escapes in hot-spot areas and their screening and evaluation have led to some *Katte*- and rhizome rot-resistant lines in the germplasm collection project of the Indian Institute of Spices Research at Kozhikode, Kerala State, under the administrative control of the Indian Council of Agricultural Research, New Delhi. Some of these lines are high yielding as well. For the production of planting material, which is an important component of cardamom production, the disease-resistant lines can be made use of. At the same time, the agronomically superior lines have to be subjected to molecular-breeding programs for the production of transgenics that will incorporate *Katte* virus resistance either through coat protein-mediated resistance or otherwise.

Another area of global importance is the emergence of organic farming; indeed, organically produced crops are in far greater demand than ever before. The spices present an excellent opportunity in this area, and cardamom is no exception. A substantial amount of research and development has to go into achieving the targets in this area. Another area that may yet emerge is the production of transgenic cardamom plants that are capable of surviving biotic and abiotic stresses. Cutting-edge technology in the area of cardamom production along these lines is nonexistent as of now, but future requirements would demand such efforts, although, at this stage, one cannot hazard a guess as to what the future holds.

During the past decade and a half, world production and demand have increased in parallel, leaving very little leftover production from the previous season. In fact, during that period, both production and demand increased by about 250%. It is also during that period that the cardamom industry grew manyfold, benefiting all those engaged in production, processing, and marketing. It is hazardous to predict the future of any nonessential agricultural produce with a limited market, and cardamom comes under this category, although it is the second most prized spice in the world. However, it appears that cardamom has a somewhat bright future, although any substantial increase in production by any cardamom-producing country in a short span of time might upset the supply–demand equation. Such a situation might lead to a steep drop in prices, resulting in cardamom planters abandoning the crop. A situation akin to this was experienced by Guatemala in 1997–1998, when its production was at its peak, and the Middle East market for cardamom crashed because of the decline in the oil boom.

In spite of the fact that both the aroma and flavor of cardamom are acceptable in the manufacture of many foods, consumption in the developed world, primarily in the United States, Japan, the European Union, Australia, and New Zealand, is low. At present, the use of cardamom as raw material for the manufacture of processed foods, as cardamom oil, and as oleoresin will be in the range of 1500t per annum in the developed world. This is primarily because no agency is making a concerted effort to boost cardamom use in the aforesaid countries. Because cardamom is produced in the developing world and the cardamom industry supports many small cardamom farers, it is desirable to have a concerted public relationship drive in cardamom use. There are a few reputable spice importers' associations, such as the American Spice Trade Association, European Spice Association, and All Nippon Spice Association, that can be contacted for promotional purposes. As with pimento and black pepper, for which producers and importers have joined hands, an effort can also be initiated in the case of cardamom.

Cardamom consumption is scant everywhere but in India. Thus, for sound development of the agricultural industry, the best safeguard is the domestic market, since international markets can fluctuate from competition between producing countries, as has been the case with India and Guatemala. Promoting cardamom consumption in both traditional and nontraditional countries of consumption must be initiated. The phenomenon of cheap Guatemalan cardamom flooding the Indian market, because of both the liberalization of the economy and clandestine import through the Nepalese border, has shattered the Indian cardamom market. The lesson that must be learned from this episode is that, unless farmers are enabled to produce competitively and diversify cardamom use, the future for cardamom will be bleak.

2.13 Large Cardamom (*Amomum Subulatum Roxb.*)

Large cardamom, also known as Nepal cardamom (*Amomum subulatum Roxb.*), is a spice that is cultivated in the sub-Himalayan region of northeastern India, especially Sikkim, since time immemorial. In the past, the aboriginal inhabitants of Sikkim, called *Lepchas,* collected capsules of large cardamom from the natural forests, but later on these forests passed into village ownership and the villages started to cultivate large cardamom. The presence of wild species, locally known as *churumpa,* and the variability within the cultivated species support the view of its origin in Sikkim (Subba, 1984). Subsequently, cultivation spread to northern Uttar Pradesh, northeastern states of India (Arunachal Pradesh, Mizoram, and Manipur), Nepal, and Bhutan. Sikkim is the largest producer of large cardamom, and annual production in India is about 3500–4000t of the cured product. Average productivity is 100–150 kg ha^{-1}, but in well-maintained plantations the productivity can go up to 1000–2000 kg ha^{-1}. Nepal and Bhutan are the other two countries cultivating this crop, with an annual production of about 1500t each. Large cardamom is used in *Ayurvedic* medicines in India, as mentioned by the great Indian sage and medical practitioner *Susruta* in the sixth century B.C., and was also known to the Greeks and Romans as *Amomum* (Ridley, 1912). Large cardamom contains ca. 1.98–2.67% of volatile oil and is used mainly to flavor food products (Gupta et al., 1984). The seeds also possess certain medicinal properties, such as carminative, stomachic, diuretic, cardiac (as a stimulant), and antiemetic properties, and are a remedy for throat and respiratory problems (Singh, 1978). As a marketable commodity, large cardamom is sold mainly in northern India. Over the past few years, it also has been exported, and in 1997–1998 India earned about US$3 million as export earnings, when the country exported 1784t. Pakistan, Singapore, Hong Kong, Malaysia, the United Kingdom, and the Middle East countries are the major importers of large cardamom.

2.13.1 Habit and Habitat

Amomum subulatum is a perennial herb that belongs to the family Zingiberaceae under the order Scitaminae. The plant consists of subterranean rhizomes and several leafy aerial shoots and/or tillers. The number of such rhizomatous leafy shoots varies between 15 and 140 in a single plant or clump of plants. The height of leafy shoots ranges from 1.7 to 2.6m, depending on the cultivar; the plant possesses 9–13 leaves in each tiller. The leaves are distichous, simple, linear, and lanceolate, glabrous on both sides with a prominent mid rib. Inflorescence is a condensed spike on a short peduncle. Flowers are bracteate, bisexual, zygomorphic, epigynous, and cuspinated. The yellowish perianth is differentiated into calyx, corolla, and anther crest. Each spike contains about 10–15 fruits (capsules) and, rarely, up to 20–25 capsules, depending on cultivars. Flowering season begins early at lower altitudes, with peak flowering during March–April, whereas it starts at higher altitudes in May, with a peak during June–July. Harvesting begins during August–September at lower altitudes and in October–December at higher altitudes (Gupta and John, 1987). The fruit is an oval-shaped capsule, trilocular with many seeds. The capsule wall is echinated, reddish brown to dark pink. Seeds are white when immature and become dark gray to black toward maturity. The capsules formed at the basal portion of spike are bigger and bolder than the others (Rao et al., 1993).

Large cardamom is grown in cold and humid conditions under shade trees at altitudes varying from 800 to 2000 m amsl with an average precipitation of 3000–3500 mm, spread over about 200 days. The temperature ranges from 6 °C in December–January to 30 °C in June–July (Singh, 1988). Frost, hailstorm, and snowfall are the major deleterious factors affecting large cardamom. The crop grows well in moist, but well drained, loose soil. The depth of soil varies from a few centimeters to several meters, depending on the topography and soil formation. The soil is acidic and rich in organic matter (Mukherji, 1968).

2.13.2 Cultivars

The commercially grown cultivars of large cardamom belong to the species *Amomum subulatum* Roxburgh. Out of a total of 150 species of Amomum occurring in the tropics of the world, only about 8 are considered to be native to the eastern sub-Himalayan region: *Amomum subulatum* Roxb., *Amomum costatum* Benth., *Amomum linguiformae* Benth., *Amomum pauciflorum* Baker., *Amomum corynostachyum* Baker., *Amomum dealbatum* Roxb. (*Amomum sericeum* Roxb.), *Amomum kingii* Roxb., and *Amomum aromaticum* Roxb. (Hooker, 1886). Later, 18 species of *Amomum* were reported from the northeastern Himalayan regions (Anon., 1950). In the Indian subcontinent itself, there is another center of diversity in the Western Ghats region in southwest India. Gamble (1925) has reported six species from this region.

There are five main cultivars of large cardamom: *Ramsey, Sawney, Golsey, Varlangey* (*Bharlangey*), and *Bebo* (Gyatso et al., 1980). They are all well known. Some of their subcultivars (*Ramnag, Ramla,* Madhusey, *Mongney,* and so on) are also seen in cultivation in small areas of Sikkim State in northeastern India. Another cultivar, *Seremna* or *Lepbrakey* (a *Golsey* type), is gaining in importance and is spreading to more areas at lower altitudes (Upadhyaya and Ghosh, 1983).

2.13.2.1 Ramsey

The name *Ramsey* was derived from two *Bhutia* (the language spoken in Bhutan and one of the languages spoken in northeastern India) words: *Ram*, meaning "mother," and *Sey* meaning "gold" (yellow colored). This cultivar is well suited for higher altitudes, even above 1500 m on steep slopes. Grown-up clumps 8–10 years of age possess 60–140 tillers. The color of the tillers is maroon or maroonish green. The second fortnight of May is the peak flowering season. Capsules are small, of average length around 2.27 and 2.5 cm wide, with 30–35 capsules in a spike, each containing 16–30 seeds. The harvest is during October–November. Peak bearing capsules are noticed in alternate years, and the tendency to produce a heavy crop one year and a light or no crop the next year is generally referred to as "alternate bearing," a phenomenon common in mango as well. The *Ramsey* cultivar is more susceptible to the viral diseases *Foorkey* and *Chirke*, especially if planted at lower altitudes. It occupies a major area of large cardamom in Sikkim and the Darjeeling district of the state of West Bengal in India. Two strains of this cultivar, namely, *Kopringe* and *Garadey*, from Darjeeling district and having stripes on the leaf sheath, are reported to be tolerant to *chirke* virus (Karibasappa et al., 1987).

2.13.2.2 Sawney

This cultivar obtained its name from the Nepalese word *Sawan*, which signifies the month of August, by when the crop is ready to be harvested. The cultivar grows well at low to middle altitudes. It is widely adaptable and can also grow at high altitudes, even as high as 1300–1500 m. It is robust in nature and consists of 60–90 tillers in each clump. Its tillers are similar in color to that of *Ramsey*. Each productive tiller produces two spikes, on average. The average length and diameter of a spike are 6 and 11 cm, respectively. Flowers are longer (6.23 mm) and yellow in color, with pink veins. The second half of May is the peak flowering season (Rao et al., 1993). Capsules are bigger and bold, and the number of seeds in each capsule are more (35) than in *Ramsey*. Harvest begins in September–October and may extend up to

November in high-altitude areas. *Sawney* is susceptible to both *Chirke* and *Foorkey* viral diseases. Cultivars, such as *Red Sawney* and *Green Sawney*, received their names from the color of the capsule. *Mongney,* a strain found in the south and west districts of Sikkim, is a nonrobust type with small round capsules resembling those of *Ramsey.*

2.13.2.3 Golsey (Dzoungu Golsey)

The name *Golsey* is derived from Hindi (the widely spoken North Indian language) and *Bhutia* (the language spoken in Bhutan), with the root *Gol* meaning "round" and *Sey* meaning "gold" (yellow colored). *Golsey* is suitable for low altitudes, below 1300 m amsl, especially in the Dzongu area of North Sikkim. Unlike other cultivars, *Golsey* plants are not robust. They have 20–25 straight tillers with erect leaves. Alternate, prominent veins are extended to the edges of the leaves (Biswas et al., 1986). Unlike the tillers of *Ramsey* and *Sawney, Golsey* tillers are green. Each productive tiller produces two spikes, on average. Flowers are bright yellow. On average, each spike is 5.3 cm long and 9.5 cm in diameter and contains an average of seven capsules. Capsules are big and bold, 2.46 cm long and 3.92 cm wide, and contain 60–62 seeds. This cultivar becomes ready for harvest in August–September. *Golsey* is tolerant to the two viral diseases *Chirke* and *Foorkey* and also to leaf streak diseases. The cultivar is known for its consistent performance, although it does not produce a heavy yield. Many local cultivars are known in different locations, such as *Ramnag* in North Sikkim. The word *Ramnag* can be split into its root, *Ram*, meaning "mother," and *Nag*, meaning black, referring to the dark pink-colored capsules. *Seto-Golsey* is from the west district of Sikkim and has robust leafy stems or tillers and green capsules. *Madhusey*, with elliptic and pink-colored capsules, has a robust leafy stem with sweet seeds, compared with other cultivars (Rao et al., 1993).

2.13.2.4 Ramla

Ramla plants are tall and vigorous, like *Ramsey*, and have capsule characters, like *Dzongu Golsey*. Tillers are pink colored, like those of *Ramsey* and *Sawney.* Cultivation is restricted to a few middle-altitude plantations in North Sikkim. The capsules are dark pink, with 25–38 seeds per capsule. *Ramla* appears to be a natural hybrid between *Dzongu Golsey* and *Ramsey.* It is susceptible to *Chirke* and *Foorkey* viral diseases.

2.13.2.5 Varlangey

The Varlangey cultivar grows in low-, medium-, and high-altitude areas in South Regu (East Sikkim) and at high altitudes at Gotak (Kalimpong subdivision in Darjeeling district of the state of West Bengal, India). Its yield performance is exceptionally high at the higher altitudes—that is, 1500 m amsl. It is a robust type, and the total number of tillers ranges from 60 to 150. The color of the tillers resembles that of *Ramsey*—that is, maroon to maroonish green toward the collar zone. The girth of the tillers is more than that of *Ramsey* tillers. Each productive tiller produces almost three spikes, on average, with an average of 20 capsules per spike. The capsules are bigger than those of *Ramsey* and are bold with 50–65 seeds. Harvest begins in the last week of October. This cultivar is also susceptible to both of the viral diseases *Chirke* and *Foorkey*.

2.13.2.6 Bebo

The Bebo cultivar is grown in the Basar area of Arunachal Pradesh. The plant has unique features of rhizome and tillering. The rhizome rises above the ground level, with roots penetrating deep into the soil, and the young tillers are covered under thick leafy sheath. Bebo is supposed to be tolerant to *Foorkey* viral disease. The spikes have a relatively long peduncle (10–15 cm), and the capsules are bold, red brown, or light brown; seeds contain a low level of essential oil (2% v/w) (Dubey and Singh, 1990).

2.13.2.7 Seremna (Sharmney or Lepbrakey)

The Seremna cultivar is grown in small pockets at Hee-Gaon in west Sikkim at low altitude and is known for its high-yield potential. Plant features are similar to those of *Dzongu Golsey*, but the leaves are mostly drooping—hence the name *Sharmney*. (The root of the word *Sharm* in Hindi means "modesty in a female with bowed head.") The total number of tillers ranges from 30 to 49, and the plant is not robust. On average, two to three spikes emerge from each productive tiller, with an average of 10.5 capsules per spike, each having 65–70 seeds.

Comparative morphological characters of the four most important cultivars, namely, *Ramla, Ramsey, Sawney*, and *Golsey*, are given in Tables 2.60 and 2.61.

2.13.3 Plant Propagation

Large cardamom is propagated through seeds, rhizomes (sucker multiplication), and tissue culture techniques. Cultivars suitable for specific areas, altitudes, agroclimatic conditions, and mother plant or clump of known performance are selected for the collection of seeds, rhizomes, and vegetative buds.

2.13.3.1 Nursery Practices

Large cardamom is propagated through seeds. A healthy plantation, free from viral disease in particular, is selected for seed capsules. Normally, gardens yielding 1000 kg ha^{-1} or more during the previous 3 years are selected. Higher (taller) spike-bearing (reproductive) tillers per plant (bush), a greater number of spikes, capsules that are bold, a greater number of seeds per capsule, and so on, are some of the criteria on the basis of which seed capsules are selected from the plot. Spikes are harvested at maturity, and seed capsules are collected from the lowest two circles in the spike. After dehusking, the seeds are washed well in water to remove any mucilage that is covering them, mixed with wood ash, and dried under shade. The dried seeds are then treated with 25% nitric acid for 10 min, a practice that is known to induce the seeds to germinate early and that is said to yield a greater percentage of

Table 2.60 Growth Performance of *Ramla, Ramsey, Sawney,* and *Golsey* Cultivars of Large Cardamom (Average of 3 Years)

Plant Character	Ramla	Ramsey	Sawney	Golsey	CD Rate (p = 0.05)
Plant height (cm)	200.83	192.08	196.00	190.05	6.54
Number of tillers per plant	59.08	42.00	40.50	39.80	12.67
Number of spikes per plant	40.25	36.00	35.00	30.00	4.70
Spike length (cm)	7.06	7.00	6.40	6.50	0.49
Spike breadth (cm)	8.60	7.00	6.00	7.20	1.47
Number of capsules per spike	16.00	12.00	13.00	14.00	1.12
Fresh capsule yield per plant (g)	375.00	185.00	190.00	216.00	14.00
Dry capsule yield per plant (g)	70.00	47.00	48.00	52.00	15.90

Note: CD, critical difference.

Table 2.61 Capsule Characteristics of Different Cultivars of Large Cardamom (Average of 3 Years)

Capsule Characteristic	Ramla	Ramsey	Sawney	Golsey	CD Rate (p = 0.05)
Fresh weight per capsule (g)	4.00	3.50	4.00	4.50	0.56
Dry weight per capsule (g)	0.90	0.75	0.85	1.00	0.14
Moisture (%)	13.00	14.00	13.00	15.00	1.54
Number of seeds per capsule	38.00	36.00	35.00	40.00	3.28
Volatile oil (%)	2.67	2.50	2.00	1.98	0.48

Note: CD, critical difference.

germination (Gupta, 1989). The acid-treated seeds are washed thoroughly in running water to remove the acid residue and are surface dried under shade. The seeds are sown immediately after such treatment.

2.13.3.2 Nursery Site Selection

An open area with gentle slope and having a facility for irrigation is selected for establishment of the nursery. A large-cardamom nursery is raised in two stages: the primary nursery and the secondary nursery. Seedlings raised in the former by seeds are transplanted into the latter or into polybags.

Primary Nursery

Seeds are sown in September–October or early November (prewinter) or in February–March (postwinter). September sowing results in quicker and better germination.

Seedbed is prepared in a well-drained area. Soil is cut to a depth of 30 cm and exposed to sunlight for a week. A bed of 15 to 25 cm in height, 90 cm in width, and a convenient length is prepared, into which well-decomposed cattle manure or compost is filled. Seeds at the rate of 100 g for a bed size of 1 m × 3 m are sown in furrows along the width and are covered with a thin layer of soil. The space between furrows is maintained at 10 cm. After sowing, the beds are covered with thick mulch of paddy straw or dry grass and are watered regularly to keep the seedbed moist. The seedbeds are examined about a month after sowing, and once germination starts after this period, the following operations are conducted:

1. Overhead shading is supplied by a bamboo mat.
2. Mulch is removed from the bed, cut into small pieces, and spread in between seed rows.
3. Seedbeds are watered regularly and kept moist.

Once seedlings in the primary seedbed reach the 3–4-leaf stage, which occurs by February–March when seeds are sown in September–October, or in April–May when seeds are sown in February–March, they are transplanted into polybags or into secondary nursery beds (Gupta, 1989).

Polybag Nursery

Topsoil in virgin land or forest area that is rich in leaf mold is collected and mixed with well-decomposed cattle manure to get a good potting mixture. A mixture of 5:1 of topsoil to cattle manure is prepared and filled in polybags 8 in. × 8 in. with perforations at the base for drainage. Polybags are arranged under overhead shaded shed. Primary seedlings are transplanted in polybags (one seedling per bag) during February–March or April–May. Polybag seedlings are watered regularly with a rose can to keep the soil moist. Care must be taken to cover the collar region and the exposed roots of seedlings with a thin layer of topsoil, which aids anchorage and tillering. Excessive watering must be avoided. Polybag seedlings attain a height of 30–40 cm, with two to four tillers, by July–August if transplanting is done in February–March. These seedlings are planted in the main plantation in July–August. The polybags are removed, and seedlings with soil ball intact are planted. Sometimes the seedlings are maintained in the polybags until April of the next year and are planted in May–June (Anon., 1998).

Secondary Nursery

Seedlings from the primary nursery are sometimes transplanted in beds. Beds of a size and nature similar to those of primary beds are prepared, and seedlings at the 3–4-leaf stage are transplanted in March, April, or May, maintaining a spacing of 15 cm between seedlings. A layer of well-decomposed cattle manure is applied and incorporated into the soil. Watering is done at regular intervals to keep the soil moist. The entire secondary nursery is maintained under overhead shade (preferably black agro-net), and seedlings are maintained for 10–12 months. The expected growth of seedlings is about 45–60 cm in height with 5–10 tillers each, and these seedlings are transplanted in June–July in the main field (Anon., 1998a).

Propagation through Rhizome

High-yielding, disease-free planting materials are selected for multiplication. Trenches 2ft wide, 2ft deep, and of a convenient length are made across slopes. Trenches are filled with topsoil, leaf mold, and decomposed leaf litter. Rhizomes with one mature tiller and two young shoots or vegetative buds are planted at a spacing of 3ft in the trenches in June–July. Thick mulching with dry leaf or grass is applied at the base of the rhizome, and watering is done regularly to keep the soil moist. Once fresh vegetative buds appear, well-decomposed cattle manure is applied 1ft around the rhizome and incorporated into the soil. A rhizome multiplication plot is maintained with 50% shade, either under shade trees or under agro-net shade. When rhizomes are planted in June–July, about 15–20 tillers are produced from each rhizome within 6–10 months. Each such clump is split into units of two to three tillers, which are used either for planting in the main field in June–July or for further multiplication (Anon., 1998a).

Micropropagation

Large cardamom can be multiplied on a large scale through micropropagation. Protocols for micropropagation were developed by the Indian Institute of Spices Research (Nirmal Babu et al., 1997a; Sajina et al., 1997a). Auxiliary buds 0.5–2cm in length from promising, virus-free mother plants are used as explants. The explants are thoroughly washed in clean running water and then in a detergent solution, after which they are treated in 0.15% mercuric chloride solution for 2min and then passed through absolute alcohol for 30s. The treated explants are cultured in a modified MS medium, solidified with agar, and with the following adjuvant:

Step 1: For initial bud development and growth of the bud *in vitro* (culture period, 6–8 weeks): kinetin 3–5mg^{-1} + IBA 1–2mg^{-1} + sucrose 20gliter^{-1}
Step 2: For proliferation of the auxiliary bud rhizome (6–8 weeks): BAP 2mgliter^{-1} + NAA 3–5mgliter^{-1} + sucrose 20gliter^{-1}
Step 3: For rooting and establishment of plantlets (6–8 weeks): IBA 1–2mgliter^{-1} + KN 3–5gliter^{-1} + sucrose 20gliter^{-1}.

For complete details of the procedure, see Nirmal Babu et al. (1997a) and Sajina et al. (1997b).

2.13.3.3 Plantation Management

Soil Condition, Preparation of Land, and Shade Development

Large cardamom is grown in loamy forest soils having depths varying from a few centimeters to several meters. These soils are of medium fertility (Bhutia et al., 1985; Biswas et al., 1986). Because the terrain varies from gentle to deep slope, chances of water stagnation are meager. (Water stagnation is highly deleterious to the crop; see Singh et al., 1998.) In general, large cardamom is cultivated on hill slopes and often in terraced lands, which were earlier put to the paddy crop, after raising adequate shade trees. Large cardamom is a shade-loving plant and grows tall under dense shade (60–70% of full daylight interception) to light shade (about 30% full daylight interception), as reported by Singh et al. (1989). The daylight intensity required for optimum growth of large cardamom ranges from 5000 to 20,000 lux. Therefore, in virgin forests, it is necessary to clean undergrowth. Overhead shade regulation is essential in areas that have insufficient shade, so shade saplings of different shade trees are planted in June–July. The most commonly used shade trees are *Utis (Alnus nepalensis,* 600–2000 m amsl), *Panisaj (Terminalia myriocarpa,* 400–1000 m amsl), *Malato (Macaranga denticulata,* 670–1515 m amsl), among others. It is advisable to plant more than one of the commonly used shade trees that are native to the locality. In the case of fallow land, *Utis* is the first choice, as it is quick growing, is capable of fixing atmospheric nitrogen, and has faster rates of nutrient recycling. Monoculture of *Utis* can fix as much as 29–117 kg ha^{-1} of atmospheric nitrogen. The yield of large cardamom has been found to increase as much as 2.2 times under a canopy of *Utis.* Sharma and Ambasht (1988) have found large-cardamom-based agroforestry under the influence of *Utis* more productive, with a faster rate of nutrient recycling. Depending on altitude, planting shade trees in rows at a distance of 9–10 m is ideal. Before planting the sapling, the course and direction of the Sun and the slope of the hill must be taken into consideration. Usually, the tree rows are run along the southwest direction inside the plantation.

Planting

For planting of large cardamom, pits are opened at a spacing that is suitable for the variety or cultivar. In the case of robust varieties such as *Sawney, Varlangey,* and *Ramsey,* the spacing followed is 150 cm × 150 cm, while a spacing of 120 cm × 120 cm is used for *Golsey* (Dzongu). Pits are opened in April–May. The size of the pits is usually 30 cm × 30 cm × 30 cm. After the first showers, and at least 15–20 days before planting, pits are filled with topsoil, decomposed cattle manure, compost, or leaf mold, along with 100 g of rock phosphate. Ideally, planting is best done in June–August, but it depends on the rains. Staking is essential to provide good anchorage.

Mulching

The plant base should be mulched with dried leaves, weeds, and trash. Mulching must be done for fresh plants, as well as existing ones, in October–November. This practice helps conserve soil moisture and recycle plant nutrients.

2.13.4 Plant Nutrition

Most of the nutrients are removed by the vegetative growth in large cardamom; much less is removed by capsules and spikes. Robust varieties such as *Ramsey, Sawney,* and so on remove twice the quantity of nutrients, compared with nonrobust varieties such as *Dzongu Golsey.* To produce 100 kg of dry large cardamom, the robust types remove 10.33 kg of nitrogen, 1.95 kg of phosphorus, 26.24 kg of potassium, 19.10 kg of calcium, and 11.9 kg of magnesium. The nonrobust type *Dzongu Golsey* removes, by comparison, 5.74 kg of nitrogen, 0.99 kg of phosphorus, 3.54 kg of potassium, 9.18 kg of calcium, and 5.86 kg of magnesium. Old leafy shoots removed during harvest are used as soil mulch. Because large cardamom is grown under forest cover, neither manuring nor applying fertilizer is usually practiced. A low-volume and less nutrient-exhausting crop, large cardamom has a degree of sustenance in terms of nutrient cycling. However, to obtain high yields, applying chemical fertilizer is a must.

Application of nitrogen–phosphorus–potassium fertilizer in three splits—in April–May after the first summer showers, followed by a second application in June and the last in September–October before the monsoon ceases—is recommended. Fertilizer must be applied in circular bands at a distance of 30–45 cm from the clump, with mild forking. A fertilizer schedule of 18.4 g, 6 g, and 18.6 g of nitrogen, phosphorus, and potassium, respectively, per clump produced a large number of spikes per clump in *Sawney*. However, *Golsey* did not respond to fertilizer application. Foliar application of urea (0.5–1%), DAP (0.5%), and muriate of potash (0.5–1%) in February, April, and October enhanced yield in *Ramsey*.

Large-cardamom cultivation is an ecofriendly organic way of farming with minimal inputs and with the least deleterious interactions with the soil system.

2.13.4.1 Weeding

About 51 species of weeds have been recorded in large-cardamom plantations (Anon., 1984). Depending on the intensity of weed growth, two to three rounds of weeding are required per year. First weeding is done in February–April, before the first application of fertilizer and just before flowering. Second and third weedings are done before harvest in August–September–October, along with the removal of dried leaves, unproductive tillers, and so on. The weeded materials are used for mulching.

2.13.4.2 Shade Regulation

Most shade tree species are deciduous in nature; hence, frequent shade regulation is not required. However, during the early years of shade establishment and at 2- to 3-year intervals, the undergrowing side branches are cut to encourage straight growth and to allow the branches to spread at least 3–4 m above ground level so that moderate shade is maintained (Gupta, 1986)

2.13.4.3 Irrigation

When the plantation is irrigated in summer months, yields are better, as the cardamom plant is extremely susceptible to drought and water stress. Irrigation during November to March sustains good yield. In plantations, surface irrigation is generally utilized.

2.13.4.4 Roguing and Gap Filling

One of the main reasons for poor productivity (yields as low as 50–150 kg ha^{-1}) in large cardamom is that most of the plantations have become senile and unproductive. The two viral diseases *Chirke* and *Foorkey* have been a cause of low yields. Hence, regular roguing of diseased and senile plants and filling the gaps with disease-free seedlings is a must to maintain productivity.

2.13.5 Crop Improvement

2.13.5.1 Flowering and Pollination

Large cardamom is essentially cross-pollinated; hence, insect pollination is the rule. Flower morphology is adapted for such a mode of pollination. Each spike consists of about 40–50 flowers, which open in an acropetal sequence over a period of about 15–25 days. Flower opening commences between 3 and 4 A.M., and anthers dehisce almost instantly, whereas stigma receptivity lasts 24 h. During rain-free days, stigma receptivity lasts up to 36 h (Gupta and John, 1987).

Bumblebees (*Bombus* sp.) are the main pollinators, although a variety of honeybees and other insects also pollinate flowers (Varma, 1987). The maximum foraging activity of the bees takes place between 6 and 7 A.M., and their size is compatible with the flower size, facilitating pollination. Rainy days are not conducive to pollination.

2.13.5.2 Genetic Investigations

Karibasappa et al. (1989) carried out coefficient-of-variation investigations on mature tillers per clump, panicles per clump, and capsule–panicle ratio in *Sawney, Pink Golsey, Ramsey, Ramnag,* and *Madhusey,* and the investigations indicated high heritability coupled with genetic advance for characters such as length of mature tiller, panicles per clump, panicle weight, and capsule yield. The capsule yield was directly correlated with girth, panicle weight, panicle per clump, maturity of tillers, and capsule–panicle ratio. Karibasappa et al. also conducted correlation studies, which indicated that the mature-seed index, total soluble sugars of seed mucilage, and test weight of 1000 seeds were positively correlated with oleoresin content and negatively correlated with cineol content.

2.13.5.3 Clonal Selection

The germplasm collection of large cardamom has been explored by the Indian Cardamom Research Institute, Regional Research Station, Gangtok, in Sikkim State, and a gene bank of 180 accessions has been established at Pangthang. Rao et al.(1990) reported a promising selection of *Barlanga* cultivar from high altitudes. The cultivar has desirable attributes, such as a high ratio of mature tillers to productive spikes (1:3.6) and capsules having a bold size (50–80 seeds per capsule). On the basis of a preliminary evaluation, four selections, namely, SBLC-5, SBLC-42, SBLC-42, and SBLC-47 A, have been identified by the Institute as having high-yield potential. These selections are multiplied in large numbers by micropropagation and are distributed among the farmers for cultivation.

2.13.6 Insect Pest Management

More than 22 insect pests are known to be associated with large cardamom, and only a few of them cause substantial damage to the crop (Azad Thakur and Sachan, 1987;

Bhowmik, 1962). to the material that follows describes some of the more important insect pests.

2.13.6.1 Leaf Caterpillar

The leaf-eating caterpillar (*Artona chorista* Jordon, Lepidoptera: Zygaenidae) is a major pest of large cardamom in Sikkim and West Bengal State (Singh and Varadarasan, 1998). Its outbreak was recorded in 1978 in Sikkim, where about 2000 acres of large-cardamom plantations were severely defoliated (Subba, 1984). The leaf caterpillar was first recorded as *Clelea plumbiola* Hamson on large cardamom by Bhowmik (1962). *Artona chorista* occurs sporadically in epidemic form in Sikkim and West Bengal every year. Usually, the incidence of pests is higher in June–July and October–March. Severe damage was recorded in Lower Dzongu, Phodong, Ramthung Basit (North Sikkim), Soreng, Hee, Chako (West Sikkim), and Kewizing (South Sikkim).

2.13.6.2 Nature and Extent of Damage

Leaf caterpillars are monophagous and highly host specific. They are gregarious in nature, and 60–200 caterpillars are found on each leaf. They feed on chlorophyll under the leaf surface, leaving transparent epidermis and veins (skeletonization). The damaged portion becomes brownish and hence can easily be identified. Yield is adversely affected through defoliation. The area of a medium-sized cardamom leaf is 160–170 cm^2, and a mature larva consumes about 2.12 cm^2 each day (Singh and Varadarasan, 1998).

Infestations are managed through mechanical, chemical, and biological means. Mechanical means are physical collection and destruction. For chemical control, insecticides, such as Qinalphos or Endosulfan (0.05%), are effective. The danger of chemical control is that the chemicals also eliminate natural predators of the caterpillars. Because large cardamom is chiefly a naturally occurring vegetation, the use of toxic insecticides is best avoided. For biological control, a species of predatory pentatomid bug has been found effective. The bug kills one to three larvae per day by sucking their body fluid.

2.13.6.3 Hairy Caterpillar

The hairy caterpillars are a group of defoliators that infest large cardamom. A severe infestation affects yields adversely. Among the hairy caterpillars, *Eupterote* sp. is the predominant one, of which the major species are *Eupterote fabia* Cramer and *Eupterote* sp. (Lepidoptera: Eupterotidae). The caterpillar infests plantations during the monsoon period, from August onward, and the infestation can last up to December. *Eupterote fabia* and other species are sporadic and polyphagous, and feed on leaves of cardamom, causing defoliation. Sometimes, *Eupterote fabia* causes severe defoliation.

The adult moth is large (10.8 cm across the wing) and yellow in color. The female moth lays eggs on the undersides of leaves in clusters of 20–140. The hatching period lasts 19–21 days. The mature larva measures 7.4 cm long. Larval development is completed in 83–97 days, and the adult emerges in 120–180 days (Azad Thakur and Sachan, 1987). The hairy caterpillar is a minor pest of large cardamom.

2.13.6.4 Aphids

Aphids do more damage as a vector rather than as a pest. They are the transmitters of the viruses that cause *Chirke* and *Foorkey* diseases. Records show that the aphid population is high during summer months at lower altitudes. The following are the major species:

1. *Pentalonia nigronervosa* f. *caladii* (Groot) (Hemiptera: Aphididae)
2. *Micromyzus kalimpongensis* (Hemiptera: Aphididae)
3. *Rophalosiphum maidis* (Fitch) (Hemiptera: Aphididae)
4. *Rophalosiphum padi* (Lin.) (Hemiptera: Aphididae).

Pentalonia nigronervosa f. *caladii* and *Micromyzus kalimpongensis* are known to be the vectors of *Foorkey*, or yellow virus disease. The aphids colonize at the base (rhizome) of the clump, and if their population grows too much, they move to the aerial portion of the clump. Two to six aphids per tiller were recorded from *Foorkey*-infected plants during summer months. *P. nigronervosa* f. *caladii* is also reported as the vector of *Katte* viral disease in small cardamom. These aphids are dark brown in color, are small, and measure 1–1.15 mm in length. They remain mostly inside the soil, close to rhizomes, and suck the sap from the pseudostem. The alate (winged) and apterous (wingless) forms complete their life cycle in 20–30 days.

Maize aphids, *R. maidis* and *R. padi*, have been reported on the lower surfaces of the leaves of large cardamom, congregating near the midrib and veins. These aphids are known to be the vector of another viral disease: mosaic streak, or *Chirke* (Raychaudhary and Chatterjee, 1965).

The removal and destruction of diseased plants is helpful in controlling the further spread of disease and in the reduction of the aphid population. Spraying of 0.03% mimethoate or phosphomidon after the removal of *Foorkey*- and *Chirke*-infected clumps in March–April gives adequate control of the aphids.

2.13.6.5 Shoot Fly

The shoot fly, *Merochlorops dimorphus* Cherian (Diptera: Chlopidae), recorded as a major pest of large cardamom, damages young shoots. Low to moderate damage is recorded in large-cardamom plantations in Sikkim and West Bengal. In the main field, more damage is recorded at higher altitudes than in lower altitudes. Another shoot fly, *Bradysia* sp. (Diptera: Sciaridae), has been reported to be damaging to large cardamom (Kumar and Chaterjee, 1993). As much as 54% of new shoots are infected. Infestation leads to browning of the tip of the shoot, and later the shoot completely dries up, causing the symptom "dead heart." A single pale glossy-white larva bores the young shoots and feeds on the central core of the pseudostem from top to bottom, resulting in the death of the shoot. The best management of the insect pest is to remove the infested young shoot at ground level and destroy it.

2.13.6.6 Stem Borer (Glypheterix *sp. Lepidoptera: Glyphiperidae)*

The stem borer is a major and specific pest of large cardamom. Its infestation intensity varies. Azad Thakur and Sachan (1987) have reported 19% infestation in 1978–1979, but infestation can sometimes be far more severe. The infestation occurs from March to November. The larvae feed on the central portion of the shoot, and as a result, the terminal leaf of the plant dries up. This symptom is known as "dead heart," as in the case of shoot fly infestation.

2.13.6.7 White Grubs (Holotrichia *sp. Coleoptera: Melonthidae)*

The white grub is polyphagous and infests the roots and rhizomes of large cardamom. The infested plants show yellowing of leaves and symptoms of withering. The grubs are "C" shaped with a brown head. An infestation was recorded at Panthang (East Sikkim) and Kabi (North Sikkim) in September–December 2005. *Holotrichia* sp. is a minor pest; hence, no control measures are adopted.

The Leaf-folding caterpillar (*Cotesia euthaliae* Bhatnagar. Hymenoptera: Braconidae) was recorded as a larval parasitoid. It inflicts minor damage. The caterpillar pupates inside the folded leaf during winter. Mechanical control consisting of collecting infested leaves along with the caterpillars is the best manner of controlling the pest.

2.13.6.8 Minor Pests

Mealy Bugs

These pests infect underground rhizomes, feeding on the roots and the rhizomes themselves and causing yellowing of the plant in summer. The pest is recorded at Neem (East Sikkim), Tarku (South Sikkim), Chawang (North Sikkim), and Singling (West Sikkim) and is usually found in March–October.

Leaf Thrips

Heliothrips haemorrhoidalis (Thysanoptera: Thripidae) is a minor pest on leaves, infesting their undersurfaces and sucking the sap. The damage is more on seedlings and is recorded throughout the year. *Rhipiphorothrips cruentatus* Hood (Thripidae: Thysananoptera) is also reported to be a minor pest of large cardamom on seedlings (Azad Thakur and Sachan, 1987). The leaf thrips population subsists more on the lower leaf surface than the upper surface. The infested leaves turn brown and wither gradually. For management, infested leaves are removed and destroyed. If an infestation is severe, it can be controlled by spraying Monocrotophos or Quinalphos at 0.025% (Singh et al., 1994).

Lacewing Bug

Stephanitis typical (Distant) (Hemiptera:Tigidae) is a minor sucking pest on large-cardamom leaves. A severe infestation was recorded in 1997 in North Sikkim, where about 1000 plants in an isolated patch were damaged. The infested area was open, without any shade trees. The damage was recorded in the main field, where shade was thin during the pre- and postmonsoon period. The infestation is recognizable even from a distance, owing to the grayish-yellow feeding spots on the leaves. The bugs suck the sap on the lower surfaces of the leaves. In the case of a severe infestation, plant growth and yield are adversely affected (Singh et al., 1994).

Grasshoppers

These insects infest the plants in nurseries as well as in the main field. Both nymphs and adults feed on leaves. The major species are *Mazarredia* sp. Bolivar (Tettigonidae) and *Chrotogonus* sp. (Acrididae).

Bagworm

Acanthopsyche sp. (Lepidoptera: Psychidae) is a minor pest. Its larvae cause small holes in leaves. Damage is negligible, but the insect is seen in the plantation throughout the year.

Fruit Borer

The grub of the scolytid beetle *(Synoxy* sp.) makes a hole in the immature capsule, feeds on the seeds inside, and pupates inside the capsule. The pest has been recorded at Hee Gaon (West Sikkim).

Scale Insects

These insects colonize near the mid vein on the lower surface of the leaf. They suck the leaf's sap, resulting in brownish spots on leaves. A minor infestation of scale insects was recorded in seedlings in Cardamom Nursery, Mallipayong (South Sikkim).

Rhizome Weevil

This weevil is brownish in color, and the adult is about 1.5 cm. The insect feeds on the rhizome by boring a tunnel into it. It was recorded at Kabi (North Sikkim) in the month of April 2005.

Leaf Beetle

The adult, *Lema* sp. (Coleoptera: Chrysomelidae) of this insect is greenish brown and 8–10 mm in size. The leaf beetle makes irregular holes in tender leaves.

Green Beetle

Basilepta femorata Jacoby (Coleoptera: Eumolpidae) has been recorded in April and October 2005 on the leaves. The exact nature of damage is yet unknown. The grub is a major pest on *E. cardamomum* Maton (Varadarasan et al., 1991). Azad Thakur and Sachan (1987) reported that adult beetles also are very destructive as they nibble and eat away fresh leaf buds.

Nematodes

The dangerous root-knot nematode, *Meloidogyne incognita,* has been found to infect seedlings in the nurseries and plants in the main field, causing considerable damage in both cases. The infected seedlings or plants show stunted growth, and the leaves become narrow and acquire a rosette form due to reduction in internodal length. The root system shows excessive branching with galls. Deep digging and exposure of the soil to sunlight (solarization) before preparation of the nursery may reduce the intensity of nematodes. Using the same site to maintain the nursery must be avoided, lest the infection persist. Farmers are encouraged to change nurseries every year and to raise the seedlings in polybags containing a good potting mixture. This practice not only reduces the nematode population, but also prevents the movement of nematodes through the soil from one location to another.

2.13.6.9 Other Pests

Rodents, squirrels, and wild cats damage the fruits before harvest. The black cat, a nocturnal mammal, is known to cause heavy losses, as they are voracious feeders of near-mature capsules.

2.13.6.10 Storage Pests

A reddish-brown caterpillar of cardamom bores into the capsule and feeds on the mucilaginous seed coat. The hard seeds are unaffected. The appearance of the capsules and the quality of seeds can be adversely affected.

2.13.7 Diseases

The large-cardamom plant is susceptible to a number of diseases, which are mainly viral and fungal in origin. The viral diseases are the most severe; the fungal diseases are not the major problem. The two viral diseases causing major problems are *Chirke* and *Foorkey*. Among the fungal diseases, flower rot, clump rot, leaf streak, and wilt are known to cause damage to the plant and ultimately reduce crop yield.

2.13.7.1 Chirke *Disease*

Chirke disease is a mosaic disease causing pale streaks on the leaves. The streaks turn brown and cause leaves and plants to dry and wither. The flowering in diseased plants is extensively reduced, and only 1–5 flowers develop in one inflorescence, as against 16–20 in an inflorescence of healthy plants (Raychaudhary and Chatterjee, 1965). By the end of the third year after planting, the loss can be as much as 85%. The cultivar *Kopringe* is resistant to *Chirke* disease, whereas the perennial weed *Acorus calamus* L. was found to be highly susceptible (Raychaudhary and Ganguly, 1965). The disease is readily transmitted by mechanical sap inoculation, and in the field it is spread by aphids, such as *Rhopalosiphum maidis* Fitch., within a short acquisition feeding period of 5 min.

Primary spread of the disease from one area to another is through infected rhizomes. Further spread in the field is by aphids (Raychaudhary and Chatterjee, 1965). A serological method that gives results rapidly was developed to locate *Chirke*-diseased plants under field conditions in the manner described by Bradley and Munro (Ganguly, 1966).

2.13.7.2 Foorkey *Disease*

Foorkey disease is characterized by dwarf tillers with small, slightly curled pale-green leaves. The virus (spherical particles 37 μ in diameter; Ahlawat et al., 1981a,b) induces a remarkable reduction in the size of leafy shoots and of leaves of the infected plants and stimulates the proliferation of a large number of stunted shoots arising from the rhizome. The spikes or inflorescence is transformed into leafy vegetative parts, and the formation of fruit formation is altogether suppressed. The diseased plants remain unproductive and gradually degenerate. Symptoms of *Foorkey* disease appear on both seedlings and fully grown plants (Varma and Capoor, 1964). Unlike *Chirke* disease, *Foorkey* disease is transmitted, not through the sap, but by aphids, such as *Pentalonia nigronervosa* Cog. and *Micromyzus kalimpongensis* Basu (Basu and Ganguly, 1968). The primary spread of the disease from one area to another is through infected rhizomes, and further spread within the plantation is through aphids. Infected rhizomes can be killed by injecting Agroxone-40.

2.13.8 *Management of* Chirke *and* Foorkey *Diseases*

The following strategies considerably minimize infestations by the causative viruses (Chattopadhyay and Bhomik, 1965):

1. Treatment of Infected Plants
   a. Diseased plants are regularly rouged.
   b. Diseased plants are uprooted and destroyed when detected.
   c. Uprooted plants must be taken to an isolated place, chopped into small pieces, and buried in deep pits for quick decomposition.
2. Prophylactic Measures
   a. Use healthy and disease-free planting material, preferably seedlings.
   b. Avoid replanting suckers from areas that are prevalent in disease.
   c. Avoid raising nurseries in the vicinity of infected plantations.

2.13.8.1 *Leaf Streak Disease*

Leaf streak disease is a fungal disease caused by *Pestalotiopsis royenae* (D. Sacc) Steyaert., is a serious disease among foliar diseases, and is prevalent round the year. The chief symptom of the disease is the formation of numerous translucent streaks on young leaves along the veins. The infection starts from the emerging folded leaves, and infected leaves eventually dry up, causing a loss of the green part of the leaf, which leads to a loss of photosynthetic surface and, eventually, a loss of yield. *Dzongu Golsey* is found to be more susceptible to leaf streak than other cultivars (Srivastava, 1991). Two schedules per year of three rounds of 0.2% spray of copper oxychloride at fortnightly intervals, one in February–March and the other in September–October, can control the spread of the disease.

2.13.8.2 *Flower Rot*

Flower rot is a fungal disease caused by *Fusarium* and *Rhizoctonia* sp.When infection takes place before or at the time of fertilization, the infected flowers turn dark brown and fail to develop into capsules. If infection occurs after flowering or during fruit set, the infected fruit or capsule loses color and odor (Srivastava, 1991). The disease can be managed by avoiding (1) an accumulation of leaf mass or mulch over the inflorescence or spike during the monsoon season and (2) soil spills over the spikes.

2.13.8.3 Wilt

Wilt, a fungal disease caused by *Fusarium oxysporum*, is prevalent in swampy and open areas. An early symptom is chlorosis of the older leaves, commencing from the petiole region and progressing inward toward the young leaves. As the infection progresses, the pseudostems turn rotten, blocking the vascular bundles. Finally, the pseudostem collapses and the plant dries up. Drenching 0.5% Dithane M-45 or Thiram will help check the spread of the disease in nurseries as well as in the main field. Planting in swampy or dry areas is best avoided.

2.13.9 Harvesting and Postharvest Technology

The first crop comes to harvest about 2–3 years after the planting of seedlings or suckers. However, stabilized yields are obtained only from the fourth year onward and continue during the next 10–12 years. Harvesting starts in August–September at low altitudes and continues until December at high altitudes. Usually, harvesting is done in one round; hence, the harvested produce often contains capsules of varying maturity. Harvesting can be commenced when the seeds in the topmost capsules in the spike attain a dark-gray color. A special type of knife, known locally as *Elaichi chhuri* (meaning "cardamom knife") is used for the harvest. The stalk of the spike is cut close to the leafy shoot. After harvest, individual capsules are separated manually and then are cured to reduce their moisture content to 10–12%. The traditional curing known as the *Bhatti* curing system (direct heat drying) is followed. Large cardamom is also cured by the flue pipe curing system, involving indirect heating.

2.13.9.1 The Bhatti *System*

The local name for kiln is *Bhatti*. A *Bhatti* consists of a platform fabricated from bamboo mats or wire mesh, laid over a four-walled structure made of stone pieces with a V-shaped opening in the front for feeding firewood. Capsules are spread over the platform and are dried by direct heat generated from the burning of the firewood. To cure 100 kg of dry capsules, about 70 kg of firewood is required in this traditional kiln (Sundriyal et al., 1994). Both green and dry wood can be used, and as a result, a huge volume of smoke is generated that passes through the cardamom (Singh, 1978). Depending on the thickness of the cardamom spread, it takes 60–72 h to cure (John and Mathew, 1979). The color of cardamom cured in this system is dark brown to black. If smoke percolates through the cardamom spread, the original color is lost and the seeds have a smoky smell, which only will fetch a lower price in the market (Karibasappa, 1987).

2.13.9.2 Portable Curing Chamber

This is a prototype of the "Copra Dryer" (for the drying of dehusked coconut kernel) developed by the Central Plantation Crops Research Institute, Kasaragod, Kerala State, another research institute under the administrative umbrella of the Indian Council of Agricultural Research, New Delhi. The unit consists of an air-heating chamber and a furnace cylinder with a chimney. The whole unit is fitted in a detachable angular iron frame. The air-heating chamber is enclosed by asbestos sheets on all four sides, leaving the top open for drying. Some space is left below the walls for air to enter into the chamber. Fresh cardamom is spread on the platform (made of wire mesh size 3–5 cm) to a thickness of about a 15- to 20-cm layer. Firewood is burned inside the furnace cylinder. Air around the furnace gets heated up quickly and is convected upward, after which it passes through the produce, thereby drying it. The temperature of the heating chamber is regulated by the rate of burning of the firewood and also by regulating the chimney valves. It takes about 20 h to curing about 50 kg of raw samples (Annamalai et al., 1988).

2.13.9.3 Flue Pipe Curing System

The Spices Board of India has introduced a novel method of curing called the "flue pipe curing system" in place of the *Bhatti* system. The method was described earlier in Chapter 8, on the harvesting and processing of cardamom. The whole process of curing takes about 28–29 h. Cardamom capsules thus cured are immediately collected and rubbed in trays or processed in cardamom-polishing machine to remove the tail (Anon., 1998). Clean produce is then packed in polyethylene-lined gunny bags and stored in wooden boxes. The cardamom cured in this manner gives, on average, 25% of the freshly harvested produce (John and Mathew, 1979). Cardamom dried by this method has the following advantages over cardamom dried in the *Bhatti* system (Karibasappa, 1987):

1. The original pink color and sweet camphor aroma and flavor are retained.
2. The cured cardamom fetches a better market price.
3. Both curing expenses and firewood consumption are low.
4. Total curing takes only 28–29 h.
5. Uniform drying is ensured.

2.13.10 Natural Convection Dryer

A dryer similar to the one just described has been designed by the Central Food Technological Research Institute, Mysore, Karnataka State (Joseph et al., 1996). The dryer consists of a furnace, flue ducts, a wire mesh tray for charging capsules, and supporting structures. The furnace is fabricated from 8-mm-thick MS sheets. A brick lining inside the furnace provides insulation. Flue ducts are made of 1.6 mm GI sheets, and the ducts are arranged in two tiers, one over the other, with sufficient space in between. Two crimped steel wire mesh trays with a border are placed one over the other. Firewood is burned to generate heat for drying. The hot flue gases passing through the ducts set up convection currents in the air between the duct wall and wire mesh trays. The convection currents pass upward through the mesh and the bed of cardamom capsules on the mesh is subjected to drying. The dryer has a thermal efficiency of 5.6, better than that of the conventional flue curing kilns. It can dry 300 kg of large cardamom at a time to a level of 10% moisture in 12 h. This method is much more efficient than the conventional drying system (Joseph et al., 1996).

2.13.10.1 Gasifier Curing System

The gasifier curing system, an upgrade of the *Bhatti* system, was developed by Tata (the leading industrial house of India, with a great legacy behind it) Energy Research Institute (TERI) in Delhi. The solid fuel, firewood, is converted into gaseous fuel by partial combustion through gasification and a thermochemical reaction. A mixture of producer gas consisting of carbon monoxide, carbon dioxide, methane, and elemental nitrogen is obtained. The mixture is combustible and is used to provide burning. Through an updraft type of biomass gasifier, air enters the gasifier from the bottom and producer gas is taken out from the top to cure the cardamom. A prototype unit, which is fitted in the existing *Bhatti* curing system, has been successfully field-tested in Sikkim and an improved large cardamom with a better appearance and a more volatile oil content is obtained (Anon., 1998b).

The gasifier system is more advantageous than the traditional *Bhatti* system because the uncontrolled burning of firewood logs in the *Bhatti* system results in the loss of volatile oil; besides, exposure to smoke imparts a smoky smell to the volatile oil. There is also charring of capsules due to localized overheating, but the controlled burning in a gasifier system helps to retain more of the volatile oil. Also, the quality of the cured cardamom is better without the smoky smell, because of the clean burning of the gaseous fuel. Finally, the cured capsules retain the natural pink color of the harvested produce (Anon., 1998b).

2.13.11 Chemical Composition

On analysis, dried large-cardamom seeds were found to contain the following chemical composition and properties:

Moisture: 8.5%
Protein: 6%
Volatile oil: 2.8%
Crude fiber: 22%
Starch: 43.2%
Ether extract: 5.3%
Alcohol extract: 7%
One hundred grams of dried seeds contain
Calcium: 666.6 mg
Magnesium: 412.5 mg
Phosphorus: 61 mg
Fluoride: 14.4 ppm.

The seeds contain the glycosides petunidin 3,5-diglucoside, leucocyanidin 3-*O*-β-D-glucopyranoside, and subilin, an aurone glucoside. Cardamomum—a chalcone, alpinetin—and a flavanone are also found in the seeds (Shankaracharya et al., 1990). On steam distillation, the powdered seeds yield 1–3.5% of a dark-brown, mobile essential oil. An investigation has shown that the volatile oil content is 2.44% in *Sawney,* 2.42% in Pink *Golsey,* 2.25% in *Ramnag,* and 1.66% in *Ramsey* (Shankaracharya et al., 1990). The oil has the following properties: specific gravity (at 20 °C), 0.9142; refractory index (at 26 °C), 1.46; optical Rotation, –18° 3; acid value, 2.9; saponification value, 14.53; and saponification value after acetylation, 40.2. Large-cardamom oil is characterized by a flat cineol odor, a harsh aroma, and inferior flavor, as against the warm, spicy, aromatic odor of ordinary cardamom (Table 2.62). Large-cardamom oil is almost devoid of α-terpinyl acetate and is rich in 1,8-cineole (Balakrishnan et al., Gurudutt et al., 1996).

Table 2.62 Composition of Volatile Oil of Large Cardamom and Ordinary Cardamom

Constituent	Large Cardamom (Range in %)	True Cardamom (Range in %)
α-Terpinene	0.5–11.13	0.37–2.5
α-Pinene	2.0–3.11	1.10–13.00
β-Pinene	2.4–3.67	0.2–4.9
Sabinene	0.2–9.10	2.5–4.9
Camphene	0.44	0.02–0.13
ν-Terpinene	0.2–16.2	0.04–11.2
Limonene	6.38–10.3	0.12–2.1
p-Cymene	0.20–0.30	0.40–0.70
1,8 - cineole	63.3–75.27	23.4–51.30
Linalool	0.41	2.1–4.5
Geraniol	0.12	0.25–0.38
α-Terpineol	4.9–7.2	0.86–1.90
Terpinen-4-ol	1.42–2.0	0.14–15.3
Nerlidol	0.12–1.0	0.23–1.60
Nerlacetate	0.14	0.02–0.09
α-Terpinyl acetate	5.10	34.60–52.5
α-Bisabolene	1.3–3.6	0.07–0.83
β-Terpineol	0.8	0.70–2.10

Source: Shankaracharya et al. (1990), Balakrishnan et al. (1984).

2.13.12 Properties and Uses

Large cardamom is used as an ingredient, as well as a flavoring agent with masala and curry powders; in flavoring dishes, cakes, and pastries; as a masticatory; and for medicinal purposes. The seeds are used for chewing along with betel quid (a combination of beetle leaf, areca nut, and lime, with or without tobacco). In the Gulf countries, large cardamom is used as a cheaper substitute for spicing tea in place of ordinary cardamom. In the Indian systems of medicine, namely, *Ayurveda* and *Unani,* large cardamom is used as a preventive, as well as a curative, for throat trouble, lung congestion, inflammation of the eyelids, digestive disorders, and even in the treatment of pulmonary tuberculosis (Kirtikar and Basu, 1952). The seeds are fragrant adjuncts for other stimulants, bitters, and purgatives. The seeds have a sharp, good taste and are a tonic for the heart and liver. The pericarp is reported to be good for alleviating headache, and it is said to heal stomatitis (Anon., 1950). A decoction of the seeds is used as a gargle in afflictions of the teeth and gums. Large-cardamom seeds are used with melon seeds as a diuretic in treating kidney stones. Large-cardamom seeds promote the elimination of bile and are useful in liver congestion. They are also used in the treatment of gonorrhea. In large doses with quinine, they are used in neuralgia. The seed oil is applied to the eyes to prevent inflammation.

Cardamom can be directly used in the preparation of pickles, meat, and vegetable dishes and in the preparation of *pulao*, a rice-based fragrant preparation that is very popular in North Indian cuisine. Large cardamom also finds use in the industrial sector, to flavor toothpaste, sweets, soft drinks, toffees, milk, and alcoholic beverages. The ripe fruits are eaten raw by the people of Sikkim and Darjeeling and are considered a delicacy (Gyatso et al., 1980).

2.13.13 Conclusions

Large cardamom is a crop of the northeastern Himalayan tracts, the largest producer being Sikkim, in India. In South Asian countries, it is used extensively as a spice and as a substitute for true cardamom in the Middle East regions. Large cardamom is also important in tribal and indigenous medicine.

There is but meager information on research and development of this crop. Efforts have yet to be initiated to develop superior genotypes combining a high yield and quality. A search for aroma quality is essential to locating lines of superior flavor and quality composition. More important is the management of diseases and pests, especially developing resistant or tolerant lines against the two serious viral diseases known as *Chirke* and *Foorkey.* The rich genetic diversity found in the center of its origin (Sikkim and the adjoining areas) has to be screened to locate natural resistance against pests and diseases, as well as high yielders. A search for types that are adaptable to a lower elevation will be a boon if found: crops can then be cultivated in the lower hills. A good tissue culture protocol for large cardamom is available, and large-scale multiplication of some of the elite lines will provide disease-free superior planting material to the growers. Appropriate research is also needed to provide and popularize an efficient drying technology that will be suitable to the location and acceptable to small and medium-scale growers.

2.14 False Cardamom

In addition to true cardamom and large cardamom, all other crops that produce aromatic seeds are lumped together as false cardamom. They do not have much commercial importance, except for *Aframomum corrorima (Amomum melegueta,* the Korarima cardamom or "grains of paradise"), which is cultivated on a small scale in some West African countries. Still, most of these false cardamoms are important locally as spices and flavorants and as remedies for various ailments. Detailed studies of these species are lacking, although a few have been subjected to chemical analysis. This chapter briefly considers false cardamoms.

2.14.1 *Elettaria*

Sri Lankan wild cardamom *E. ensal* (Gaertn). Abheywickrme *(E. major* Thawaites)

Elettaria is the Sri Lankan wild cardamom, morphologically similar to true cardamom, but a more robust plant, bearing erect panicles and much-elongated fruits (3–5 cm). Some controversy continues to exist regarding the taxonomic status of this species. Burtt (1980), as well as Burtt and Smith (1983), did not treat it as a separate species, but included it under *E. cardamomum* only. Abheywickrama (1959) treated it as a separate species because of its more robust nature, patently different fruit size, and, of course, different chemical composition. Burtt is of the opinion that these characters are insufficient to separate the Sri Lankan wild cardamom into different species. His chemical analysis of the seeds showed that the two most important constituents of cardamom oil—1,8-cineole and α-terpinyl acetate—are present only in traces in Sri Lankan wild cardamom. Bernhard et al. (1971) carried out comparative chemical analyses of the different cardamoms, including Sri Lankan wild cardamom, to detect the various constituents of the seeds (Photo 2.8).

Photo 2.8 Wild cardamom.

Malaysian and Indonesian *Elettaria*

Elettaria longituba (Ridl.) Holtt., in Gard. Bull. Sing., 13, 238, 1950.

Elettaria longituba is a perennial, vigorous herb that is endemic to Malaysia and that seems to be conspecific with *E. aquatilis,* reported from Sumatra. It is a large species, and its flowers appear singly or in long intervals, with only a few flowers occurring per cincinnus. Flowering stolons (panicle) are 3 to 4m long, and the anthers dehisce by pores without a hairy flap. The fruit is long, globose, or pyriform, smooth or slightly ribbed. There is no reported use for the plant or its fruit (Holttum, 1950).

While studying the Zingiberaceae of Indonesia, Sakai and Nagamasu (2000) listed seven species of *Elettaria* from the region:

Elettaria rubida R. M. Sm. (described in *Botany Journal Linnaeus Society*, 85, 66, 1982) has a red inflorescence with orange flowers. The anther is ecristate, and the leaves are obovate with an attenuated base.

Elettaria stolonifera (K. Schum.) Sakai and Nagamasu (described in *Edinburgh Journal of Botany*, 57, 227–243, 2000). Plants are 0.8–1.5m tall and trail on the ground. Inflorescence is 25–60cm long, flowers are white, and anthers are 3mm long and dehisce in the upper half only. The anther crest is three lobed, with fruit unknown.

Elettaria kapitensis Sakai and Nagamasu (described in *Edinburgh Journal of Botany*, 57, 227–243, 2000). The plant is about 0.7m tall, inflorescence is 40 to 60cm long, flowers are white, the calyx is 17mm long, and the corolla tube and calyx are free above the ovary. The anthers dehisce throughout their length, and the anther crest is about 2mm long. The fruit is unknown.

Elettaria surculosa (K. Schum.) B. L. Burt and R. M. Smith (described in Notes of the Royal Botanical Garden, Edinburgh 31, 312, 1972. This species is identical to Elettaria multiflora R. M. Sm. described in Notes of the Royal Botanical Garden, Edinburgh 43, 452, 1986). Plants are about 1.2–2.0m tall, inflorescence creeps on the ground and is up to 2m long. It has white flowers, and the anthers dehisce by small pores covered with a flap with long hairs up to 2mm long. The fruit is about 40mm × 16mm long, ellipsoid, sparsely pubescent, and reddish brown in color.

Elettaria linearicrista Sakai and Nagamasu (described in *Edinburgh Journal of Botany*, 57, 227–243, 2000). Plants are about 0.6–1.5m tall, the leaves are narrowly oblong, inflorescence is about 40cm long, the axis is densely pubescent and somewhat erect, the flowers are white, the anthers dehisce by longitudinally elongated pores at the upper middle of the thecae, and the fruit is unknown.

Elettaria longpilosa Sakai and Nagamasu (described in *Edinburgh Journal of Botany*, 57, 227–243, 2000). Plants are 1–1.3m tall, the leaves are densely pubescent on the lower surface, inflorescence is about 50cm long and creeping just below the ground, the flowers are white, the anthers' thecae dehisce by a small pore just above the midpoint, the pore is covered with a flap with long hairs, the anther crest is deeply three lobed, and the fruit is unknown.

Elettaria brachycalyx Sakai and Nagamasu (described in *Edinburgh Journal of Botany*, 57, 227–243, 2000). The plants are about 1.0m tall, inflorescence is about 60cm long with white flowers, the calyx is about 6mm long and fissured for two-third of its length, and the anther's thecae dehisce by pores covered with a flap of long hairs, the anther crest is about 1mm long and three lobed, and the fruit is unknown.

2.14.1.1 Aframomum sp.

Korarima cardamom [*Aframomum corrorima* (Braun). Syn., *Aframomum melegueta* (Roscoe) Schum (*Meleguetta*) pepper, grains of paradise, or alligator pepper]. Korarima cardamom grows in the wild and is also cultivated sporadically in Ethiopia, Nigeria, and nearby regions. The species is endemic to this African region. It is a perennial aromatic herb with strong, fibrous, subterranean scaly rhizomes and a leafy stem of about 1–2 m high. The plants grow naturally at 1700–2000 m amsl in Ethiopia and in other tropical West African countries. They flower from January to September, and the fruits mature 2–3 months after flowering. The species is propagated by both seeds and rhizome parts. No cultivation practices have been recorded. The dried products are usually of poor quality due to improper drying, and that is usually the case with many rural regions of Africa. Dried fruits are either sold in local markets or exported to other African countries, with small quantities exported to Saudi Arabia, Iran, and other Gulf countries. A rust disease caused by *Puccinia aframom* Hans food is common to this plant.

The seeds contain 1–2% essential oil with a typical odor, and the species is called "nutmeg cardamom." Ajaiyeoba and Ekundayo (1999) carried out a gas chromatography–mass spectrography investigation on hydrodistilled oil from samples purchased from the local market in Nigeria. The samples were found to contain 27 compounds, of which α-humulene and β-caryophyllene made up the most, 82.6% of the oil.

Menut et al. (1991) compared the chemical composition of *Aframomum* and *Amomum* and came up with the following composition:

Type 1: *Aframomum angustifolium, Aframomum corrorima, Aframomum mala*, and *Aframomum muricatum*, which contain roughly equal amounts of hydrocarbons and oxygen-containing compounds.

Type 2: *Aframomum compactum, Aframomum kravanh,* and *Aframomum subulatum*, which contain oxygen-containing monoterpenes with 1,8-cineloe as the major constituent.

Type 3: *Amomum globosum* (from Thailand), *Amomum villosum, Amomum giganteum*, which contain oxygen-containing monoterpenes different from 1,8-cineloe.

Type 4: *Amomum globossum* (from China), which contains sesquiterpenes, α-humulene, and farnasol.

Type 5: *Amomum ptychloimatum*, which contains aliphatic compounds.

Uses

Korarima cardamom was once important in trade with Europe and part of the West African coast, which was then known as the "Grain Coast" (now Liberia). It was in demand by the local people as a spice and stimulating carminative and for a variety of other ailments. The fruit pulp around the seed is eaten especially before maturity and is chewed as a stimulant. The seeds are used as a spice to flavor all types of dishes in tropical African countries. The seeds are also used to flavor tea and coffee, as in Arab counties, and in some special kinds of breads. In Ghana, the seeds are used in enema preparations; in Lagos and Sierra Leone, they are used against fever and throat inflammations.

For external use, the seeds are crushed and rubbed on the body as a counter-irritant or applied as a paste to relieve headache, etc. (Dalziel, 1937). A whole-plant decoction is taken as a febrifuge. The roots have a cardamom-like taste, and a decoction is used to relieve constipation. The roots are regarded as a vermifuge against tapeworm, which infests humans through ill-cooked meat. The juice of the young leaves acts as a styptic (Dalziel, 1937). The juice was used in United Kingdom and United States in the preparation of medicines for cattle and also in the preparation of spiced wine (called hippocras), flavored with korarima cardamom, cinnamon, and ginger (Bentley and Trimen, 1880). The gastroprotective and antiulcer properties of an ethanol extract of korarima cardamom seeds were investigated in rats by Rafatullah et al.(1995), using pyloric legation, hypothermic restraint stress, indomethacin (indometacin), cystamine, and narcotizing agents. Galal (1996) studied the antimicrobial activity of 6-paradol and related compounds present in korarima cardamom. These compounds were active against *Mycobacterium chelonei, Mycobacterium intracellulare, Mycobacterium smegmatis*, and *Mycobacterium xenopi*. Escoubas et al. (1995) have shown that an extract of the seeds exhibited antifeedant activity against termites.

Aframomum daniellii Schum: This perennial herb occurring in West Africa grows to about 75 cm and has leaves 25–30 cm long. There are many red flowers in the inflorescence, and the capsule is red and smooth, with smooth seeds that are a shining olive brown with a white margin. The pulp of the seed is agreeably acidic and is eaten by the natives to refresh their breath because it has a turpentine-like taste. *Aframomum daniellii* relieves thirst during fever and inhibits the growth of several kinds of fungi and bacteria.

Aframomum granum-paradisi K. Schum (black *Amomum*): This species occurs in Ghana, Gambia, Sierra Leone, and southern Nigeria. The seeds have a shiny surface, are aromatic, and possess a camphoraceous flavor. The seeds are used together with shea butter (from *Vittellaria paradoxa* Gaertner, Sapotaceae) to make body pomade. The natives use the whole plant medicinally, as with *Aframomum corrorima.* The leaves are cooked with food to impart flavor. The plant has been wrongly identified as grain of paradise.

Aframomum latifolium K. Schum (large or grape seeded *Amomum*): This species is indigenous to Sierra Leone. The seeds resemble grape pits and are smooth, shiny, and weakly aromatic. The ripe fruit is acidic and is eaten by the natives to refresh the breath; it is also believed to be a good reliever of fatigue and is used in fever (Dalziel, 1937). A decoction of the whole plant is mixed with the leaves of *Morinda lucida* (Rubiaceae) and used by the Susus tribe in a daily bath to restore strength after fever and to relieve fatigue.

Aframomum biauriculatum K. Schum: Chisowa et al. (1998) analyzed the rhizome of this species and identified 29 compounds in the rhizome oil and 17 compounds in the leaf oil. More than 58% of the leaf oil is β-pinere.

Aframomum sanguineum K. Schum: This species is a native of West Africa. The seeds of the plant are used in cosmetics, to flavor tea and rice, and in traditional medicine to prevent throat infections and to treat stomachache, dysentery, and snakebite. Among the most important constituents in the essential oils, it is 1,8-cineole which is more than 38% of the total (Hari et al., 1994).

2.14.1.2 Amomum *sp.*

Bengal Cardamom Amomum Aromaticum Roxb (Jalpaiguri Cardamom)

This is a perennial herb that grows to a height of about 1 m and is indigenous to Assam, Meghalaya, Sikkim, North Bihar, and northern West Bengal, extending to Nagaland and Uttar Pradesh. The leaves are oblong-lanceolate, and the plant has pale-yellow flowers and rugose capsules. On steam distillation, seeds yield 1–2% of essential oil, which has a strong camphoraceous and cineol-like odor and taste and contains large quantities of cineole. The seeds, which have medicinal properties, are used as a spice in place of *Amomum subulatum,* the large cardamom.

Amomum Dealbatum Roxb

This is a herbaceous perennial with large oblong-lanceolate leaves and a globose, short, peduncled spike. The plant occurs in the Khasi hills of the eastern Himalayas. On distillation, the leaves give an essential oil (0.018%) whose chief constituents are 1β-pinene and, to a lesser extent, α-pinene. The seeds are feebly aromatic in taste and odor, and they are used as a substitute for large cardamom.

Amomum Longiligulare Wu

This is a species that occurs in Vietnam and is used locally for medicinal purposes as well as to flavor food. It is used as a medicine to treat dyspepsia, vomiting, diarrhea, and dysentery. Of the different constituents, the seed oil contains nearly 50% camphor (Dung and Thin, 1992).

Amomum tsao-ko Crevost Lemarim (Vietnam cardamom)

This herb stands 2–3 m high and has oblong-to-lanceolate leaves. The capsules are 2.5 cm long and oblong. The plant is cultivated in South China and Vietnam, especially in the forest regions of the Hoang Lien Son Mountain. The seeds contain 1–2% of essential oil of which the main constituent (nearly 35%) is 1,8-cineole. Nearly 80% of the chemical constituents have been reported in the seed oil (Lizhu-Qiang et al., 1998). The seeds are used as a spice in Vietnam and China as a substitute for true cardamom. The species is also used as a local remedy for many ailments, including dyspepsia, diarrhea, and indigestion. Seven antioxidants have been found in the plant (Fang Jen and De, 1996).

Amomum Krevanh

Indigenous to Thailand, where it is used in traditional medicine, this species is often called Thai cardamom. Kamchonwongpaisan et al. (1995) have isolated a novel diterpene peroxide, the structure of which they then elucidated. The compound exhibited potent activity against *Plasmodium falciparum.*

Amomum Xanthiodis Wall (Tavoy cardamom)

This plant is distributed in the forests of the Indo–Burma, Malayan region. It is a perennial herb with a leafy stem reaching about 1.5–2 m. Spikes are globose and shortly peduncled, with few flowers. Fruits (the capsule) are oblong and pale brown.

The plant, which is anodyne, is good at controlling diarrhea, dysentery, cramps, hiccups, nausea, splenitis, and stomachache. It also can be used as a carminative and sedative. The seeds contain a number of chemical constituents, including D-camphor, linalool, nerolidol, and terpene, and are effective in relieving painful urination and regulating bowel movements. For this use, powdered seed mixed with butter is administered. The Chinese and Burmese use this species as an important condiment (Kirtikar and Basu, 1952).

Amomum Compactum Soland (Round Cardamom) Syn. Amomum Kepulaga Sprague and Burkill

Round cardamom, also known as Siam cardamom, is reported to be antitoxic, antiemetic, carminative, and stomachic. In China, it is a folk remedy for ague, cachexia, cancer, catarrh, cold, coughs, cramps, dyspepsia, gout, heartburn, nausea, hepatitis, ophthalmia, rheumatism, and vomiting. It is rarely used alone, more frequently in combination with other plant products. Mixed with fresh egg yolk, it is given during parturition. Elsewhere, the plant is used in folk remedies for indurations of the liver and uterus and against cancer (Perry and Metzger, 1980).

Amomum Constalum Benth

Endemic to the Eastern Himalayan forests, this is a perennial herb with a stout, leafy stem that is 1.5–2 m long. Spikes are peduncled and 5–7.5 cm long. Seeds are aromatic. In Chinese medicine, the seeds are employed for ailments of the stomach and for asthma, pulmonary afflictions, and general debility. Tribal members use the seeds to relieve stomachache.

Amomum Pavieanum Pierre and Gagnep

An infusion of this plant, endemic to Southeast Asia, is used in the treatment of diarrhea and general debility following certain dysenteries. The rhizome is used as a seasoning agent in the eastern and southern parts of Thailand (Perry and Metzger, 1980). The oil yield (0.23% v/w) is colorless, with an anise-like odor (Scheffer et al., 1988). Gas chromatographic analysis has identified 41 compounds in the rhizome oil. The dominant (91.6%) compound of the oil is methylchavicol. Only four other components amounted to more than 1%: α-pinene, camphene, β-pinene, and camphor. The composition of *Amomum pavieanum* is quite distinct from that of other species, being marked by the absence of 1,8-cineole as the dominant component. There appears to be no other *Amomum* sp. having an oil composition resembling that of *Amomum pavieanum* (Scheffer et al., 1988).

Amomum Acre Valeton

This is an Indonesian species found mainly in the Sulawesi region. The fruit and inner part of the petioles are used for pickling, often in vinegar, and subsequently are used as a spice and a flavorant. Very young stem is used directly. The plant itself is poorly understood. Two forms—*Rombo* and *Kautopi*—have been distinguished (De Guzman and Siemonsma 1999). There is only scarce information on the chemical composition of the seed oil, etc.

Amomum Ochreum Ridley

This species occurs in Peninsular Malaysia. The seeds are used locally as a substitute for cardamom. The plant is 3–4 m tall and produces inflorescences (spikes) from the base. The flowers have an obovate three-lobed labellum that isoorange yellow with red veins and spots. The anther appendage is transversely oblong and faintly red spotted (De Guzman and Siemonsma 1999).

Amomum Testaceum Ridley

This species occurs in Thailand, Vietnam, Peninsular Malaysia, and Borneo, and is only occasionally cultivated. The seeds are aromatic and locally used the way true cardamom is used. The plant is a rhizomatous herb with leafy shoots about 3 m tall and inflorescence an oblongoid spike emerging from the rhizome. The flowers have an obovate labellum that is white with a broad, dull-yellow patch toward the apex and a paler yellow median band flanked by purple lines. The stamens have an 8-mm-long filament and a three-lobed anther appendage (De Guzman and Siemonsma 1999). The fruit is slightly ribbed, pinkish, and slightly hairy. The seeds are brown, aril thin, and aromatic, and their chemical composition is unknown.

Amomum Xanthophlebium Baker (Syn. Amomum Stenoglossum Baker)

This plant occurs in Peninsular Malaysia and Borneo. The flowers are used to flavor curries. The plant is a rhizomatous herb, with a leaf stem up to 4–5 m tall, and inflorescence consists of ellipsoidal spikes arising from the rhizome. The flowers have an obovate labellum with crinkled edges and are white suffused with red stripes and spots. The fruit is obovoid, smooth, suppressed, silky, and hairy (De Guzman and Siemonsma 1999). The flowers are used by locals to flavor various dishes. The seeds are used by tribal members in local medicinal preparations. There is no information on the chemical composition.

The following species of Amomum occur in the Western Ghat forests of southern India: *Amomum involucratum* Benth, *Amomum hypoleucum* Thw, *Amomum cannicarpum* (Wight) Benth. ex Baker, *Amomum muricatum* Beddome, *Amomum ghaticum* Bhatt, *Amomum masticatorium* Thwaites, and *Amomum pterocarpum* Thwaites. There is no information about their use in local or tribal medicine. No study has been undertaken with regard to the chemical composition of the seeds of these species.

2.15 Specification for Cardamom

2.15.1 Requirements

2.15.1.1 Cardamom with Capsules

The cardamoms shall be nearly ripe fruits of *Elettaria cardamomum* (L.) Maton in the form of capsules that have been dried. The capsules will be of a color ranging from light green to brown, cream, and white, and will be global in size or three cornered and having a ribbed appearance. The capsules may be clipped and their pedicels removed. The capsules shall be well formed and with sound cardamom seeds inside. The capsules may also be bleached. The cardamoms may be graded on the basis of color, clipping, size, mass per liter, bleaching or otherwise, proportions of extraneous matter, and place of origin.

The Cardamom Seeds

Capsules may be decorticated and the separated seeds packed for trade purposes. The cardamom seeds may be graded on the basis of mass per liter and extraneous matter.

Taste and Aroma or Flavor

The taste and aroma or flavor of cardamom capsules and seeds shall be characteristic and fresh. The material shall be free from foreign taste and aroma or flavor, including rancidity and mustiness. Cardamom capsules and seeds shall also be free from visible moulds and insect infestation.

Mass per Liter

The mass per liter of cardamom capsules and seeds shall be determined in accordance with the method given in 4 of Indian Standards 1797–1973.

Moisture content: The moisture content in all grades of cardamom except the grade "Bleached" or "Half Bleached" shall not exceed 10% when determined in accordance with the method given in 10 of Indian Standards 1797–1973. The moisture content in the grade "Bleached" or "Half Bleached" shall not exceed 13% when determined by the same method.

Volatile Oil Content

The volatile oil content of different grades of cardamom capsules shall not be less than 3.5% on a capsule basis, and not be less than 4% on a seed basis, when determined in accordance with the method given in 15 of Indian Standards 1797–1973.

2.15.1.2 Extraneous Matter

The proportion of calyx pieces, stalk bits, and other extraneous matter in cardamom in capsules and seeds shall not be more than 5% and 2% by mass, respectively, when determined in accordance with the method given in 5 of Indian Standards 1797–1973.

Empty and Malformed Capsules

The proportion of shriveled capsules that are not fully developed shall not be more than 7% by mass and shall be determined after separating them in accordance with the method given in 5 of Indian Standards 1797–1973.

Blacks and Splits

Blacks are capsules having a visible blackish to black color, and splits are those capsules that are open at the corners for more than half the length. The proportion of blacks and splits shall not be more than 15% by count. For this purpose,

100 capsules shall be taken from the sample and the number of blacks and splits separated and counted.

Light Seeds

The proportion of light seeds in cardamom seeds shall not be more than 5% by mass when separated in accordance with the method given in 5 of Indian Standards 1797–1973. Light seeds shall be seeds that are brown or red, broken, or immature and shriveled.

Grades

The cardamoms in capsules and seeds may also be graded before packing. There shall be 25 grades of cardamom with capsules of three cardamom seeds. The designations of the grades and their requirements are given in Table 2.63.

2.15.1.3 Packing and Marking

Packing

Capsules: Cardamom capsules shall be packed in clean, sound, and dry tinplate containers or wooden cases suitably lined with polythene or waterproof paper or kraft paper or in new jute bags lined with polythene waterproof paper.

Seeds: Cardamom seeds shall be packed in clean and dry tinplate containers or wooden cases lined with polyethylene or waterproof paper or kraft paper.

Marking

The following particulars shall be marked or labeled on each container:

1. Name of the material and the trade name or brand name, if any
2. Name and address of the manufacturer or the packer
3. Batch or code number
4. Net mass, in metric units
5. Grade of the material (if graded)
6. Country of origin.

Table 2.63 Grade Designations of Cardamom Seeds and Their Requirements

Grade Designation	Trade Name	Extraneous Matter, by Mass (%)	Light Seeds, by Mass (%)	Mass(g liter^{-1} min^{-1})	General Characteristics
CS1	Prime	0.5	3.0	675	Decorticated and dry seeds of any variety of *Elettaria cardamomum*
CS2	Shipment	1.0	5.0	660	
CS3	Brokens	2.0	–	–	

Sampling: Representative samples of the material shall be drawn and tested for conformity to this specification as prescribed in 3 of the Indian Standards 1797–1973.

2.15.1.4 Specification for Large Cardamom

Scope: This standard specifies the requirements and methods of sampling and testing for large cardamom, in capsules and seeds of *Amomum subulatum* Roxb., and other related species.

Reference:

The following Indian Standards (IS) are necessary adjuncts to this standard:

IS No: 1070: 1977 Title

1797: 1985 Water for general laboratory use

13, 145: 1991 Method of sampling for spices and condiments.

2.15.1.5 Requirements

Description

Large cardamom in capsules: Large cardamom is the dried, nearly ripe fruit of *Amomum subulatum* Roxb. and other related species. The capsules shall be of a color ranging from brown to pink, ovoid, and more or less triangular shaped with a ribbed appearance. The capsules may be clipped, and their pedicels removed. The capsules shall be well formed with sound seeds inside.

Large cardamom seeds: The cardamom capsules may also be decorticated and the seeds separately packed for trade purposes.

Odor and taste: The odor and taste of large-cardamom capsules and seeds shall be characteristic and fresh. The capsules shall be free from foreign odor and taste.

Freedom from insects, molds: Large-cardamom capsules and seeds shall be free from living insects and mold and shall be practically free from dead insects and rodent contamination visible to the naked eye with such magnification as may be necessary. If the magnification exceeds 10 times, this fact shall be stated in the test report.

Extraneous matter: Large-cardamom capsules and seeds shall be free from visible dirt or dust. The proportion of pieces of calyx and stalk and other extraneous matter shall not be more than 5% when determined by the method specified in 4 of IS 1797: 1985.

Empty and shriveled capsules: The proportion of immature and shriveled capsules shall not be more than 7% when determined after separating them in accordance with the method given in 4 of IS 1797: 1985.

Light seeds: The proportion of light seeds in large-cardamom seeds shall not exceed 5% when determined in accordance with the method given in 4 of IS 1997: 1985.

Mass per liter: The mass of large-cardamom capsules and any seeds contained shall be determined according to the method given in 3 of IS 1797: 1985.

Chemical requirements: Large-cardamom capsules and seeds shall also comply with the requirements given in Table 2.64.

Table 2.64 Grades and Specifications for Cardamom

Grade Designation	Trade Name	Extraneous Matter, by Mass (max. %)	Empty and Extraneous Malformed Matter, by Capsules, by Mass (max. %) Count (max. %)	Unclipped Capsules, by Count (max. %)	Immature and Shriveled Capsules, by Weight (max. %)	Blacks and Splits, by Count (max. %)	Size (DIA) of Hole of the Sieve on which Retained (mm)	Mass Minimum (g liter^{-1})	Color	General Characteristics
Alleppey AG, green										
AGEB	Cardamom extra bold	Nil	2.0	Nil	2.0	Nil	7.0	435	Deep green	Kiln dried, three cornered, and having a ribbed appearance
AGB	Cardamom bold	Nil	2.0	Nil	2.0	Nil	6.0	415	Green or light green, creamy	
AGS	Cardamom superior	Nil	3.0	Nil	5.0	Nil	5.0	385		
AGS 1	Shipment green 1	Nil	5.0	Nil	7.0	10.0	4.0	350		
AGS 2	Shipment green 2	Nil	5.0	Nil	7.0	12.0	4.0	320		
AGL	Light green	Nil	–		–	15.0	3.5	260		
Coorg CG, green										
CGEB	Coorg extra bold	Nil	Nil	Nil	Nil	Nil	8.0	450	Golden light greenish or brownish to brown	Global shape
CGB	Bold	Nil	2.0	Nil	3.0	Nil	7.5	435		

CG 1	Superior	Nil	3.0	Nil	5.0	Nil	6.5	415		Skin ribbed or smooth; pedicel separated
CG 2	Superior coorg green or mota green	Nil	5.0	3.0	7.0	Nil	6.0	385		
CG 3	Shipment	Nil	5.0	5.0	7.0	10.0	5.0	350		
CG 4	Light	Nil	–	–	–	15.0	3.5	280		
Bleached or Half Bleached										
BL 1		Nil	Nil	Nil	Nil	Nil	8.5	340	Pale creamy dull white	Fully developed capsules are bleached, global, or three cornered with ribbed or smooth skin
BL 2		Nil	Nil	Nil	Nil	Nil	7.0	340		
BL 3		Nil	Nil	Nil	Nil	Nil	5.0	340		
Bleached White										
BW 1	Mysore/ Mangalore bleachable cardamom	Nil	1.0	Nil	Nil	Nil	7.0	460	White, light green, or light gray	Fully developed capsules suitable for bleaching
BW 2	Mysore/ Mangalore bleachable cardamom	Nil	1.0	Nil	Nil	Nil	7.0	460		
BW 3	Mysore/ Mangalore bleachable bulk cardamom	Nil	2.0	Nil	Nil	Nil	4.3	435	–	

(Continued)

Table 2.64 (Continued)

Grade Designation	Trade Name	Extraneous Matter, by Mass (max. %)	Empty and Extraneous Malformed Matter, by Capsules, by Mass (max. %) Count (max. %)	Unclipped Capsules, by Count (max. %)	Immature and Shriveled Capsules, by Weight (max. %)	Blacks and Splits, by Count (max. %)	Size (DIA) of Hole of the Sieve on which Retained (mm)	Mass Minimum ($g liter^{-1}$)	Color	General Characteristics
BW 4	Mysore/ Mangalore bleachable bulk cardamom, unclipped	Nil	2.0	Nil	Nil	Nil	4.3	435	–	
Mixed			2.0	–					–	
MEB	Mixed extra bold	–	2.0	–	2.0	Nil	7.0	435	–	
MB	Mixed bold	–	2.0	–	2.0	Nil	6.0	415	–	
MS	Mixed superior	–	3.0	–	5.0	Nil	5.0	385	–	Dried and mixed capsules of different varieties of *Elettaria cardamomum*
MS 1	Mixed shipment I	–	5.0	–	7.0	10.0	4.0	350	–	
MS 2	Mixed shipment II	–	5.0	–	7.0	12.0	4.0	320	–	
ML	Mixed light	–	–	–	–	15.0	3.5	260	–	

[a]The determination of moisture content and total ash shall be made on the whole capsules.
[b]The determination of volatile oil shall be made on the seeds obtained after separation of the skin and decortication.

Table 2.65 Chemical Requirements

Serial No.	Characteristics	Requirement	Methods of Test; Reference to CI of IS 1797:1985
1	Moisture content, percent (mm^{-1}), max.	12.0	9
2	Volatile oil content (ml 100 g^{-1}) on dry basis, minimum	1.0	15

Packing

Cardamom capsules shall be packed in clean, sound, and dry tinplate containers or in suitably lined wooden cases or suitably lined new jute bags. Lining materials may be, for example, waterproof paper, kraft paper, or plastic material.

Cardamom seeds shall be packed in clean and dry tinplate containers or in suitably lined wooden cases—for example, waterproof paper, kraft paper, or plastic material of food-grade quality.

Marking

The following particulars shall be marked or labeled on each container:

1. Name of the material, and trade name or brand name, if any
2. Name and address of the manufacturer or packer
3. Batch or code number
4. Net mass
5. Year of harvest
6. Sampling.

Representative samples of large cardamom in capsules and seeds shall be drawn by the method prescribed in IS 13,145: 1991.

Tests

Tests shall be carried out in accordance with 3.4, 3.6, 3.7, 3.8, and 3.9 and column 4 of Table 2.65.

Quality of Reagents

Unless specified otherwise, pure chemicals and distilled water shall be employed in tests. (Note: "Pure chemicals" means chemicals that do not contain impurities that affect the results of testing and analysis.)

References

Abheywickrama, B.A., 1959. A provisional checklist of the flowering plants of Ceylon. Ceylon J. Sci. (Biol. Sec.) 2, 119–240.

Abraham, C.C., 1975. Insect pests of stored spices and their control. Arecanut. Spices Bull. 7, 4–6.

Abraham, P., 1956. Spices as intercrops in coconut and arecanut gardens. Arecanut J. 7, 56–58.

Abraham, P., 1957. "Karuna" is excellent as shade tree for cardamom. Indian Farming 8 (2), 34–38.

Abraham, P., 1958. New knowledge for cardamom growers. Indian Farming 8 (2), 34–38.

Abraham, P., 1965. The Cardamom in India. ICAR Publication, New Delhi. pp. 1–46

Abraham, P., Thulasidas, G., 1958. South Indian Cardamom and their Agricultural Value. Indian Council of Agricultural Research Bulletin, vol. 79. ICAR, New Delhi, pp. 1–27.

Abraham, V.A., Gopinathan, K.V., Padmanabhan, V., Saranappa, Deforestation and change of micro–macro climate conditions. Cardamom J. 12 (18), 3–7.

Adams, F., 1971. Ionic concentrations and activities in soil solutions. Soil Sci. Soc. Am. Proc. 35, 420–426.

Adams, F., 1974. Soil solution. In: Carson, E.W. (Ed.), The Plant Root Environment (pp. 441–481). University Press of Virginia, Charlottesville.

Agnihothrudu, V., 1968. Description of the fungus *Phaeodactylium venketesanum* on cardamom. Proc. Indian Acad. Sci. Sect. B 68, 206–209.

Agnihothrudu, V., 1969. A leaf disease of cardamom from Kerala with a note on fungi found on cardamom and allied genera all over the world. Cardamom News Annu., 35–40.

Agnihothrudu, V., 1974. Is there a bacterial disease in cardamom? Cardamom News 6, 5.

Agnihothrudu, V., 1987. Diseases of small and large cardamom. Rev. Trop. Plant. Pathol. 4, 127–147.

Ahlawat, Y.S., Raychaudhuri, S.P., Yora, K., Dot, Y., 1981. Electron microscopy of the virus causing Foorkey disease of large cardamom, *Amomum subulatum* Roxb. Natl. Acad. Sci. Lett. 4 (4), 165.

Aiyappa, K.M., Nanjappa, P.P., 1967. Highlights on Cardamom Research: Its Problems and Prospects. Cardamom Industry. Cardamom Board, Cochin, Kerala.

Ajaiyeoba, E.O., Ekundayo, O., 1999. Essential oil constituents of *Aframomum melegueta* (Roscoe) K. Schum. Seeds (alligator pepper) from Nigeria. Flavour Fragr. J. 14, 109–111.

Alagianagalingam, M.N., Kandaswamy, T.K., 1981. Control of capsule rot and rhizome rot of cardamom (*Elettaria cardamomum* Maton.). Madras Agric. J. 68, 564–567.

Ali, M.I.M., 1982. Field evaluation of fungicides against leaf blotch disease of cardamom. Pesticides 11, 38–39.

Ali, S.S., 1983. Nematode problems in cardamom and their control measures. Sixth Workshop of All India Co-ordinated Spices and Cashewnut Improvement Project, Calicut, Kerala, India, November 10–13, 1983.

Ali, S.S., 1984. Effect of three systemic nematicides against root knot nematodes in a primary nursery of cardamom. In: First International Congress on Nematology, Ontario, Canada, August 1984 (abstract).

Ali, S.S., Venugopal, M.N., 1993. Prevalence of damping off and rhizome rot disease in nematode infested cardamom nurseries in Karnataka. Curr. Nematol. 4 (1), 19–24.

Aloskar, L.V., Kakkar, K.K., Chakre, O.J., 1992. Second Supplement to Glossary of Indian Medicinal Plants, Part 2. Publication and Information Directorate, New Delhi. p. 289.

Al-Zuhair, H., El-Sayeh, B., Ameen, H.A., Al-Shoora, H., 1996. Pharmacological studies of cardamom oil in animals. Pharmacol. Res. 34, 79–82.

Anand, T., Govindaraju, C., Sudharshan, M.R., Srinivasulu, P., 1998. Epidemiology of vein-clearing of small cardamom. Indian J. Virol. 14, 105–109.

Anandaraman, S., Reineccius, G.A., 1980. Microencapsulation of flavours. Food 1 (9), 14.

Annamalai, S.J.K., Patel, R.T., John, T.D., 1988. Improved curing method for large cardamom. Spice India 1 (4), 5.

Annual Report for 1996–1997. Indian Institute of Spices Research (IISR), Calicut, India (1996–1997).

Anon., 1950. Amomum. Garden Bull. 13, 192–214.

Anon. (1952a). The Wealth of India Council of Scientific and Industrial Research, New Delhi, India.

Anon., 1952. *Elettaria cardamomum* Maton. (Zingiberaceae). The Wealth of India. CSIR, New Delhi, India.

Anon., 1955. Final Report of the ICAR. Scheme for Scientific Aid to Cardamom Industry in South India Madras State (from October 1944 to March 1954).

Anon., 1958. South Indian cardamoms: their evolution and natural relationships Tech. Bull. 57. ICAR, New Delhi.

Anon., 1970. Cardamom culture. Part I.. Cardamom News 4 (6), 2–3.

Anon., 1972. Eighth Annual Report. University of Agricultural Sciences, Bangalore. p. 191

Anon., 1975. C.C. Spice open new mill. Food Process. Ind. 44 (529), 36.

Anon., 1976. Cardamom in Karnataka, UAS Tech. Series No. 14. University of Agricultural Sciences, Hebbal, Bangalore. p. 5.

Anon., 1977. Pulverizing system. Food Process 38 (3), 108.

Anon., 1978. New clone developed. Farmers Friend 3 (2), 4–5.

Anon., 1979. Propagation of Cardamom (Nursery Practices of Cardamom). University of Agricultural Sciences, Regional Research Station, Mudigere. p. 5.

Anon., 1980. Pests and Diseases in the Main Plantation in Cardamom Package of Practices. Cardamom Board, Cochin, Kerala, India.

Anon., 1982. Condiments and spices. In: Handbook of Agriculture. ICAR, New Delhi, India.

Anon., 1984. Large Cardamom: Package of Practices. Cardamom Board, Gangtok, Sikkim.

Anon., 1985. Cardamom Package of Practices, Pamphlet No. 9. Central Plantation Crops Research Institute, Kasaragod, India. p. 30.

Anon., 1985. Cardamom Package of Practices, Pamphlet No. 28. Central Plantation Crops Research Institute, Kasaragod, India. p. 20.

Anon., 1986. Annual Report 1986. ICRI, Myladumpara. pp. 51–53

Anon., 1986. Annual Report. ICRI, Spices Board, India.

Anon., 1986. Cardamom Cultivation. Cardamom Board, Cochin. p. 11.

Anon., 1987. Annual Report. ICRI, Spices Board, India.

Anon., 1988. Economics of Cardamom Cultivation for Small Growers, Spices Boards, Cochin, Kerala, India.

Anon., 1989. Cardamom Package of Practices. National Research Center for Spices, Calicut, Kerala, India.

Anon., 1989. Bi-annual Report 1987–1989. Indian Cardamom Research Institute, Myladumpara. pp. 41–47.

Anon., 1989. Annual Report for 1988–1989. National Research Center for Spices, Calicut, Kerala, India. pp. 37–38.

Anon., 1990. Cardamom Nursery Management Extension Folder No. 1. Spices Board, Indian Cardamom Research Institute, Myladumpara. p. 7.

Anon., 1991. Post Harvest Technology of Cardamom: Project Report. Tea Research Sub Station, United Planters Association of South India (UPASI), Vandiperiyar, Kerala, India.
Anon., 1993. Consolidated Report on Rootgrubs. Indian Cardamom Research Institute, Spices Board, p. 116.
Anon., 1993. Nutritional Management for Cardamom. Technical Guide for Cardamom Planters. Spices Board, Cochin.
Anon., 1997. Cardamom: Package of Practices. Spices Board, Kochi, p. 10.
Anon., 1998a. Final report of ICAR ad-hoc scheme "Evaluation of crop response to application of micronutrients to small cardamom (*Elettaria cardamom* Maton.)," p. 10.
Anon., 1998. Design, Development and Field-testing of an Advanced Cardamom-curing Prototype for Sikkim. Project Report. Tata Energy Research Institute, New Delhi, India.
Argersinger Jr., W.J., Davidson, A.W., Bonner, O.D., 1950. Thermodynamics and ion exchange phenomena. Trans. Kans. Acad. Sci. 53, 404–410.
Ayyar, T.V.R., 1935. A new species of *Thysanoptera* from S. India (*Taeniothrips cardamomi* sp. nov.). Bull. Entomol. Res. 26, 357–358.
Azad Thakur, N.S., Sachan, J.N., 1987. Insect pests of large cardamom (*Amomum subulatum* Roxb.) in Sikkim. Bull. Entomol. 28, 46–58.
Badei, A.Z.M., El-Akel, A.T.M., Morsi, H.H.H., 1991. Evaluation of chemical, physical and antimicrobial properties of cardamom essential oil. Bull. Faculty Agric. University of Cairo 42 (1), 183–197.
Badei, A.Z.M., Morsi, H.H.H., El-Akel, A.T.M., 1991. Chemical composition and antioxidant properties of cardamom essential oil. Bull. Faculty Agric. University of Cairo 41 (1), 199–215.
Bajaj, Y.P.S., Furmanova, M., Olszowska, O., 1988. Biotechnology of the Micro-Propagation of Medicinal and Aromatic Plants,, vol. I. Springler-Verlag, Berlin. pp. 60–103.
Bakan, J.A., 1973. Microencapsulation of foods and related products. Food Technol. 27 (11), 33.
Bakan, J.A., 1978. Microencapsulation. In: Peterson, M.S., Johnson, R. (Eds.), Encyclopedia of Food Science. Avi. Pub. Co. Inc., Westport Conn., USA.
Balakrishnan, K.V., George, K.M., Mathulla, T., Narayana Pillai, O.G., Chandran, C.V., Verghese, J., 1984. Studies in Cardamom. 1. Focus on oil of *Amomum subulatum* Roxb. Indian Spices 21 (1), 9–12.
Balasimha, D., 1989. Light penetration patterns through arecanut canopy and leaf physiological characteristics of intercrops. J. Plantation Crops (Suppl), 61–67.
Balu, A., 1991. Studies on storage pests of cardamom. Third Annual Research Council Meeting Agenda, Indian Cardamom Research Institute, p. 27. Spices Board, India.
Balu, A., Gopakumar, B., Chandrasekar, S.S., 1991. Third Annual Research Council Meeting Agenda. Indian Cardamom Research Institute, Spices Board, India, pp. 25–26.
Bambawale, P.M., 1980. Economics of sprinkler irrigation system. Cardamom 12 (12), 21–27.
Banerjee, S., Sharma, R., Kale, R.K., Rao, A.R., 1994. Influence of certain essential oils on carcinogen-metabolising enzymes. Nutr. Cancer 21, 263–269.
Barber, S.A., 1974. Influence of plant root on ion movement in soils. In: Carson, E.W. (Ed.), The Plant Root and its Environment (pp. 525–564). University of Virginia, Charlottesville.
Barber, S.A., 1984. Soil Nutrient Bioavailability: A Mechanistic Approach.. Wiley, New York.
Basu, A.N., Ganguly, B., 1968. A note on the transmission of "foorkey disease" of large cardamom by the aphid, *Micromyzus kalimpongensis* Basu. Indian Phytopathol 21, 127.
Bavappa, K.V.A., 1986. Research at CPCRI. Tech. Bull., CPCRI, Kasaragod, India. p. 34
Beckett, P.H.T., 1971. Potassium potentials: a review. In: Potash Review. Int. Potash. Inst. Berne, Switzerland, pp. 1–41.

Beeson, C.F.C., 1941. The Ecology and Control of Forest Insects of India and Neighbouring Countries. F.R.I., Dehradun, India. p. 1007.

Belavadi, V.V., Parvathi, C., 1998. Estimation of honey bee colonies required for effective pollination in cardamom. In: Proceedings of the National Symposium on Diversity of Social Insects and other Anthropods and the Functioning Ecosystems, III Congress of IUSSI. Mudigere, p. 30.

Belavadi, V.V., Chandrappa, H.M., Shadakshari, Y.G., Parvathi, C., 1993. Isolation distance for seed gardens of cardamom (*Elettaria cardamomum* Maton.). In: Veeresh, G.K., Umashanker, R., Ganeshan, K.N. (Eds.), Pollination in Tropics (pp. 241–243).

Belavadi, V.V., Venkateshalu, V.V., Vivek, H.R., 1997. Significance of style in cardamom corolla tubes for honeybee pollination. Curr. Sci. 73, 287–290.

Belavadi, V.V., Parvathi, C., Raju, B., 1998. Optimal foraging by honey bees on cardamom. In: Proc. XIII Plantation Crops Symposium, December 16–18, Coimbatore, India (abstract).

Bentley, R., Trimen, H., 1880. Medicinal Plants, vol. 1V (reprint). J.A. Churchill, London. pp. 267–268.

Bernhard, R.A., Wijesekera, R.O.B., Chichester, C.O., 1971. Terpenoids of cardamom oil and their comparative distribution among varieties. Phytochemistry 10, 177–184.

Bhandari, N., 1989. Spices beauty and body care. In: Strategies for Export Development of Spices. Spices Board, Cochin, pp. 54–56.

Bhat, K.S., 1974. Intensified inter/mixed cropping in areca garden: the need of the day. Arecanut Spices Bull. 5, 67–69.

Bhat, K.S., Leela, M., 1968. Cultural requirement of arecanut. Indian Farming 18 (4), 8–9.

Bhowmik, T.P., 1962. Insect pests of large cardamom and their control in West Bengal. Indian J. Entmol. 24, 283–286.

Bhutia, D.T., Gupta, R.K., Biswas, A.K., 1985. Fertility status of the soils of Sikkim. Sikkim Science Society Newsletter 5.

Biswas, A.K., Gupta, R.K., Biswas, A.K., 1986. Characteristics of different plant parts of large cardamom. Cardamom 19 (3), 7–10.

Blancow, N.W., 1972. Martindale: The Extra Pharmacopoeia.. The Pharmaceutical Press, London, UK.

Boguslawski, E.V., Lach, G., 1971. Die K-Nachlieferung des Bodens im Pflanzenex-periment im Vergleich mit dem austauschbaren Kalium. Z. Acker. Pflanzenbau 134, 135–164.

Bossen, I., 1982. Plantation and labour force dissemination in Guatemala. Curr. Anthropol. 23, 3.

Bramley, P.M., 1997. Isoprenoid metabolism. In: Dey, P.M., Harborne, J.B. (Eds.), Plant Biochemistry (pp. 417–434). Academic Press, California.

British Pharmaceutical Codex (1963). Pharmaceutical Press, London, UK. p. 139.

British Pharmacopoeia (1980). vol. 2. HMSO Cambridge, UK, p. 834.

British Pharmacopoeia (1993). General Medical Council, London, UK, p. 138.

Brouder, S.M., Cassman, K.G., 1994. Evaluation of a mechanistic model of potassium uptake by cotton in vermiculitic soil. Soil Sci. Soc. Am. J. 58, 1174–1183.

Brouwmeester, H.J., Gershenzon, J., Konings, M.C.J.M., Croteau, R., 1998. Biosynthesis of the monoterpenes limonene and carvone in the fruit of caaway. I. Demonstration of enzyme activities and their changes with development. Plant Physiol. 117, 901–912.

Burtt, B.L., 1980. Cardamom and other Zingiberaceae in Hortus Malabaricus. In: Manilal, K.S. (Ed.), Botany and History of Hortus Malabaricus (pp. 139–148). Oxford/IBH, New Delhi.

Burtt, B.L., Smith, R.M., 1983.Dassanayake, M.D. A Revised Hand Book to the Flora of Ceylon, vol. 1V. Amerind Pub., New Delhi.

Capoor, S.P., 1967. Katte disease of cardamom. In: Seminar on Cardamom. Cardamom Board, Cochin, Kerala, India, p. 25.
Capoor, S.P., 1969. Katte disease of cardamom. Cardamom News 3, 2–5.
Chandran, K., Raja, P., Joseph, D., Suryanarayana, M.C., 1983. Studies on the role of honeybees in the pollination of cardamom. In: Proc. 2nd Int. Conf. Api. Trop. Climate, pp. 497–504.
Chandrappa, H.M., Shadakshari, Y.G., Sudharsan, M.R, Raju, B., 1997. Preliminary yield trial of tissue cultured cardamom selections. In: Edison, S., Ramana, K.V., Sasikumar, B., Nirmal Babu, K., Eapen, S.J. (Eds.), Biotechnology of Spices, Medicinal and Aromatic Plants (pp. 102–105). Indian Society of Spices, Calicut, India.
Chandrappa, H.M., Shadakshari, Y.G., Dushyandhakumar, B.M., Edison, S., Shivashankar, K.T., 1998. Breeding studies in cardamom (*Elettaria cardamomum* Maton.). In: Mathew, N.M., Jacob, C.K. (Eds.), Developments in Plantation Crops Research (pp. 20–27). Allied Pub., New Delhi.
Charles, J.K., 1986. Productivity in cardamom: an insight. Cardamom J. 19 (12), 13–20.
Chattopadhyay, S.B., 1967. Diseases of Plants Yielding Drugs, Dyes and Spices. Indian Council of Agriculture Research, New Delhi, India. p. 100.
Chattopadhyay, S.B., Bhomik, T.P., 1965. Control of "foorkey" disease of large cardamom in West Bengal. Indian J. Agric. Sci. 35, 272–275.
Chaudhary, R., Chandel, K.P.S., 1995. Studies on germination and cryopreservation of cardamom (*Elettaria cardamomum* Maton.) seeds. Seed Sci. Biotechnol. 23 (235–240).
Cherian, A., 1977. Environmental ecology: an important factor in cardamom cultivation. Cardamom J 9 (1), 9–11.
Cherian, A., 1979. Produce better cardamom seedlings for ensuring high yield. Cardamom J. 11 (1), 3–7.
Cheriyan, M.C., Kylasam, M.S., 1941. Preliminary studies on the cardamom thrips (*Taeniothrips cardamomi* Ramk.) and its control. Madras Agric. J. 29, 355–359.
Chisowa, E.H., Hall, D.R., Farman, D.I., 1998. Volatile constituents of leaf and rhizome oils of *Aframomum biauriculatum* K. Schum. J. Essent. Oil Res. 10, 447–449.
Chowdhary, S., 1948. Notes on fungi from Assam. Lloydia 21, 152–156.
Claassen, N., 1990. Die Aufnahme von Nahrstoffen aus dem Boden durch die hohere Pflanze als Ergebnis von Verfugbarkeit und Aneignungsvermogen. Severin-Verlag, Gottingen, Germany.
Claassen, N., Barber, S.A., 1976. Simulation model for nutrient uptake from soil by growing plant root system. Agron. J. 68, 961–964.
Claassen, N., Hendriks, K., Jungk, A., 1981. Rubidium-Verarmung des wurzelnahen Bodens durch Maispflanzen. Bodenk 144, 533–545.
Claassen, N., Syring, K.M., Jungk, A., 1986. Verification of a mathematical model by simulating potassium uptake from soil. Plant Soil 95, 209–220.
Clark, G., Stuart, C., Easton, M.D., 2000. Eucalyptol. Perfumer Flavourist 25, 6–16.
Clevenger, J.F., 1934. Volatile oil from cardamom seed. J. Assoc. Agric. Chem. 17, 283.
CPCRI (1980). Annual Report. Central Plantation Crops Research Institute, Kasaragod, Kerala, India. pp. 121–122.
CPCRI (1984). Annual Report. Central Plantation Crops Research Institute, Kasaragod, Kerala, India. p. 27.
CPCRI (1985). Annual Report. Central Plantation Crops Research Institute, Kasaragod, Kerala, India. pp. 118–122.
Croteau, R., Sood, V.K., 1985. Metabolism of monoterpenes: evidence for the function of monoterpene catabolism in peppermint (*Mentha piperita*) rhizomes. Plant Physiol. 77, 801–806.

Curtin, D., Smillie, G.W., 1983. Soil solution composition as affected by timing and incubation. Soil Sci. Soc. Am. J. 47, 701–707.

Dalziel, J.M., 1937. The Useful Plants of West Tropical Africa. Crown Agents for Overseas Dev., London. pp. 470–472.

Dandin, S.B., Madhusoodanan, K.J., George, K.V., 1981. Cardamom descriptor. In: Proc. 1Vth Sym. Plant Crops (Placrosym–1V), CPCRI, Kasaragod, India, pp. 401–406.

David, B.V., Sundarraj, 1993. Studies on *Dialeurodes* of India. *Kanakarajella* gen. Nov. J. Entomol. Res. 17, 253.

David, B.V., Narayanaswami, P.S., Murugesan, M., 1964. Bionomics and control of the castor shoot and capsule borer *Dichocrocis punctiferalis* Guen. in Madras State.

De Gues, J.G., 1973. Fertilizer Guide for the Tropics and Subtropics, Second ed. Zurich Centre d'Etude de l'Azote.

De Guzman, C.C., Siemonsma, J.S. (Eds.),, 1999. Plant Resources of South-East Asia. No. 13. Spices. Backhuys Pub., Leiden.

Deshpande, R.S., Kulkarni, D.S., 1973. Deficiency symptoms in cardamom (*Elettaria cardamomum*). Mysore J. Agric. Sci. 7, 246–249.

Deshpande, R.S., Kulkarni, S.D., Viswanath, S., Suryanarayana Reddy, B.G., 1971. Influence of lime and nitrogenous fertilizers on soil biology. Mysore J. Agric. Sci. 5 (1), 77–81.

Desrosier, N., 1978. Reitz Master Food Guide. AVI Pub. Co., USA.

Dhanalakshmy, C., Leelavathy, K.M., 1976. Leaf spot of cardamom caused by *Phaeotrichoconis crotalariae*. Plant Dis. Reporter 60, 188.

Dhanapal, K., Thomas, J., 1996. Evaluation of *Trichoderma* isolates against rot pathogens of cardamom. In: Manibhushan Rao, K., Mahadevan, A. (Eds.), Recent Trends in Biocontrol of Plant Pathogens (pp. 65–67). Today and Tomorrow Publishers, New Delhi, India.

Dimitman, J.E., 1981. An aphid transmitted virus of cardamom in Guatemala (abstract). Phytopathology 71, 104–105.

Dimitman, J.E., Flores, A., Nickoloff, J.A., 1984. Cardamom mosaic: a member of the poty virus group in Guatemala (abstract). Phytopathology 74, 844.

Dubey, A.K., Singh, K.A., 1990. Large cardamom: a spice crop of India. Indian Farming 49 (2), 17–18.

Dung, N.X., Thin, N.N., 1992. Some important medicinal and aromatic plants from VietnamRaychaudhuri, S.P. Recent Advances in Medicinal, Aromatic and Spice Crops, vol. 2. Today and Tomorrow Publication, New Delhi.

During, C., Duganzich, D.M., 1979. Simple empirical intensity and buffering capacity measurements to predict potassium uptake by white clover. Plant Soil 51, 167–176.

Eapen, S.J., Venugopal, M.N., 1995. Field evaluation of *Trichoderma* sp. and *Paecilomyces lilacinus* for control of root knot nematodes and fungal disease of cardamom nurseries. Indian J. Nematol. 25, 115–116. (abstract).

Ellis, M.B., 1971. Demataceous Hyphomycetes. Commonwealth Mycological Institute, Kew, Surrey, UK. p. 608.

El-Tabir, K.E.H., Shoch, M., Al-Shora, M., 1997. Exploration of some pharmacological activities of cardamom seed (*Elettartia cardamomum*) volatile oil. Saudi Pharm. J. 5 (2–3), 96–102.

Escoubas, P., Lajide, L., Mizutani, J., 1995. Termite antifedent activity in *Aframomum melegueta*. Phytochemistry 40, 1097–1099.

Fang Jen, W.U., 1996. Isolation, purification and identification of antioxidative components from fruits of Amomum tsao-ko. J. Chin. Agric. Chem. Soc. 34, 438–451.

Fohse, D., Claassen, N., Jungk, A., 1991. Phosphorus efficiency of plants II. Significance of root radius, root hairs, and cation–anion balance for phosphorus influx in seven plant species. Plant Soil 132, 261–272.

Francis, M.J.O., O'Connell, M., 1969. The incorporation of mevalonic acid into rose petal monoterpenes. Phytochemistry 8, 1705–1708.

Gaertner (1791). Quoted from Abheywickrama (1959).

Galal, A.M., 1996. Antimicrobial activity of 6-paradol and related components. Int. J. Pharmacol. 34, 64–69.

Gamble, J.S., 1925. The flora of the Presidency of Madras. Bot. Sur. India, Calcutta (Reprint). Ganapathy, T.R., Bapat, V.A., Rao, P.S. (1994). *In vitro* development of encapsulated shoot tips of cardamom. Biotechnol. Technol. 8, 239–244.

Ganguly, B., 1966. A rapid test for detecting Chirke affected large cardamom (*Amomumsubulatum* Roxb.) plants in field. Sci. Culture 32 (2), 95–96.

Gapon, E.N., 1933. Theory of exchangeable adsorption in soil. J. Gen. Chem. (USSR) 3, 144–163. (abstract in Chem. Abs. 28, 41–49, 1934).

Geetha, S.P., Manjula, C., Sajina, A., 1995. *In vitro* conservation of genetic resources of spices. Proceedings of the Kerala Science Congress, Palakkad. pp. 12–16.

Geetha, S.P., Nirmal Babu, K., Rema, J., Ravindran, P.N., Peter, K.V., 2000. Isolation of protoplasts from cardamom (*Elettaria cardamomum* Maton.) and ginger (*Zingiber officinale* Rosc.). J. Spices Aromatic Crops 9, 23–30.

George, C.K., 1990. Production and export of cardamom in Guatemala. Spice India 3 (9), 2–6.

George, C.K., John, K., 1998. Future of Cardamom Industry in India. Spice India 11 (4), 20–24.

George, K.V., 1967. Katte disease threatens the future of cardamom plantation industry. Cardamom News 1, 1–3.

George, K.V., 1971. Research of cardamom: some suggestions. Cardamom News 5, 2–3.

George, K.V., 1976. Production constraints in cardamom. Cardamom News 8 (3), 9–16.

George, K.V., Dandin, S.B., Madhusoodanan, K.J., Koshy, J., 1981. Natural variations in the yield parameters of cardamom (*Elettaria cardamocmum*). In: Visveshwara, S. (Ed.), Proc. PLACROSYM 1981 (pp. 216–231,). ISPS, Kasaragod, India.

George, M., Jaysankar, N.P., 1977. Control of Chenthal (bacterial blight) disease of cardamom with penicillin. Curr. Sci. 46, 237.

George, M., Jaysankar, N.P., 1979. Distribution and factors influencing chenthal disease of cardamom. In: Proc. PLACROSYM-II. CPCRI, Kasaragod. pp. 343–347.

George, M., Joseph, T., Potty, V.P., Jaysankar, N.P., 1976. A bacterial blight disease of cardamom. J. Plantation Crops 4, 23–24.

George, M.V., Mohammed Syed, A.A., Korikanthimath, V.S., 1984. *Diospyros ebenum* Koenig, an ideal shade tree for cardamom. J. Plantation Crops 12 (2), 160–163.

Gerhardt, U., 1972. Changes in spice constituents due to the influence of various factors. Felischwirtschaft 52 (1), 77–80.

Gershenzon, J., McConkey, M.E., Corteau, R.B., 2000. Regulation of monoterpene accumulation in leaves of peppermint. Plant Physiol. 122, 205–213.

Goldman, A., 1949. How spice oleoresins are made? Am. Perfum. Essent. Oil Rec. 53, 320–323.

Gonsalves, D., Trujillo, E., Hoch, H.C., 1986. Purification and some properties of a virus associated with cardamom mosaic, a new member of the poty virus group. Plant Disease 70, 65–69.

Gopakumar, B., Kumaresan, D., 1991. Evaluation of certain insecticides against cardamom whitefly, *Dialeurodes cardamomi* (David and Subr). Pestology 15, 4–5.

Gopakumar, B., Singh, J., 1994. Evaluation of neem-based insecticides against cardamom thrips *Sciothrips cardamomi* (Ramk.). In: International Symposium on Allelopathy in Sustainable Agriculture, Forestry and Environment. Delhi, India, p. 124 (abstract).

Gopakumar, B., Kumaresan, D., Varadarasan, S., 1988. Whitefly management in cardamom plantation. Cardamom 20, 5–6.
Gopakumar, B., Kumaresan, D., Varadarasan, S., 1989. Note on the incidence of *Meta-podistis polychrysa* Meyrick (Lepidoptera: Gtyphipterigidae) on small cardamom (*Elettaria cardamomum* Maton.). Entomon 14, 170.
Gopal, R., Chandraswamy, D.W., Nayar, N.K., 1990. Correlation and path analysis in cardamom. Indian J. Agric. Sci. 60, 240–242.
Gopal, R., Chandraswamy, D.W., Nayar, N.K., 1992. Genetic basis of yield and yield components in cardamom. J. Plantation Crops 20 (Suppl), 230–232.
Gopalakrishnan, N., 1994. Studies on the storage quality of carbon dioxide-extracted cardamom and clove oils. J. Agric. Food Chem. 42, 796–798.
Gopalakrishnan, N., Narayanan, C.S., 1991. Supercritical CO_2 extraction of cardamom. J. Agric. Food Chem. 39, 1976–1978.
Gopalakrishnan, M., Luxmi Varma, R., Padmakumari, K.P., Beena, S., Howa, U., Narayanan, C.S., 1991. Studies on cryogenic grinding of cardamom. Ind. Perfumer 35 (1), 1–7.
Gopalakrishnan, M., Narayanan, C.S., Kumaresan, D., Bhaskaran, P., 1989. Physico-chemical changes in cardamom infested with thrips. Spice India 2 (12), 18–19.
Gopalakrishnan, M., Narayanan, C.S., Grenz, M., 1990. Non saponifiable lipid constituents of cardamom. J. Agric. Food Chem. 38 (12), 2133–2136.
Gopalakrishnan, N., Laxmivarma, R., Padmakumari, K.P., Symon, B., Umma, H., Narayanan, C.S., 1990. Studies on cryogenic grinding of cardamom. Spice India 3 (7), 7–13.
Govindarajan, V.S., Narasimhan, S., Raghuveer, K.G., Lewis, Y.S., 1982. Cardamom: production, technology, chemistry and quality. CRC Crit. Rev. Food Sci. Technol. 16 (3), 229–326.
Govindarajan, V.S., Shanti, N., Raghuveer, K.G., Lewis, Y.S., 1982. Cardamom: production, technology, chemistry and quality. CRC Crit. Rev. Food Sci. Nutr. (Florida) 16, 326.
Govindaraju, C., Venugopal, M.N., Sudharshan, M.R., 1994. An appraisal of "Kokke Kandu": a new viral disease of cardamom and Katte (mosaic) disease of Karnataka. J. Plantation Crops 22, 57–59.
Govindaraju, C., Thomas, J., Sudharsan, M.R., 1996. "Chenthal" disease of cardamom caused by *Colletotrichum gloeosporioides* Penz and its management. In: Mathew, N.M., Jacob, C.K. (Eds.), Developments in Plantation Crop Research (pp. 255–259). Allied Publishers, New Delhi, India.
Griebel, C., Hess, G., 1940. "Die Haltbarkeit abgepackter gemahlener Gewurze.". Zeits. Untersuch der Lebensmittel. 79, 184–191.
Grimme, H., 1974. Potassium release in relation to crop production. In: International Potash Institute (ed.) und Potassium Research and Agricultural Production, pp. 131–136. Int. Potash. Inst., Berne, Switzerland.
Guenther, E., 1952. Cardamom. In: The Essential, Oils, vol. V. Robert E. Krieger Publishing, New York, p. 85.
Guenther, E., 1975. The Cardamom Oils. In: The Essential Oils, vol. V. Robort E. Krieger Publishing Company, New York, pp. 85–106.
Gupta, P.N., 1986. Shade regulation in large cardamom. Cardamom 19 (8), 5–7.
Gupta, P.N., Nagvi, A.N., Mistra, L.N., Sen, T., Nigam, M.C., 1984. Gas chromatographic evaluation of the essential oils of different strains of *Amomum subulatum* Roxb. growing wild in Sikkim. Sonderdruck ans Parfumeric and Kodmetik 65, 528–529.
Gupta, U., 1989. Studies on germination of seeds of large cardamom. Spice India 2 (3), 14.
Gupta, U., John, T.D., 1987. Floral biology of large cardamom. Cardamom 20 (5), 8–15.
Gurudutt, K.N., Naik, J.P., Srinivas, P., Ravindranath, B., 1996. Volatile constituents of large cardamom (*Amomum subulatum*). Flavour Fragr. J. 1, 7–9.

Gurumurthy, B.R., Hegde, M.V., 1987. Effect of temperature on germination of cardamom (*Elettaria cardamomum* (L.) Maton). J. Plantation Crops 15, 5–8.

Gurumurthy, B.R., Nataraj, S.P., Pattanshetti, S.P., 1985. Improved methods of drying. Cardamom 18 (9), 3–7.

Gyatso, K., Tshering, P., Basnet, B.S., 1980. Large cardamom of Sikkim. Dept. Agric. Govt. Sikkim 2 (4), 91–95.

Hari, L., Bukuru, J., Pooter, H.L. de, 1994. The volatile fraction of *Aframomum sanguineum* K. Schum. from Burudi. J. Essent. Oil Res. 6, 395–398.

Hashim, B., Aboobaker, V.S., Madhubala, R.K., Rao, A.R., 1994. Modulatory effects of essential oils from spices on the formation of DNA adducts by aflatoxin B 1 *in vitro*. Nutr. Cancer 21, 169–175.

Heath, H.B., 1978. Flavour Technology: Profiles, Products, Applications, AVI Publishing Co., Inc., CT, USA. p. 113

Hendriks, L., Claassen, N., Jungk, A., 1981. Phosphatverarmung des wurzelnahen Bodens und Phosphataufnahme von Mmais und Raps. Z. Pflanzenern. Bodenk 144, 486–499.

Hinsinger, P., Jaillard, B., Dufey, J.E., 1992. Rapid weathering of a trioctahedral mica by roots of ryagrass. Soil Sci. Soc. Am. J. 56, 977–982.

Hirasa, K., Takemasa, M., 1998. Spices Science and Technology. Marcel Dekker, New York, USA.

Hoagland, R., Martin, J.C., 1933. Absorption of potassium by plants in relation to replaceable, non-replaceable and solution potassium. Soil Sci 36, 1–34.

Holttum, R.E., 1950. The Zingiberaceae of the Malay Peninsula. Garden's Bull. Singapore 13, 236–239.

Huang, Y.B., Hsu, I.R., Wu, P.C., Ko, H.M., Tsai, Y.H., 1993. Crude Drug enhancement of percutaneous absorption of indomethacine *in vitro* and *in vivo* penetration. Kao Hsiung J. Haresh Ko Tsa Chih 9, 392–400.

Huang, Y., Lam, S.L., Ho, S.H., 2000. Bioactivities of essential oil from *Elettaria cardamomum* (L.) maton to *Sitophilus zeamais* Mostschulsky and *Tribolium castaneum*. Herbst. J. Stored Products Res. 36, 167–171.

IISR (1995). Annual Report for 1994–1995. Indian Institute of Spices Research, Calicut, Kerala, India. pp. 5–6.

IISR (1996). Annual Report for 1995–1996. Indian Institute of Spices Research, Calicut, Kerala, India. pp. 66–67.

IISR (NRCS) (1997–1998). Annual Report (1997–1998). Indian Institute of Spices Research, Calicut, India.

ITC (1977). Spices—A Survey of the World Market, vols. 1 and 2. International Trade Centre, UNCTAD/GATT, Geneva.

ITC/SEPC (1978). Market Survey of Consumer Packed Spices in Selected Countries. International Trade Centre, UNCTAD/GATT, Geneva.

Ikeda, R.M., Stanley, W.L., Vannier, S.H., Spitler, E.M., 1962. Monoterpene hydrocarbon composition of some essential oils. J. Food Sci. 27, 455.

Ito, Y., Miura, H., Miyaga, K., 1962. Cited from Hirasa and Takemasa (1998).

Jackson, B.P., Snowdon, D.W., 1990. Atlas of Microscopy of Medicinal Plants. Culinary Herbs and Spices. Belhaven Press, UK.

Jacob, S.A., 1981. Biology of *Dichocrocispunctiferalis* Guen. on turmeric. J. Plantation Crops 9, 119–123.

Jacob, T., Usha, R., 2001. 3'-Terminal sequence analysis of the RNA genome of the Indian isolate of cardamom mosaic virus, a new member of the genus *Madura* virus of Potyviridae. Virus Genes 23, 81–88.

Jacob, T., Jebasingh, Venugopal, M.N., Usha, R., 2002. High genetic diversity in the coat protein and the 3'-untranslated regions among the geographical isolates of cardamom mosaic virus from South India, in press.

Jain, P., Khanna, N.K., Trehan, N., Godhwani, J.L., 1994. Anti-inflammatory effects of an Ayurvedic preparation, Brahmi rasayan, in rodents. Ind. J. Expl. Biol. 32, 633–636.

Jardine, P.M., Sparks, D.L., 1984. Potassium-calcium exchange I a multi-reactive soil system II. Thermodynamics. Soil Sci. Soc. Am. J. 48, 45–50.

John, J.M., Mathew, P.G., 1977. Sprinkler irrigation in cardamom—a sure step to increased production. Cardamom 9 (6), 17–21.

John, J.M., Mathew, P.G., 1979. Large cardamom in India. Cardamom 3 (10), 13–20.

John, M., 1968. Hints for raising cardamom nurseries. Cardamom News 2 (6), 4–5.

Joseph, J., Raghavan, B., Nangundaiah, G., Shankaranarayana, M.L., 1996. Curing of large cardamom (*Amomum subulatum* Roxb.): fabrication of a new dryer and a comparative study of its performance with existing. J. Spices Aromatic Crops 5, 105–110.

Joseph, K.J., 1978. Multicropping concept in Forestry. In Proceedings of the First Annual Symposium on Plantation Crops (PLACROSYM–1) (E.V. Nelliat, Ed.), pp. 441–443. CPCRI, Kasaragod.

Joseph, K.J., 1985. Marketing and price formation of cardamom in Kerala. M. Phil. Thesis. Jawaharlal Nehru University, New Delhi, India.

Joseph, K.J., 1986. Spotted locusts (*Aularches milians* L.) and its integrated management. Cardamom J 19 (17), 5.

Joseph, K.J., Narendran, T.C., Joy, P.J., 1973. Studies on Oriental *Brachymeria* (Chalacidoidae) Report. PL 480 Research Project, Taxonomic Studies on the Oriental Species of *Brachymeria* (Hymenoptera: Chalacididae). University of Calicut, Calicut, Kerala.

Kachru, K.P., Gupta, R.K., 1993. Drying of spices: status and challenges. In: Proceedings of the National Seminar on Post Harvest Technology of Spices, ISS, Calicut, pp. 15–27.

Kalra, C.L., Seligal, R.C., Manan, J.K., Kulkarni, S.G., Berry, S.K., 1991. Studies on preparation, packaging and quality standards of ground spice mixes. Part 1 Garam masala. Beverages Food World 18, 21–24.

Kamchonwongpaisan, S., Nilanonta, C., Tarnchompoo, B., Thebtaranonth, C., Thebataranonth, Y., Yuthavong, Y., Kongsaeree, P., Clardy, J., 1995. An antimalarial peroxide from *Amomum kevanh* Pierre. Tetrahedron Lett. 36, 1821–1824.

Karibasappa, G.S., 1987. Post harvest studies in large cardamom (*Amomum subulatum* Roxb.). Sikkim. Sci. Soc. News Lett. 6 (3), 2–10.

Karibasappa, G.S., Dhiman, K.R., Biswas, A.K., Rai, R.N., 1987. Variability and association among quantitative characters and path analysis in large cardamom. Indian J. Agric. Sci. 57, 884–888.

Karibasappa, G.S., Dhiman, K.R., Rai, R.N., 1989. Half sib progeny analysis for variability and association among the capsule characters and path studies on oleoresin and its cineole content in large cardamom. Indian J. Agric. Sci. 53, 621–625.

Kasi, R.S., Iyengar, K.G., 1961. Cardamom propagation. Spices Bull 1 (3), 10–11.

Kasturi, T.R., Iyer, B.H., 1955. Fixed oil from *Elettaria cardamomum* seeds. J. Indian Inst. Sci. 37A, 106.

Keerthisinghe, G., Mengel, K., 1979. Phosphatpufferung verschiedener Boden und ihre Veranderung infolge PhosphataltERung. Mitt. Dtsch. Bodenk. Ges. 29, 217–230.

Khader, K.B.A., Antony, K.J., 1968. Intercropping: a paying proposition for areca growers. What crops to grow. Indian Farming 18 (4), 14–15.

Khader, K.B.A., Sayed, A.A.M., 1977. Fertilizing cardamom: its importance. Cardamom 9 (1–2), 13–14.

Kirtikar, R.P., Basu, B.D., 1952. Indian Medicinal Plants, vol. 4. LM Balu Pub., Allahabad (Rep.).
Koller, W.D., 1976. The importance of temperature on storage of ground natural spices. Z. Lebensm. Unters. Forsch. 160, 143–147.
Kologi, S.D., 1977. Cardamom Research at Mudigere. In: Towards Higher Yield in Cardamom. CPCRI, Kasaragod, India, pp. 28–29.
Korikanthimath, V.S., 1980. Nursery studies in mulches for use in cardamom primary nursery. Annual Report. Central Plantation Crops Research Institute, Kasaragod, Kerala, India.
Korikanthimath, V.S., 1982. Nursery studies in cardamom. Annual Report. Central Plantation Crops Research Institute, Kasaragod, Kerala, India.
Korikanthimath, V.S., 1983. Seminar on Production and Prospects of Cardamom in India. University of Agricultural Sciences, Dharwad, Karnataka, India.
Korikanthimath, V.S., 1983. Nursery studies in cardamom. Annual Report. Central Plantation Crops Research Institute, Kasaragod, Kerala, India.
Korikanthimath, V.S., 1983. Practical Planting and Shade Management in Cardamom. Planters Chronicle, UPASI, Coonoor. pp. 405–406.
Korikanthimath, V.S., 1986. Systems of planting-cum-fertilizer levels in cardamom under rain-fed condition. Annual Report. National Research Center for Spices, pp. 30–31.
Korikanthimath, V.S., 1987. Studies on analysis of rainfall and its impact on cardamom Proceedings of the National Seminar on Agrometeorology of Plantation Crops. Regional Agricultural Research Station, Kerala Agricultural University, Pilicode, Kerala, India. March 12–13, 1987.
Korikanthimath, V.S., 1989. Systems of planting-cum-fertilizer levels in cardamom under rainfed condition. Ann. Rep. National Research Center for Spices, 1988–1989, Calicut, Kerala, India.
Korikanthimath, V.S., 1990. Efficient management of natural resources for cardamom (*Elettaria cardamomum* Maton.) production International Symposium on Natural Resources management for Sustainable Agriculture. Indian Society of Agronomy, New Delhi, India, February 6–10, 1990.
Korikanthimath, V.S., 1991. Shade management for high productivity in cardamom (*Elettaria cardamomum* Maton.). Spice India 4 (2), 15–21.
Korikanthimath, V.S., 1992. High production technology in cardamom. In: Sarma, Y.R., Devasahayam, S., Anandaraj, M. (Eds.), Black Pepper and Cardamom: Problems and Prospects (pp. 20–30). Indian Society for Spices, Calicut, Kerala, India.
Korikanthimath, V.S., 1992. Large scale multiplication of cardamom: the NRCS experiment. In: Sarma, Y.R., Devasahayam, S., Anandaraj, M. (Eds.), Black Pepper and Cardamom: Problems and Prospects (pp. 65–67). Indian Society for Spices, Calicut, Kerala, India.
Korikanthimath, V.S., 1999. Rapid clonal multiplication of elite cardamom selections for generating planting material, yield upgradation and its economics. J. Plantation Crops 27, 45–53.
Korikanthimath, V.S., Mulge, R., 1998. Assessment of elite cardamom lines for dry matter distribution and harvest index. J. Med. Aromatic Plants 20, 28–31.
Korikanthimath, V.S., Naidu, R., 1986. Influence of stage of harvest on the recovery percentage of cardamom. Cardamom J 19 (11), 5–8.
Korikanthimath, V.S., Rao, G.S., 1993. Leaf area determination in cardamom (*Elettaria cardamomum*). J. Plantation Crops 11, 151–153.
Korikanthimath, V.S., Venugopal, M.N., 1989. High Production Technology in Cardamom, Tech. Bull.. National Research Center for Spices, Calicut, Kerala, India.
Korikanthimath, V.S., Kiresur, V., Hiremath, G.M., Hegde, R., Mulge, R., Hosmani, M.M., 1988. Economics of mixed cropping of coconut with cardamom. Crop Res 15 (23), 188–195.

Korikanthimath, V.S., Mulge, R., Hegde, R., Hosmani, M.M., 1988. Crop combinations and yield pattern in coffee mixed cropped with cardamom. J. Plantation Crops 26 (1), 41–49.

Korikanthimath, V.S., Venugopal, M.N., Naidu, R., 1989. Cardamom production: success story. Spices India 119, 19–24.

Korikanthimath, V.S., Ankegowda, S.J., Yadkumar, N., Hegde, R., Hosmani, M.M., 2000. Microclimatic and photosynthetic characteristics in arecanut and cardamom mixed cropping systems. J. Spices Aromatic Crops 9, 61–63.

Korikanthimath, V.S., Hegde, R., Gayathri, A., 2000b. Investigations on cardamom based cropping systems. Report of the Ad-hoc Scheme. IISR, Calicut.

Kovar, J.L., Barber, S.A., 1988. Phosphorus supply characteristics of 33 soils as influenced by seven rates of phosphorus addition. Soil Sci. Soc. Am. J. 52, 160–165.

Krishna, K.V.S., 1968. Cultivation of cardamom: selection of plants suitable for seed purpose. Cardamom News 2, 1–2.

Krishna, K.V.S., 1997. Cardamom plantations in Papua new Guinea. Spice India 10 (7), 23–24.

Krishnamoorthy, M.N., Natarajan, C.P., 1976. Preliminary studies on bleaching cardamoms. Cardamom J 8 (8), 17–19.

Krishnamurthy, K., Khan, M.M., Avadhani, K.K., Venkatesh, J., Siddaramaiah, A.L., Chakravarthy, A.K., Gurumurthy, S.R., 1989a. Three Decades of Cardamom Research at Regional Research Station, Mudigere (1958–1988), Technical Bulletin No. 2. p. 94. Regional Research Station, Mudigere, Karnataka, India.

Krishnamurthy, K., Khan, M.M., Avadhani, K.K., Venkatesh, J., Siddaramaiah, A.L., Chakravarthy, A.K., Gurumurthy, B.R., 1989. The Decades of Cardamom Research at R. R. S. Mudigere, Tech. Bull. University of Agricultural Sciences, Bangalore, India.

Krishnan, R.P., Guha, P.C., 1950. Mysore cardamom oil. Curr. Sci. 19, 157.

Kubo, I., Himejima, M., Murari, H., 1991. Antimicrobial activity of flavour components of Cardamom. J. Agric. Food Chem. 39, 1984–1986.

Kuchenbuch, R., Jungk, A., 1984. Wirkung der Kaliumdungung auf die. Kalium verfügbarkeit in der Rhizosphäre von Raps. Z. Pflanzenernahr. Bodenkd 147, 435–448.

Kuhlmann, H., Wehrmann, J., 1984. Kali-Dungeempfehlung auf der Grundlage von 81K-Dungeversuchen zu Gertreide und Zuckerruben. Z. Pflanzenernahr. Bodenkd 147, 349–360.

Kulandaivelu, G., Ravindran, K.C., 1982. Physiological changes in cardamom genotypes exposed to warm climate. J. Plantation Crops 20 (Suppl), 294–296.

Kulkarni, D.S., Kulkarni, S.V., Suryanarayana Reddy, B.G., Pattanshetty, H.V., 1971. Nutrient uptake by cardamom (*Elettaria cardamomum* Maton.), vol. 1. Proceedings of the International Symposium on Soil Fertility Evaluation, New Delhi, India, pp. 293–296.

Kumar, A., Chatterjee, S.V., 1993. Record of new pest *Bradysia* sp. and its predator, *Phaonia simulans* on large cardamom. J. Appl. Zool. Res. 3 (1), 103–104.

Kumar, K.B., Kumar, P.P., Balachandran, S.M., Iyer, R.D., 1985. Development of clonal plantlets in cardamom (*Elettaria cardamomum*). J. Plantation Crops 13, 31–34.

Kumar, M.D., Santhaveerabhadraiah, S.M., Ravishankar, C.R., 2000. Effect of fertilizer levels on the yield of small cardamom under natural shade. In: Spices and Aromatic Plants. Indian Society of Spices, IISR, Calicut, India. pp. 179–180.

Kumaresan, D., 1982. Efficacy of modern synthetic insecticides against cardamom thrips. Pesticides 16, 26–27.

Kumaresan, D., 1983. Field evaluation of insecticides for the control of cardamom thrips. South Indian Hort 31, 151–152.

Kumaresan, D., 1988. Cardamom pests. In: Pest and Disease Management (A Guide to Planters). Indian Cardamom Research Institute, Spices Board, India, p. 40.

Kumaresan, D., Varadarasan, S., 1987. Review and current status of research on insect pest control in cardamom cropping systems. In: Workshop on Insect Pest Management of Coffee, Tea, Cardamom Cropping Systems. Central Coffee Research Station, Chickmagalur, Karnataka, India, January 23–24, 1987.

Kumaresan, D., Regupathy, A., Bhaskaran, P., 1988. Pests of Spices,. Rajalakshmi Publications, Nagercoil, Tamil Nadu, India. p. 241.

Kumaresan, D., Regupathy, A., George, K.V., 1978. Control of cardamom stem borer *Dichocrocis punctiferalis* Guen. J. Plantation Crops 6, 85–86.

Kumaresan, D., Varadarasan, S., Gopakumar, B., 1989. General accomplishments towards better pest management in cardamom. Spice India 2, 5–8.

Kurup, K.R., 1978. Trickle irrigation. Cardamom 10 (6), 11–16.

Kuruvilla, K.M., Madhusoodanan, K.J., 1988. Effective pollination for better fruit set in cardamom. Spice India 1 (6), 19–21.

Kuruvilla, K.M., Sudharshan, M.R., Madhusoodanan, K.J., Priyadarshan, P.M., Radhakrishnan, V.V., Naidu, R., 1992. Phenology of tiller and panicle in cardamom (*Elettaria cardamomum* Maton.). J. Plantation crops 20 (Suppl), 162–165.

Kuruvilla, K.M., Sudharshan, M.R., Madhusoodanan, K.J., Priyadarshan, P.M., Radhakrishnan, V.V., Naidu, R., 1992. Phenology of tiller and panicle in cardamom (*Elettaria cardamomum* Maton.). J. Plantation Crops 20 (Suppl), 162–165.

Kuttappa, K.M., 1969. Capsule shedding in cardamom. Cardamom News 3 (5), 2–3.

Kuttappa, K.M., 1969. Cardamom: digging. Cardamom News 2 (7), 3–5.

Lawrence, B.M., 1970. Terpenoids in two *Amomum* species. Phytochemistry 9, 665.

Lawrence, B.M., 1978. Major tropical spices: sardamom (*Elettaria cardamom*). Essential Oils, 105–155.

Lewis, D.G., Quirk, J.P., 1967. Phosphate diffusion in soil and uptake by plants III. ^{31}P-movement and uptake by plants as indicated by ^{32}P-autoradiography. Plant Soil 26, 445–453.

Lewis, Y.S., 1973. The importance of selecting the proper variety of a spice for oil and oleoresin extraction. Tropical Product Institute Conference Papers, London.

Lewis, Y.S., Nambudiri, E.S., Natarajan, C.P., 1967. Studies on some essential oils. Indian Food Packer 11 (1), 5.

Lewis, Y.S., Nambudiri, E.S., Krishnamurthy, N., 1974. Flavour quality of cardamom oils. Proc. VI. Int. Congr. Essential Oils, London.

Linschoten, J.H. Van, 1596. Voyage of John huygen Van Linschoten in India, vol. II, pp. 86–88 (quoted by Watt, 1872).

Lizhu-Qiang, Lei, Wang-Hua, D., Rong, H., Yuan-Qing, Z., 1998. Chemical constituents of the essential oil in Amomum tsao-ko from Yunnan Province. Acta Bot. Yunnancia 20, 119–122.

Lu, S., Miller, M.H., 1994. Production of phosphorus uptake by field-grown maize with the Barber-Cushman model. Soil Sci. Soc. Am. J. 58, 852–857.

Lukose, R., Saji, K.V., Venugopal, M.N., Korikanthimath, V.S., 1993. Comparative field performance of micropropagated plants of cardamom. (*Elettaria cardamomum*). Indian J. Agric. Sci. 63, 417–418.

Lumbanraja, J., Evangelou, V.P., 1992. Potassium quantity: intensity relationships in the presence and absence of NH_4 for three Kentucky soils. Soil Sci 154, 366–376.

Mabberley, D.J., 1987. The Plant Book.. Cambridge University Press, Cambridge.

Madhusoodanan, K.J., Radhakrishnan, V.V., 1996. Cardamom breeding in Kerala. Breeding of Crop Plants in Kerala. University of Kerala, Trivandrum, India. pp. 73–81.

Madhusoodanan, K.J., Sudharshan, M.R., Priyadarshan, P.M., Radhakrishnan, V.V., 1990. Small cardamom: botany and crop improvement. In: Cardamom Production Technology, ICRI, Spices Board, India. pp. 7–13.

Madhusoodanan, K.J., Kuruvilla, K.M., Potty, S.N., 1998. Cardamom hybrids for higher yield and better quality capsule. Spice India 11 (3), 6–7.

Madhusoodanan, K.J., Radhakrishnan, V.V., Kuruvilla, K.M., 1999. Genetic resources and diversity in cardamom. In: Sasikumar, B., Krishnamoorthy, B., Rema, J., Ravindran, P.N., Peter, K.V. (Eds.), Biodiversity, Conservation, and Utilization of Spices, Medicinal and Aromatic Plants (pp. 68–72). IISR, Calicut.

Mahabala, G.S., Bisaliah, S., Chengappa, P.G., 1991. Resource use efficiency and age-return relationship in cardamom plantations. J. Plantation Crops 20 (Suppl), 359–365.

Mahindru, S.N., 1978. Spice extraction: cardamom. Indian Chem. J. 13 (1), 28–30.

Mahindru, S.N., 1982. Spices in Indian Life. Sultanchand & Sons, New Delhi.

Manomohanan, T.P., Abi, C., 1984. *Elettaria cardamomum,* a new host for *Phytopthora palmivora* (Butler). In: Proc. PLACROSYM-VI. CPCRI, Kasaragod, pp. 133–137.

Marsh, A.C., Moss, M.K., Murphy, E.W., 1977. Composition of Foods, Spices and Herbs. Raw, Processed, Prepared. USDA, Agric. Res. Serv. Hand Book, No. 8, 2 Washington DC, USA.

Martin, H.W., Sparks, D.L., 1983. Kinetics of nonexchangeable potassium release from two coastal plain soils. Soil Sci. Soc. Am. J. 47, 883–887.

Martindale, The Extra Pharmacopoeia (1996). Royal Pharmaceutical Society, London, UK, pp. 686–2, 1681–1, 1758–1.

Mathai, C.K., 1985. Quality evaluation of the 'Agmark' grades of cardamom, *Elettaria cardamomum*. J. Sci. Food Agric. 36 (6), 450–452.

Mathew, M.J., Saju, K.A., Venugopal, M.N., 1997. Management of *Pentalonia nigronervosa f. caladii* Van der groot, vector of *Cardamom Mosaic Virus* (Katte) and Cardamom vein Clearing Virus (Kokke Kandu) through eco-friendly vector control measures (abst). In Symposium on Economically Important Plant Diseases, Bangalore, Karnataka, India, December 18–20, 1997, p. 59.

Mathew, M.J., Saju, K.A., Venugopal, M.N., 1998. Efficacy of entomogenous fungi on biological suppression of *Pentalonia nigronervosa f. caladii* Van der Groot of cardamom (*Elettaria cardamomum* Maton.). J. Spices Aromatic Crops 7, 43–46.

Mathew, M.J., Saju, K.A., Venugopal, M.N., 1999a. Effect of neem products on behaviour and mortality of cardamom aphid, *Pentalonia nigronervosa f. caladii.* Van der Groot. In: Proceedings of the National Symposium on Pest Management in Horticultural Ecosystems: Environmental Implications and Thrusts, IIHR, Bangalore, J. Plantation Crops, in press.

Mathew, M.J., Venugopal, M.N., Saju, K.A., 1999b. Field evaluation of certain biopesticides in comparison with monocrotophos against cardamom aphid, vector of katte and kokke kandu diseases of cardamom (abstract). National Symposium on Biological Control of Insects in Agriculture, Forestry, Medicine and Veterinary Science, Bharathiar University, Coimbatore, Tamil Nadu, India, January 21–22, 1999, p. 87.

Mayne, W.W., 1942. Report on Cardamom Cultivation in South India. Misc. Bull. 50. ICAR, New Delhi, India, p. 67.

Mayne, W.W. 1951a. Report on Cardamom Cultivation in South India. ICAR Tech. Bull. 50, ICAR Publication, New Delhi, p. 62.

Mayne, W.W., 1951b. Report on Cardamom Cultivation in South India. Bull. 50. ICAR Publication, New Delhi. pp. 1–53.

McCaskill, D., Croteau, R., 1995. Monoterpene and sesquiterpene biosynthesis in glandular trichomes of peppermint (Mentha × piperita) rely exclusively on plastid-derived isopentenyl diphosphate. Planta 197, 49–56.

McLean, E.O., Watson, M.E., 1985. Soil measurement of plant-available potassium. In: Munson, R.D. (Ed.), Potassium in Agriculture (pp. 277–308). ASA, CSSA, and SSA, Madison, Wisconsin, WI.

Meisheri, L.D., 1993. Dehydration of horticulture produce at room temperature (27–34 degree Celsius). National Symposium on Food Processing, New Delhi, February 16–17, 1993 (mimeographed copy).

Melgode, E., 1938. Cardamom I Ceylon-Part 1. Trop. Agric. 91, 325–328.

Mengel, K., 1985. Dynamics and availability of major nutrients in soils. Adv. Soil Sci. 2, 65–131.

Mengel, K., Kirby, E.A., 1980. Potassium in crop production. Adv. Agron. 33, 59–110.

Mengel, K., Stefens, D., 1985. Potassium uptake of ryegrass (*Lolium perenne*) and red clover (*Trifolium pratense*) as related to root parameters. Biol. Fertil. Soils 1, 53–58.

Mengel, K., Uhlenbecker, K., 1993. Determination of available interlayer potassium and its uptake by ryegrass. Soil Sci. Soc. Am. J. 57, 761–766.

Menon, M.R., Sajoo, B.V., Ramakrishnan, C.K., Ramadevi, L., 1972. A new *Phytophthora* disease of cardamom (*Elettaria cardamomum* (L.) Maton. Curr. Sci. 41, 231.

Menon, M.R., Sajoo, B.V., Ramakrishnan, C.K., Ramadevi, L., 1973. Control of *Phytophthora* diseases of cardamom. Agric. Res. J. Kerala 11, 93–94.

Menut, C., Lamaty, G., Amvam Zollo, P.H., Atogho, B.M., Abondo, R., Bessiere, J.M., 1991. Aromatic plants of tropical central Africa. V. Volatile oil components of three Zingiberaceae from Cameroon. *Aframomum melegueta* (Roscoe) K. Schum. A. danielli (Hook. f.) K. Schum. and *A. sulcatum* (Oliv. and Hanb.) K. Schum. Flavour Fragr. J. 6, 183–186.

Meyer, D., Jungk, A., 1993. A new approach to quantify the utilization of nonexchangeable soil potassium by plants. Plant Soil 149, 235–243.

Mishra, A.K., Dubey, N.K., 1990. Fungitoxicity against *Aspergillusflavus*. Econ. Bot. 44, 350–533.

Mishra, A.K., Dwivedi, S.K., Kishore, N., Dubey, N.K., 1991. Fungistatic properties of essential oils of cardamom. Int. J. Pharmacognosy 29, 259–262.

Mitsios, I.K., Rowell, D.L., 1987. Plant uptake from exchangeable and non-exchangeable potassium. I. Measuring and modeling for onion roots in a chalky boulder clay soil. J. Soil Sci. 38, 53–63.

Miyazawa, M., Kameoka, H., 1975. The constitution of the essential oil and non-volatile oil from cardamom seed. J. Jpn. Oil Chemists Soc. (Yukaguku) 24, 22.

Mohenchandran, K., 1984. Planting for plantations—a study on cardamom. Cardamom J 17, 5–8.

Mollison, J.W., 1900. Cardamom cultivation in the Bombay Presidency. Agric. Ledger 11 (quoted by Ridley, 1912).

Mukherji, M.K., 1968. Large cardamom cultivation in Darjeeling district of West Bengal. Cardamom News 2 (2), 1–8.

Muralidharan, A., 1980. Biomass productivity, plant interactions and economics of intercropping in arecanut. Ph.D. Thesis. University of Agricultural Sciences, Bangalore, India, p. 271.

Muthappa, B.N., 1965. A new species of Sphaceloma on cardamom from India. Sydowia 19, 143–145.

Nadgauda, R., Mascarenhas, A.F., Madhusoodanan, K.J., 1983. Clonal multiplication of cardamom (*E. cardamomum* Maton.) by tissue culture. J. Plantation Crops 11, 60–64.

Naidu, R., 1978. Screening of cardamom varieties against *Sphaceloma* and *Cercospora* leaf spot diseases. J. Plantation Crops 6, 48.

Naidu, R., Thomas, J., 1994. Viral diseases of cardamomChadha, K.L.Rethinam, P. Advances in Horticulture, Plantation and Spice Crops, vol. 10. Malhotra Publishing House, New Delhi, India.

Naidu, R., Venugopal, M.N., 1982. Management of "Katte" disease of small cardamom. In: Proceedings of PLACROSYM V, Indian Society for Plantation Crops, Kasaragod, Kerala, India, pp. 563–571.

Naidu, R., Venugopal, M.N., 1987. Epidemiology of katte virus of small cardamom. I. Disease incubation period and role of different host parts as a source of inoculum in relation to disease spread. In: Proceedings of PLACROSYM VI, Indian Society for Plantation Crop, Kasaragod, Kerala, India, pp. 121–127.

Naidu, R., Venugopal, M.N., 1989. Epidemiology of katte virus disease of small cardamom. II. Foci of primary disease entry, patterns and gradients of disease entry and spread. J. Plantation Crops 16 (Suppl), 267–271.

Naidu, R., Venugopal, M.N., Rajan, P., 1985. Investigations on strainal variation, epidemiology and characterization of "katte" virus agent of small cardamom. Final Report of the Research Project. Central Plantation Crops Research Institute, Kasaragod, Kerala, India.

Naik, S.N., Maheshwari, R.C., 1988. Extraction of essential oils with liquid carbon-dioxide. PAFAI J 10 (3), 18–24.

Nair, C., Zachariah, P.K., George, K.V., 1982. Control of panicle rot disease of cardamom. In: Proc. PLACROSYM -V, Kasaragod, Kerala, India, pp. 133–137.

Nair, C.K., 1988. Phosphatic fertilizers for small cardamom. In: Proceedings of the Seminar on the use of Rock Phosphate in West Coast Soils, UAS, Bangalore, India and PPCL, p. 79.

Nair, C.K., Zachariah, P.K., 1975. Suitability of Mussoriephos as a phosphatic fertilizer for cardamom. Cardamom News 7 (6), 23–24.

Nair, C.K., Srinivasan, K., Zachariah, P.K., 1978. Distribution of major nutrients in the different layers of cardamom soils. In: Proc. Ist. Annual Symp. Plantation Crops, Kasaragod, Kerala State, pp. 148–156.

Nair, C.K., Srinivasan, K., Zachariah, P.K., 1980. Soil reaction in cardamom growing soils in relation to the base status. In: Proceedings of the Seminar on Diseases of Plantation Crops and Manuring of Plantation crops, Tamil Nadu Agricultural University, Madurai, India, pp. 50–53.

Nair, C.K., Natarajan, P., Jayakumar, M., Naidu, R., 1991. Rainfall analysis of the cardamom tract. J. Plantation Crops 18 (Suppl), 184–189.

Nair, C.K., Srinivasan, K., Sivadasan, C.R., 1998. Soil acidity in cardamom plantations. Spice India 1 (6), 40–41.

Nair, K.P.P., 1984a. Towards a better approach to soil testing based on the buffer power concept. In: Proceedings of the 6th International Colloquium for the Optimization of Plant Nutrition, vol. 4, Pierre-Martin Prevel, Montpellier, France, September 2–8, 1984, pp. 1221–1228.

Nair, K.P.P., 1984. Zinc buffer power as an important criterion for a dependable assessment of plant uptake. Plant Soil 81, 209–215.

Nair, K.C., Vijayan, P.K., 1973. A study on the influences of plant hormones on the reproductive behaviour of cardamom. Agric. Res. J. Kerala 11, 85.

Nair, K.N., Narayana, D., Sivanandan, P., 1989. Ecology and Economics in Cardamom Development. Centre for Development Studies, Trivandrum, Kerala, India.

Nair, K.P.P., 1992. Measuring P buffer power to improve routine soil testing for phosphate. Eur. J. Agron. 1 (2), 79–84.

Nair, K.P.P., 1996. The buffering power of plant nutrients and effects on availability. Adv. Agron. 57, 237–287.

Nair, K.P.P., 2002. Sustaining crop production in the developing world through the nutrient buffer power concept. In: Proceeding of the 17th World Soil Science Congress, vol. 2, Bangkok, Thailand, August 14–21, 2002, p. 652.

Nair, K.P.P., 2004. The agronomy and economy of black pepper (*Piper nigrum* L.): the "King of Spices." Adv. Agron. 82, 271–389.
Nair, K.P.P., Mengel, K., 1984. Importance of phosphate buffer power for phosphate uptake by rye. Soil Sci. Soc. Am. J. 48, 92–95.
Nair, K.P.P., Nand, Ram, Sharma, P.K., 1984. Quantitative relationship between zinc transport and plant uptake. Plant Soil 81, 217–220.
Nair, K.P.P., Sadanandan, A.K., Hamza, S., Abraham, J., 1997. The importance of potassium buffer power in the growth and yield of cardamom. J. Plant Nutr. 20 (7 and 8), 987–997.
Nair, M.R.G.K., 1975. Insects and Mites of Crops in India. ICAR, New Delhi. p. 408.
Nair, M.R.G.K., 1978. Cardamom. In A Monograph on Crop Pests of Kerala and Their Control. Kerala Agricultural University, Vellanikkara, Kerala, India, pp. 65–74.
Nair, P.R.S., Unnikrishnan, G., 1997. Evaluation of medicinal values of cardamom alone and in combination with other spices in ayurvedic system of medicine-project (Part I and II), 51. Government Ayurveda College, Trivandrum, Kerala, India. p. 46.
Nair, R.R., 1979. Investigations of fungal diseases of cardamom. Ph.D. Thesis. Kerala Agricultural University, Vellanikkara, Trissur, p. 161.
Nair, R.R., Menon, M.R., 1980. Azhukal disease of cardamom. In: Nambiar, K.K.N. (Ed.), Proceedings of the Workshop on *Phytophthora* Diseases of Tropical Cultivated Plants (pp. 24–33). CPCRI, Kasaragod, India.
Nambiar, K.K.N., Sarma, Y.S.R., 1974. Chemical control of capsule rot of cardamom. J. Plantation Crops 2, 30–31.
Nambiar, K.K.N., Sarma, Y.S.R., 1976. Capsule rot of cardamom. *Phythium vexans* de bary as a causal agent. J. Plantation Crops 4, 21–22.
Nambiar, M.C., Pillai, G.B., Nambiar, K.K.N., 1975. Diseases and pests of cardamom: a resume of research in India. Pesticides Annu. 12, 122–127.
Nambudiri, E.S., Lewis, Y.S., Rajagopalan, P., Natarajan, C.P., 1968. Production of cardamom oil by distillation. Res. Ind. 13, 140.
Nambudiri, E.S., Lewis, Y.S., Krishnamurthy, N., Mathew, A.G., 1970. Oleoresin of pepper. Flavour Ind. 1, 97–99.
Nanjan, K., Muthuswami, S., Thangarajan, T., Sudararajan, T., 1981. Time of planting of cardamom at Shevaroy hills. Cardamom 13 (1), 13–15.
Narayanan, C.S., Natarajan, C.P., 1977. Improvements in the flavour quality of cardamom oil from cultivated Malabar type cardamom grown in Karnataka State. J. Food. Sci. Technol. 14, 233.
Natarajan, C.P., Kuppuswamy, S., Krishnamoorthy, M.N., 1968. Maturity, regional variations and retention of green color in cardamom. J. Food. Sci. Technol. 5 (2), 65–68.
Natarajan, P., Srinivasan, K., 1989. Effect of varying levels of N, P and K on yield attributes and yield of cardamom (*Elettaria cardamomum* Maton.). South Indian Hort. 37, 97–100.
Naves, Y.R., 1974. Technologie et chimie des parfums Naturess Paris. Masson & Cie,
Nigam, M.C., Nigam, I.C., Handa, K.L., Levi, L., 1965. Essential oils and their constituents XXVIII. Examination of oil of cardamom by gas chromatography. J. Pharm. Sci. 54 (5), 799.
Nirmal Babu, K., Geetha, S.P., Manjula, C., Ravindran, P.N., Peter, K.V., 1994. Medium term conservation of cardamom germplasm: an *in vitro* approach. In: Proc. II Asia Pac. Conf. Agric. Biotech., Madras, India, p. 51 (abstract).
Nirmal Babu, K., Ravindran, P.N., Peter, K.V., 1997. Protocols for Micropropagation of Spices and Aromatic Crops. Indian Institute of Spices Research, Calicut, Kerala, India.
Nirmal Babu, K., Ravindran, P.N., Peter, K.V., 1997. Protocols for micropropagation of spices and aromatic crops. Indian Institute of Spices Research, Calicut, India.

Nirmala Menon, A., Chacko, S., Narayanan, C.S., 1999. Free and glycosidically bound volatiles of cardamom (*Elettaria cardamomum* Maton. var. miniscula Burkill). Flavour Fragr. J. 14, 65–68.

Noleau, I., Toulemonde, B., 1987. Volatile constituents of cardamom (*Elettaria cardamomum* Maton.). Flavour Fragr. J. 2, 123–127.

NRCS (1994). Annual Report for 1993–1994. National Research Centre for Spices, Calicut, Kerala, India. pp. 4–5

Nye, P.H., 1972. Diffusion of ions and uncharged solutes in soils and clays. Adv. Agron. 31, 225–272.

Omanakutty, M., Mathew, A.G., 1985. Microencapsulation. PAFAI J 7 (2), 11.

Owen, T.C., 1901. Notes on Cardamom Cultivation

Padmini, K., Venugopal, M.N., Korikanthimath, V.S., Anke Gowda, S.J., 2000. Studies on compound panicle type in cardamom (*Elettaria cardamomum* Maton.) In: Rajkumar, R. (Ed.), Recent Advances in Plantation Crops Research (pp. 97–99). Allied Pub., New Delhi, India.

Padmini, K., Venugopal, M.N., Korikanthimath, V.S., 1999. Biodiversity and conservation of cardamom (*Elettaria cardamomum* Maton.). In: Sasikumar, B., Krishnamoorthy, B., Rema, J., Ravindran, P.N., Peter, K.V. (Eds.), Biodiversity, Conservation and Utilization of Spices, Medicinal and Aromatic Plants (pp. 73–78). IISR, Calicut, India.

Palaniappan, C., 1986. Analysis of cardamom curing in conventional chamber and Melcard drier. Cardamom 19 (11), 5–8.

Panchaksharappa, M.G., 1966. Embryological studies in some members of Zingiberaceae II. *Eleattaria cardamomum,* Hitchenia, Caulna and *Zingiber macrostachyum*. Phytomorphology 16, 412–417.

Parameshwar, N.S., Venugopal, R., 1974. Capsule setting studies in *Elettaria cardamomum* Maton. Curr. Res. 3 (5), 57–58.

Parameshwar, N.S., Haralappa, H.S., Mahesh Gowda, H.P., 1989. Cardamom yield can be increased by clonal propagation. Curr. Res. 8, 150–151.

Parameswar, N.S., 1973. Floral biology of cardamom (*Elettaria cardamomum* Maton.) Mysore J. Agric. Sci. 7, 205–213.

Parameswar, N.S., 1977. Intergeneric hybridization between cardamom and related genera. Curr. Sci. 6, 10.

Park, M., 1937. Report on the work of the Mycological Division. Admn. Rep. Div. Agric. Ceylon 1936, 1728–1735.

Parry, J.W., 1969. Spices., vol. II. Chemical Pub. Co., New York, USA.

Parvathi, C., Shadakshari, Y.G., Belavadi, V.V., Chandrappa, H.M., 1993. Foraging behaviour of honeybees on cardamom (*Elettaria cardamomum* Maton.). In: Veeresh, G.K., Umashanker, R., Ganesan, K.N. (Eds.), Pollination in Tropics (pp. 99–103). IUSSI, Bangalore.

Patel, D.V., Kuruvilla, K.M., Madhusoodanan, K.J., Potty, S.N., 1997. Regression analysis in small cardamom. In: Proc. Symp. Trop. Crop. Res. Dev, Trichur, India, in press.

Patel, D.V., Kuruvilla, K.M., Madhusoodanan, K.J., 1998. Correlation studies in small cardamom (*Elettaria cardamomum* Maton.). In: Mathew, N.M., Jacob, C.K. (Eds.), Developments in Plantation Crops Research (pp. 16–19). Allied Pub., New Delhi.

Patel, R.K., Gangrade, G.A., 1971. Note on the biology of castor capsule borer, *Dichocrocis punctiferalis*. Indian J. Agric. Sci. 41, 443–444.

Pattanshetty, H.V., 1980. Selections and cloning for high productivity in cardamom (*Elettaria cardamomum* (L.) Maton. var. Minor. Watt.) Ph.D. Thesis. University of Agricultural Sciences, Bangalore, India.

Pattanshetty, H.V., Nusrath, R., 1973. May and September are most optimum time for fertilizer application to cardamom. Curr. Res. 2 (7), 47–48.

Pattanshetty, H.V., Prasad, A.B.N., 1972. Early August is the most suitable time for planting cardamom in Mudigere area. Curr. Res. 1 (9), 60–61.

Pattanshetty, H.V., Prasad, A.B.N., 1973. September is the most suitable month for sowing cardamom seeds. Curr. Res. 2 (5), 26.

Pattanshetty, H.V., Prasad, A.B.N., 1974. Exposing the cardamom panicles from a layer of leaf mulch to open pollination by bees and thereby improving the fruit set. Curr. Res. 3 (8), 90.

Pattanshetty, H.V., Prasad, A.B.N., 1976. Blossom biology, pollination and fruit set in cardamom Chadha, K.L. Proc. Int. Symp. Subtrop. Trop. Hort, vol. 1. Today and Tomorrow's Publishers, New Delhi.

Pattanshetty, H.V., Nusrath, R., Sulikeri, G.S., Prasad, A.B.N., 1972. Effects of season-cum-depth of planting on the mortality of vegetatively propagated cardamom suckers (Rhizome sets of *Elettaria cardamomum* Maton.) Mysore J. Agric. Sci. 6 (4), 413–420.

Pattanshetty, H.V., Deshpande, R.S., Sivappa, T.G., 1973. Cardamom seedlings can be protected against damping off disease by the treatment with formaldehyde. Curr. Res. 2, 20–21.

Pattanshetty, H.V., Nusrath, R., Sulikere, G.S., Prasad, A.B.N., 1974. Suitable season and depth of planting cardamom suckers. Curr. Res. 3 (8), 84.

Pattanshetty, H.V., Rafeeq, M., Prasad, A.B.N., 1978. Influence of the length and type of storage on the germination of cardamom seeds. In: Nelliat, E.V. (Ed.), Proceedings of the First Annual Symposium on Plantation Crops (pp. 267–274). Indian Society for Plantation Crops, Central Plantation Crops Research Institute, Kasaragod.

Perry, I.M., Metzger, J., 1980. Medicinal Plants of East and South-East Asia.. MIT Press, Cambridge, USA.

Philip, T.E., 1989. Spices in cookery and preservation of food. In: Strategies for Export Development of Spices, Spices Board, Cochin, Kerala, India. pp. 40–53.

Pillai, G.B., Abraham, V.A., 1971. CPCRI Annual Report 1974, 146.

Pillai, O.G.N., Mathulla, T., George, K.M., Balakrishnan, K.V., Varghese, J., 1984. Studies in cardamom 2. An appraisal of the excellence of Indian cardamom (*Elettaria cardamomum* Maton.) Ind. Spices 21 (2), 17–25.

Pillai, P.K., Santha Kumari, S., 1965. Studies on the effect of growth regulators on fruit setting in cardamom. Agric. Res. J. Kerala 3, 5–15.

Pillai, V.G., 1953. Using long rhizomes for cardamom planting. Ind. Farming 3 (9), 12–13.

Ponnappa, K.M., Shaw, G.G., 1978. Notes on the genus Ceriospora in India. Mycologia 70, 859–862.

Ponnugangum, V.S., 1946. Management of cardamom seedlings in the nursery. Planters Chronicle 41 (6), 117–119.

Pradip Kumar, K., Mary Mathew, K., Rao, Y.S., Madhusoodanan, K.J., Potty, S.N., 1997. Rapid propagation of cardamom through *in vitro* techniques. In: Proc. 9th Kerala Sci. Congr. Trivandrum, Kerala, India, p. 185.

Pratt, P.F., 1965. PotassiumBlack, C.A. Methods of Soil Analysis, Part 2. ASA, Madison, WI. Agron. Monogr. 9.

Preisig, C.L., Moreau, R.A., 1994. Effects of potential signal transduction antagonists on phytoalexin accumulation in tobacco. Phytochemistry 36, 857–863.

Premavalli, K.S., Majumdar, T.K., Malini, S., 2000. Quality evaluation of traditional products. Garam masala and puliyogere mix masala. Indian Spices 37 (2), 10–13.

Premkumar, T., Devasahayam, S., Abdulla Koya, K.M., 1994. Pests of spice crops. In: Chadha, K.L., Rethinam, P. (Eds.), Advances in Horticulture, vol. 10. Plantation and Spice Crops. Malhotra Publishing House, New Delhi, India.

Priyadarshan, P.M., Zachariah, P.K., 1986. Studies on *in vitro* culture on cardamom (*Elettaria cardamom* Maton, Zingiberaceae): Progress and limitations. In: Proc. Int.Congr. Plant Tissue and Cell Culture,. Minnesota, USA p. 107 (abstract).

Priyadarshan, P.M., Kuruvilla, K.M., Madhusoodanan, K.J., 1988. Tissue culture technology: impacts and limitations. Spice India 1 (6), 31–35.

Priyadarshan, P.M., Kuruvilla, K.M., Madhusodanan, K.J., Naidu, R., 1992. Effect of various media protocols and genotype specificity on micropropagation of cardamom (*Elettaria cardamomum* Maton.). In: Subba Rao, N.S., Rajagopalan, C., Ramakrishna, S.V. (Eds.), New Trends in Biotechnology (pp. 109–117). Oxford and IBH Publications, New Delhi, India.

Pruthi, J.S., 1980. Spices and Condiments: Chemistry, Microbiology and Technology. Academic Press Inc., New York, USA.

Pruthy, J.S., 1979. Spices and Condiments. National Book Trust, New Delhi, India. pp. 63–68

Pruthy, J.S., 1993. Major Spices of India: Crop Management and Post Harvest Technology. ICAR, New Delhi, India. p. 514.

Pura Naik, J., 1996. Physico-chemical and technological studies on large cardamom (*A. subulatum)* for its use in foods Ph.D. Thesis. University of Mysore, Mysore, India.

Purseglove, J.W., Brown, E.G., Green, C.L., Robbins, S.R.J., 1981. Spices, vol. I. Longman, New York. pp. 174–228.

Puttarudriah, M., 1955. An epidemic outbreak of cardamom hairy caterpillar. Indian Coffee 19, 151–156.

Radha, K., Joseph, T., 1974. Investigations on the bud rot disease (*Phytophthora palmvi-vora* Butl.) of coconut. Final Report of PL 480. CPCRI, Kayamkulam, Kerala, India, 1968–1973, p. 30.

Radhakrishnan, C., 1993. Economics of pepper production in India. Report of the Research Project Sponsored by Spices Board, Government of India, Spices Board, Cochin, Kerala, India.

Rafatullah, S., Galal, A.M., Al-Yahya, M.A., Al-Said, M.S., 1995. Gastric and duodenal anti-ulcer and cytoprotective effects of Aframomum melegueta in rats. Int. J. Pharmacog. 33, 311–316.

Raghavan, B., Abraham, K.O., Shankaranarayana, M.L., 1990. Encapsulation of spice and other flavour materials. Indian Perfumer 34 (1), 75–86.

Raghavan, B., Abraham, K.O., Shankaracharya, N.B., Shankaranarayana, M.L., 1991. Cardamom: studies on quality of volatile oil and product development. Indian Spices 28 (93), 20–24.

Raghothama, K.G., 1979. Effect of mulches and irrigation on the production of suckers in cardamom (*Elettaria cardamomum* Maton.). M.Sc. (Agric.) Thesis. University of Agricultural Sciences, Bangalore, India.

Rai, S.N., 1978. Nursery and planting of some tropical evergreen and semi evergreen species. Tech. Bull. Karnataka Forest Department, p. 49.

Rajan, P., 1981. Biology of *Pentalonia nigronervosa f. caladii.* Van der Groot, vector of katte disease of cardamom. J. Plantation Crops 9, 34–41.

Rajan, P., Naidu, R., Venugopal, M.N., 1989. Effect of insecticides on transmission and acquisition of katte virus of small cardamom and their use in relation to disease spread and vector control. J. Plantation Crops 16 (Suppl), 261–266.

Rajapakse, L.S., 1979. G.L.C Study of the essential oil of wild cardamom oil of Sri Lanka. J. Sci. Food Agric. 30, 521–527.

Ramakrishnan, T.S., 1949. The occurrence of *Pythium vexans* de Bary in South India. Indian Phytopathol 2, 27–30.

Ramana, K.V., Eapen, S.J., 1992. Plant parasitic nematodes of black pepper and cardamom and their management. In: Proceedings of National Seminar on Black pepper and Cardamom, Calicut, Kerala, India, May 17–18, 1992, pp. 43–47.

Ranganathan, V., Natesan, S., 1985. Cardamom soils and manuring. Planters Chronicle 80 (7), 233–236.

Rangaswami, G., Seshadri, V.S., Lucy Chanamma, K.M., 1968. A new Cercospora leaf spot of cardamom. Curr. Sci. 37, 594–595.

Ranjithakumari, B.D., Kuriachan, P.M., Madhusoodanan, K.J., Naidu, R., 1993. Studies on light requirements of cardamom nursery. J. Plantation Crops 21 (Suppl), 360–362.

Rao, B.S., Sudborough, J.J., Watson, H.F., 1925. Notes on some Indian essential oils. J. Indian Inst. Sci. 8A, 143.

Rao, D.G., 1977. Katte disease of cardamom and its control. Indian J. Hort. 34, 184–187.

Rao, D.G., 1977. Katte disease of small cardamom. J. Plantation Crops 5, 23–27.

Rao, D.G., Naidu, R., 1973. Studies on Katte or mosaic disease of small cardamom. J. Plantation Crops 1 (Suppl), 129–136.

Rao, D.G., Naidu, R., 1974. Additional vectors of Katte disease of small cardamom. Indian J. Hort. 31, 380–381.

Rao, D.G., Naidu, R., 1974. Chemical control of nursery leaf spot disease of cardamom caused by *Phyllosticta elettariaae*. J. Plantation Crops 2, 14–16.

Rao, N.S.K., Narayana Swamy, S., Chacko, E.K., Doreswamy, M.E., 1982. Regeneration of plantlets from callus of *Elettaria cardamomum* Maton. Proc. Indian Acad. Sci. 91 (B), 37–41.

Rao, Y.S., Gupta, U., Kumar, A., Naidu, R., 1990. Phenotypic variability in large cardamom. In: Proceedings of the National Symposium on New Trends in Crop Improvement of Perennial Spices. Rubber Research Institute of India, Kottayam, Kerala, India.

Rao, Y.S., Kumar, A., Chatterjee, S., Naidu, R., George, C.K., 1993. Large cardamom (*Amomum subulatum* Roxb.): a review. J. Spices Aromatic Crops 2 (1-2), 1–15.

Ratnam, B.P., Korikanthimath, V.S., 1985. Frequency and probability of dry spells at Mercara. Geobios 12, 224–227.

Ratnavele, M.V.S., 1968. Value of soil and plant analysis. Cardamom News 2 (6), 1–4.

Ravindran, P.N., Rema, J., Nirmal Babu, K., Peter, K.V., 1997. Tissue culture and *in vitro* conservation of spices: an overview. In: Edison, S., Ramana, K.V., Sasikumar, B., Nirmal Babu, K., Eapen, S.J. (Eds.), Biotechnology of Spices, Medicinal and Aromatic Plants (pp. 1–12). Indian Society for Spices, Calicut.

Raychaudhary, S.P., Chatterjee, S.N., 1965. Transmission of chirke disease of large cardamom by aphid species. Indian J. Entomol. 27, 272–276.

Raychaudhary, S.P., Ganguly, B., 1965. Further studies on "chirke" disease of large cardamom (*Amomum subulatum* Roxb.). Indian Phytopathol 18, 373–377.

Reghunath, B.R. (1989). *In vitro* studies on the propagation of cardamom (*Elettaria cardamomum* Maton). Ph.D. Thesis. Kerala Agric. University, Trichur, India.

Reghunath, B.R., Bajaj, Y.P.S., 1992. Micropropagation of cardamom. In: Bajaj, Y.P.S. (Ed.), Biotechnology in Agriculture and Forestry: High-Tech and Micropropagation III (pp. 175–198). Springer-Verlag, Berlin.

Reghunath, B.R., Gopalakrishnan, P.K., 1991. Successful exploitation of *in vitro* culture techniques for rapid clonal multiplication and crop improvement in cardamom. Proc. Kerala Sci. Cong. Kozhikode, pp. 70–71.

Reghupathy, A., 1979. Effect of soil application of insecticides on cardamom stem borer *Dichocrocis punctiferalis* Guen. In: Venkataraman, C.S. (Ed.), Proc. PLACROSYM I. Indian Society for Plantation Crops, Kasaragod, Kerala, India.

Richard, A.B., Wijesekera, R.O.B., Chichester, C.O., 1971. Terpenoids of cardamom oil and their comparative distribution among varieties. Phytochemistry 10, 177–184.

Ridley, H.N., 1912. Spices. McMillan & Co. Ltd., London.

Rijekbusch, P.A.H., Allen, D.J., 1971. Cardamom in Tanzania. Acta Hort 21, 144–150.

Roberto, C.J., 1982. The enzyme-linked immunovascular assay (ELISA) in the diagnosis of cardamom mosaic virus. Rev. Cafetaera 219, 19–23.

Rosengarten, F., 1973. The Book of Spices. Pyramid Books, New York.

Rosengarten Jr., F., 1969. The Book of Spices. Livingston Publishing Co., Wynwood, Pensylvenia, USA.

Rubido, J.F., 1967. Prospects of cardamom growing Gautemala. Revista Cafeteria Gautemala 71, 10–13.

Russo, J.R., 1976. Cryogenic grinding. Food Eng. Int. 1 (8), 33.

Sadanandan, A.K., Korikanthimath, V.S., Hamza, S., 1990. Potassium in soils of cardamom (*Elettaria cardamomum* Maton.) plantations. Seminar on Potassium in Plantation crops, Univ. Agric. Sci. Bangalore, India.

Sadanandan, A.K., Hamza, S., Srinivasan, V., 2000. Foliar nutrient diagnostic norms for optimising cardamom production. In: Spices and Aromatic Plants. Indian Soc. for Spices, IISR, Calicut, pp. 101–104.

Sahadevan, P.C., 1965. Cardamom. Farm Bulletin No. 4. Department of Agriculture, Kerala, India. p. 90.

Sahadevan, P.C., 1965b. Cardamom. The Agricultural Information Service, Trivandrum, Kerala, India.

Saigopal, D.V.R, Naidu, R, Joseph, T, 1992. Early detection of "Katte" disease of small cardamom through Enzyme Linked Immunoabsorbent Assay (ELISA). J. Plantation Crops 20 (Suppl), 73–75.

Sajina, A., Mini, P.M., John, C.Z., Nirmal Babu, K., Ravindran, P.N., Peter, K.V., 1997. Micropropagation of large cardamom (*Amomum subulatum* Roxb.). J. Spices Aromatic Crops 6, 145–148.

Sajina, A., Minoo, D., Geetha, S.P., Samsudheen, K., Rema, J., Nirmal Babu, K., et al., 1997. Production of synthetic seeds in a few spices crops. In: Edison, S., Ramana, K.V., Sasikumar, B., Nirmal Babu, K., Eapen, S.J. (Eds.), Biotechnology of Spices, Medicinal and Aromatic Plants (pp. 65–69). Indian Society of Spices, Calicut.

Saju, K.A., Venugopal, M.N., Mathew, M.J., 1997. Effect of phytosanitation on recurrence of Vein Clearing Virus (kokke kandu) disease of small cardamom (abst.). In: National Symposium on Resurgence of Vector Borne Viral Diseases. Gujarat Agricultural University, Anand, India, August 1–3, 1997, p. 15.

Saju, K.A., Venugopal, M.N., Mathew, M.J., 1998.). Antifungal and insect repellant activities of essential oil of turmeric (*Curcuma longa* L.) Curr. Sci. 75, 660–663.

Sakai, S., Nagamasu, H., 2000. Systematic studies of Bornean Zingiberaceae. II. Elettaria of Sarawak. Edinb. J. Bot. 57, 227–243.

Saleem, C.P., 1978. Sprinkler irrigation doubles cardamom yield. Cardamom 10 (6), 17–20.

Salzer, U.J., 1975. Analytical evaluation of seasoning extracts (oleoresins) and essential oils from seasonings. Int. Flavours Food Additives 6 (3), 151.

Samraj, J., 1970. Mosaic disease of cardamom. Cardamom News 4, 12.

Sankarikutty, B., Narayanan, C.S., Rajamani, K., Sumanthikutty, M.A., Omanakutty, M., Mathew, A.G., 1982. Oils and oleoresins from major spices. J. Plantation Crops 10 (1), 1–20.

Sankarikutty, B., Sreekumar, M.M., Narayanan, C.S., Mathew, A.G., 1988. Studies on microencapsulation of cardamom oil by spray drying technique. J. Food Sci. Technol., 352–356.

Sarath Kumara, S.J., Packiyajothy, E.V., Janz, E.R, 1985. Some studies on the effect of maturity and storage on the chlorophyll content and essential oils of the cardamom fruit (*Elettaria cardamomum*). J. Sci. Food Agric. 36 (6), 491–498.

Sastri, B.N., 1952. The Wealth of India: Raw Materials, (D-E), 150–160.
Sastry, M.N.L., Hegde, R.K., 1987. Pathogen variation in *Phytophthora* species affecting plantation crops. Indian Phytopathol. 40 (3), 365–369.
Sastry, M.N.L., Hegde, R.K., 1989. Variability of *Phytophthora* species obtained from plantation crops of Karnataka. Indian Phytopathol. 42 (3), 421–425.
Sayed, A.A.M., Korikanthimath, V.S., Mathew, A.G., 1979. Evaluation of oil percentage in different varieties and types of cardamom. Cardamom II (1), 33–34.
Schachtschabel, P., 1937. Aufnahme von nicht austauschbarem Kalium durch die Pflanze. Z. Pflanzenernahr. Bodenkd 48, 107–133.
Schachtschabel, P., 1961. Fixierung und Nachlieferung von Kalium-und Ammonium-Ionen. Beurteilung und Bestimmung des Kaliumversorgungs-grades von Boden. Landwirtsch Forsch 14, 29–47.
Schefer, J.J.C., Vreeke, A., Looman, A., Mondranondra, I.O., 1988. Composition of essential oil of the rhizome of *Amomum pavieanum* Pierre & Gagnep. Flavour Fragr. J. 3, 91–93.
Schuller, H., 1969. Die CAL-Methode, eine neue methode zur Bestimmung des pflanzen verfugbaren Phosphates in Boden. Z. Pflanzenernahr. Bodenkd 123, 48–63.
Schultz, W.G., Randall, J.M., 1970. Liquid carbon dioxide for selective aroma extraction. Food Technol. 24, 1282.
Sekhar, P.S., 1959. Control of cardamom hairy caterpillar. Ind. Coffee 23, 256–268.
Selim, H.M., 1992. Modeling the transport and retention of inorganics in soils. Adv. Agron. 47, 331–384.
Selvakumaran, S., Kumaresan, D., 1993. Final Report: ICAR Research Scheme on Studies on Bio-ecology, Damage Potential and Control of the Whitefly *Dialeurodes cardamomi* David and Subr. (Aleyrodidae: Homoptera) on Small Cardamom. Indian Cardamom Research Institute, Myladumpara, Spices Board, India, p. 44.
Shakila, R.J., Vasundhara, T.S., Rao, D.V., 1996. Inhibitory effect of spices on *in vitro* histamine production and histidine decarboxylation activity of *Morgenella morganii* and on the biogenic amine formation in market samples stored at 30 degree Celsius. Lebensm Uters Forsch 203, 71–76.
Shankar, B.D., 1980. Shade trees and their role in cardamom cultivation. Cardamom 12 (3), 3–11.
Shankaracharya, N.B., Natarajan, C.P., 1971. Cardamom: chemistry, technology and uses. Indian Food Packer 25 (5), 28–36.
Shankaracharya, N.B., Raghavan, B., Abraham, K.O., Shankaranarayana, M.L., 1990. Large cardamom: chemistry, technology and uses. Spice India 3 (8), 17–25.
Sharma, E., Ambasht, R.S., 1988. Nitrogen accretion and its energetics in Himalayan alder plantations. Funct. Ecol., 229–235.
Shultz, T.H., Flath, R.A., Black, D.R., Guadagni, D.G., Schultz, W.G., Teranishi, R., 1967. Volatiles from delicious Apple essence: extraction methods. J. Food Sci. 32, 279–283.
Shu-Yan, C., Sposito, G., 1981. The thermodynamics of ternary cation exchange systems and the subregular model. Soil Sci. Soc. Am. J. 45, 1084–1089.
Siddagangaih, Krishnakumar, V., Naidu, R., 1993. Effect of chloremquat, daminozide, ethephon and maleic hydrazide on certain vegetative characters of cardamom (*Elettaria cardamomum* Maton.) seedlings. Spices Aromatic Crops 2, 53–59.
Siddappaji, C., Reddy, D.N.R.N., 1972. A note on the occurrence of the aphid *Pentalonia nigronervosa f. caladii.* van der Groot (Aphididae: Homoptera) on cardamom (*Elettaria cardamomum* Maton.). Mysore J. Agric. Sci. 6, 194–195.
Siddappaji, C., Reddy, D.N.R.N., 1972. A note o the occurrence of the aphid, *Pentalonia nigronervosa f. caladii.* Van der Groot (Aphididae: hemiptera) on cardamom (*Elettartia cardamomum* Maton.). Mysore J. Agric. Sci 6, 192–195.

Siddappaji, C., Reddy, D.N.R.N., 1972a. Some insect and mite pests of cardamom and their control. In Third International Symposium on Subtropical and Tropical Horticulture, Today and Tomorrow Printers and Publishers, New Delhi, India, p. 174.

Siddaramaiah, A.L., 1967. Hints on raising of cardamom nursery in Mysore state. Cardamom News 1 (6), 4.

Siddaramaiah, A.L., 1988. Seed rot and seedling wilt—a new disease of cardamom. Curr. Res 17, 81–85.

Siddaramaiah, A.L., 1988. Stem, leaf sheath and leaf rot diseases of cardamom caused by *Sclerotium rolfsii* from India. Curr. Res 17, 51.

Singh, G.B., 1978. Large cardamom. Cardamom 10 (5), 3–13.

Singh, G.B., Rai, R.N., Bhutia, D.T., 1989. Large cardamom (*Amomum subulatum* Roxb.) plantation: on age old agroforestry system in eastern Himalayas. Agroforestry Syst 9, 241–257.

Singh, J., 1994. Studies on *Lema* sp. Sixth Annual Research Council Meeting on Small Cardamom and Other Spices. Agenda, Indian Cardamom Research Institute, Spices Board, Myladumpara, Spices Board, India, p. 23.

Singh, J., Varadarasan, S., 1998. Biology and management of leaf caterpillar *Artona chorista* Jordon (Lepidoptera: Zygaenidae), a major pest of large cardamom (*Amomum subulatum* Roxb.). PLACROSYM–XIII, Coimbatore, pp. 46–47(abstract).

Singh, J., Gopakumar, B., Kumaresan, D., 1993. Biological studies on blue butterfly, *Jamides alecto* (Felder) (Lycaenidae: Lepidoptera), a major cardamom capsule borer in Karnataka. Entomon 18, 85–89.

Singh, J., Sudharshan, M.R., Kumaresan, D., 1994. Cardamom foliar pest, lace-wing bug (*Stephanitis typica* (Distant) (Heteroptera: Tingidae)). Ann. Entomol 12 (2), 81–84.

Singh, J., Srinivas, H.S., John, K., 1998. Large cardamom cultivation in Uttar Pradesh Hills. Spice India 11 (4), 2–4.

Singh, V.B., Singh, K., 1996. Spices,. New Age International Ltd., New Delhi, India. p. 253

Sivadasan, C.R., Nair, C.K., Srinivasan, K., Mathews, K., 1991. Zinc requirement of cardamom (*Elettaria cardamomum* Maton.). J. Plantation Crops 18 (Suppl), 171–173.

Sivanappan, P.K., 1985. Soil conservation and water management for cardamom. Cardamom 18 (6), 3–8.

Soil Science Society of America (Committee Report), 1965. Glossary of soil science terms. Soil Sci. Soc. Am. Proc. 29, 330–351.

Sparks, D.L., 1987. Potassium dynamics in soils. Adv. Soil Sci. 6, 2–63.

Sparks, D.L., 1989. Kinetics of Soil Chemical Processes.. Academic Press, San Diego, CA.

Sparks, D.L., Huang, P.M., 1985. Physical chemistry of soil potassium. In: Munson, R.D. (Ed.), Potassium in Agriculture (pp. 20–276). American Society of Agronomy, Madison, WI.

Spices Board (1990). Status Paper on Spices Spices Board, Ministry of Commerce, Government of India, Cochin, Kerala, India.

Spices Board (2000). Draft Annual Report (1999–2000). Spices Board, Ministry of Commerce, Government of India, Cochin, Kerala, India, p. 22.

Sposito, G., 1981. Cation exchange in soils. A historical and theoretical perspective. In: Dowday, R.H. (Ed.), Chemistry in the Soil Environment. Soil Sci. Soc. Am., Madison, WI

Sposito, G., 1981. The Thermodynamics of the Soil Solution.. Clarendon Press, Oxford.

Sposito, G., Holtzelaw, K.M., Johnston, C.T., Le Vesque, C.S., 1981. Thermodynamics of sodium-copper exchange on Wyoming bentonite at 198 degree K. Soil Sci. Am. J. 45, 1079–1084.

Sposito, G., Holtzelaw, K.M., Charlet, L., Jonany, C., Page, A.L., 1983. Sodium-calcium and sodium-magnesium exchange on Wyoming bentonite in perchlorate and chloride background ionic media. Soil Sci. Soc. Am. J. 47, 51–56.

Sridhar, V.V., 1988. Studies on Nilgiri Necrosis, a new virus disease of cardamom. Ph.D Thesis. Tamil Nadu Agricultural University, Coimbatore, India.
Sridhar, V.V., Muthuswamy, M., Naidu, R., 1990. Survey on the occurrence of Nilgiri necrosis: a new virus disease of cardamom. South Indian Hort 38, 163–165.
Sridhar, V.V., Muthuswamy, M., Naidu, R., 1991. Yield loss assessment in Nilgiri Necrosis infected cardamom. South Indian Hort 39, 169–170.
Srinivasan, K., 1984. Organic carbon, total and available nitrogen in relation to cardamom soils. J. Institution Chemists (India) 56, 15–16.
Srinivasan, K., 1990. Potassium status and Potassium fixing capacity of cardamom soil. J. Institution Chemists (India) 62, 245–246.
Srinivasan, K., Bidappa, C.C., 1990. Evaluation of nutrient requirement of cardamom by desorption isotherm. J. Indian Soc. Soil Sci. 38, 166–168.
Srinivasan, K., Mary, M.V., 1981. Studies on fixation of phosphorus in acid soils of cardamom. J. Institution Chemists (India) 53, 145–146.
Srinivasan, K., Roy, A.K., Sankaranarayanan, S., Zachariah, P.K., 1986. Radial distribution of available nutrients in cardamom soils. J. Institution Chemists (India) 58, 61.
Srinivasan, K., Sudarshan, M.R., Nair, C.K., Naidu, R., 1992. Study on the feasibility of cardamom. J. Plantation Crops 20 (Suppl), 53–54.
Srinivasan, K., Rama Rao, K.V.V., Naidu, R., 1993. Organic matter addition and its nutrient contribution in cardamom plantations. J. Trop. Agric (31), 119–121.
Srinivasan, K., Vadiraj, B.A., Krishnakumar, V., Naidu, R., 1993. Micronutrient status of cardamom growing soils of South India. J. Plantation Crops 21 (Suppl), 67–74.
Srinivasan, K., Krishnakumar, V., Potty, S.N., 1998. Evaluation of fertilizer application methods on growth and yield of small cardamom (*Elettaria cardamom* Mation). In: Muralidharan, N., Rajkumar, R. (Eds.), Recent Advances in Plantation Crops Research (pp. 199–202). Allied Pub., New Delhi.
Srinivasan, K., Vadiraj, B.A., Krishnakumar, V., Potty, S.N., 2000. Available sulphur status of cardamom growing soils of South India. *J. Plantation Crops*, in press.
Srivastava, K.C., Bopaiah, M.G., Ganapathy, M.M., 1968. Soil and plant analysis as a guide to fertilization in cardamom. Cardamom News 2 (3), 4–6.
Srivastava, L.S., 1991. Occurrence of spike, root and collar rot of large cardamom in Sikkim. Plant Diseases Res 6, 113–114.
Steffens, D., Mengel, K., 1979. Das Aneignungsvermogen von *Lolium perenne* im Vergleich zu *Trifolium pratense* fur Zwischenschilht-Kalium der Touminerale. Landw. Forsch. SH 36, 120–127.
Steffens, D., Mengel, K., 1981. Vergleichende Untersuchungen zwischen *Lolium perenne* und *Trifolium pratense* uber das Aneignungsvermogen von Kalium. Mitt. Dtsch. Bodenk. Ges. 32, 375–386.
Stone, G.N., Willmer, P.G., 1989. Pollination of cardamom in Papua New Guinea. J. Agric. Res. 28 (4), 228.
Subba, J.R., 1984. Agriculture in the Hills of Sikkim,. Sikkim Science Society, Gangtok, Sikkim. p. 286.
Subba Rao, G., Naidu, R., 1981. Breeding cardamom for Katte disease resistance. In: Proc. PLACROSYN 1V, Indian Society for Plantation Crops. Kasaragod, Kerala, India, 1980, pp. 320–324.
Subba Rao, M.K., 1938. Report of the mycologist 1937–1938. Admn. Rept. Tea Sci. Dept. United. Plant. Assoc., S. India, pp. 28–42.
Subba Rao, M.K., 1939. Report of the mycologist 1937–1939, 1937–1938. Adm. Rept. Tea. Res. Dept. United Plant. Assoc., S. India, pp. 28–37.

Subbaiah, M.S., 1940. Cardamom cultivation in the Bodi hills. Madras Agricultural J 28 (10), 279–288.

Subbarao, G., Korikanthimath, V.S., 1983. The influence of rainfall on the yield of cardamom (*Elettaria cardamomum* Maton.) in Coorg district. J. Plantation Crops 11 (1), 68–69.

Sudarshan, M.R., Bhatt, S.S., Narayanaswamy, M., 1997. Variability in the tissue cultured cardamom plants. In: Edison, S., Ramana, K.V., Sasikumar, B., Nirmal Babu, K., Eapen, S.J. (Eds.), Biotechnology of Spices, Medicinal and Aromatic Plants (pp. 98–101). Indian Society for Spices, Calicut.

Sudarshan, M.R., Kuruvilla, K.M., Madhusoodanan, K.J., 1988. Productivity phenomenon in cardamom and its importance in plantation management. Spice India 1 (6), 14–21.

Sudarshan, M.R., Kuruvilla, K.M., Madhusoodanan, K.J., 1991. AS key to the identification of types in cardamom. J. Plantation Crops 18, 52–55.

Sulikeri, G.S. (1986). Effect of light intensity and soil moisture on growth and yield of cardamom (*Elettaria cardamomum* Maton.). Ph.D. Thesis. Univ. Agric. Sci., Bangalore, India.

Sulikeri, G.S., Kologi, S.D., 1978. *Phyllanthes emblica* L. leaves: a suitable mulch for cardamom nursery beds. Curr. Res 7 (1), 3–4.

Sulikeri, G.S., Pattenshetty, H.V., Kologi, S.D., 1978. Influence of the level of water table on the growth and development of cardamom (*Elettaria cardamomum* Maton.). In: Nelliat, E.V. (Ed.), Proceedings of the First Symposium on Plantation Crops, Kottayam (pp. 531). CPCRI, Kasaragod, Kerala, India.

Sundriyal, R.C., Rai, S.C., Sharma, E., Rai, Y.K., 1994. Hill agroforestry, system in South Sikkim, India. Agroforestry Syst. 26, 215–235.

Suresh, M.P., 1987. Value added products: better scope with cardamom. Cardamom 20 (1), 8–11.

Survey of India's Export Potential of Spices, 1968. The Marketing Research Corporation of India. Ltd., New Delhi vol. 2A, B59.

Suseela Bhai, R., Thomas, J., Naidu, R., 1993. Biological control of "Azhukal" disease of small cardamom caused by *P. meadii* Mc. Rae. J. Plantation Crops 21 (Suppl), 134–139.

Suseela Bhai, R., Thomas, J., Naidu, R., 1994. Evaluation of carrier media for field application of *Trichoderma* sp. in cardamom growing soils. J. Plantation Crops 22 (1), 50–52.

Suseela Bhai, R., Thomas, J., Sarma, Y. S. R., 1997. Biocontrol of capsule rot of cardamom. International Conf. Integrated Plant Disease Management for Sustainable Agriculture, New Delhi, India, November 10–15, 1997.

Suseela Bhai, R., Thomas, J., Naidu, R., 1988. Anthracnose: a new disease of cardamom. Curr. Sci. 57, 1346–1347.

Tainter, D.R., Grenis, A.T., 1993. Spices and Seasonings.. VCH Pub., Inc., USA.

Thankamma, L., Pillai, P.N.R., 1973. Fruit rot and leaf rot disease of cardamom in India. F.A.O. Plant Prot. Bull 21, 83–84.

Thimmarayappa, M., Shivashankar, K.T., Shanthaveerabhadraiah, S.M., 2000. Effect of organic manure and inorganic fertilizers on growth, yield attributes and yield of cardamom (*Elettaria cardamomum* Maton.) J. Spice Aromatic Crops 9, 57–59.

Thirumalachar, M.J., 1943. A new rust disease of cardamom. Curr. Sci. 12, 231–232.

Thomas, E.K., Indiradevi, P., Jessy Thomas, K., 1990. Performance of Indian cardamom: an analytical note. Spices India 3 (6), 19–23.

Thomas, J., Vijayan, A.K., 1994. Occurrence, Severity, Etiology and Control of Rhizome Disease of Small Cardamom, Kerala Agrl University, Vellanikkara, Trichur district.

Thomas, J., Suseela Bhai, R., Dhanapal, K., Vijayan, A.K., 1997. Integrated management of rot diseases of cardamom. International Conference on Integrated Plant Disease Management for Sustainable Agriculture. November 10–15, New Delhi, India.

Thomas, J., Naidu, R., Suseela Bhai, R., 1988. Rhizome and root rot diseases of cardamom. A review. In Proceedings of the Workshop on Strategies of the Management of Root Disease Management for Sustainable Agriculture, January 14–18, 1988, pp. 38–45.

Thomas, J., Suseela Bhai, R., Naidu, R., 1989. Comparative efficacy of fungicides against *Phytophthora* rot of small cardamom. Pesticides, 40–42.

Thomas, J., Suseela Bhai, R., Naidu, R., 1991. Capsule rot disease of cardamom *Elettaria cardamomum* (Maton.) and its control. J. Plantation Crops 18 (Suppl), 264–268.

Thomas, J., Suseela Bhai, R., Naidu, R., 1991b. Management of rot diseases of cardamom through bio-agents. In: National Seminar on Biological Control in Plantation crops, RRII, Kottayam, Kerala, India, June 27–28, p. 21 (abstract).

Thomas, K.M., 1938. Detailed Administrative Report of the Government Mycologist, Madras, 1937–1938.

Thyagaraj, N.E., Chakravarthy, A.K., Rajagopal, D., Sudharsan, M.R., 1991. Bioecology of cardamom root grub *Basilepta fulvicorne* Jacoby (Eumolpinae: Chrysomelidae: Coleoptera). J. Plantation Crops 18, 316–319.

Tokita, F., Ishikawa, M., Shibuya, K., Koshimizu, M., Abe, R., 1984. Nippon Nogeikagaku Gakkaishi 58, 585. (quoted from Hirasa and Takemasa, 1998)

Tomlinson, D.L., Cox, P.G., 1987. A new disease of cardamom (*Elettaria cardamomum*) caused by *Erwinia chrysanthemi* in Papua New Guinea. Plant Pathol 36, 79–83.

Trease, G.E., Evans, W.C., 1983. Pharmacognosy, twelth ed. Baillere Tindall, London.

Tsuda, Y., Kiuchi, F., 1989. Quoted by Hirasa and Takemasa. Koshinryo no Kinou to Seibun, Kouseikan, Tokyo, Japan 1998.

Turner, G., Gershenzon, J., Nielsen, E.E., Froehlich, J.E., Croteau, R., 1999. Limonene synthase, the enzyme responsible for monoterpene biosynthesis in peppermint is localized in leucoplasts of oil gland secretory cells. Plant Physiol. 120, 879–886.

Upadhyaya, R.C., Ghosh, S.P., 1983. Wild cardamom of Arunachal Pradesh. Indian Hort 27 (4), 25–27.

Uppal, B.N., Varma, P.M., Capoor, S.P., 1945. A mosaic disease of cardamom. Curr. Sci. 14, 208–209.

Vadiraj, B.A., Srinivasan, K., Potty, S.N., 1998. Major nutrient status of cardamom soils of Kerala. J. Plantation Crops 26 (2), 159–161.

Vanselow, A.P., 1932. Equilibria of the base-exchange reactions of bentonites, permutities, soil colloids and zeolites. Soil Sci 33, 95–113.

Varadarasan, S., 1995. Biological control of insect pests of cardamom. In: Ananthakrishnan, T.N. (Ed.), Biological Control of Social Forest and Plantation Crops Insects (pp. 109–119).

Varadarasan, S., Kumaresan, D., Gopakumar, B., 1988. Occurrence of root grubs as a pest of cardamom (*Elettaria cardamomum* Maton.) Curr. Sci. 57, 36.

Varadarasan, S., Kumaresan, D., Gopakumar, B., 1989. Bioecology and management of cardamom shoot/panicle/capsule borer *Conogethes punctiferalis* Guen. Spice India 2, 15–17.

Varadarasan, S., Kumaresan, D., Gopakumar, B., 1990. Bi-annual Report 1987–1988 and 1988–1989. Indian Cardamom Research Institute, Myladumpara, Spices Board, India. pp. 65–66.

Varadarasan, S., Kumaresan, D., Gopakumar, B., 1991. Dynamics of the life cycle of cardamom shoot borer, *Conogethes punctiferalis* Guen. J. Plantation Crops 18 (Suppl), 302–304.

Varadarasan, S., Manimegalai, R., Sivasubramonian, T., Kumaresan, D., 1990. Integrated management of cardamom root grubs. Spice India 3, 9–11.

Varadarasan, S., Sivasubramonian, T., Manimegalai, R., 1991. Control of root grub: trial cum demonstration. Spice India 4, 7–8.

Varkey, G.A., Gopalakrishnan, M., Mathew, A.G., 1980. Drying of cardamom. Workshop on Agricultural Engineering and Technology in Kerala. Agricultural University, Trichur, Kerala, India.

Varma, P.M., Capoor, S.P., 1953. Marble disease of cardamom. Indian Farming 3, 22–23.

Varma, P.M., Capoor, S.P., 1958. Mosaic disease of cardamom and its transmission by the banana aphid *Pentalonia nigronervosa* Coq. Indian J. Agric. Sci 28, 97–107.

Varma, P.M., 1962. The banana aphid (*Pentalonia nigronervosa* Coq.) and transmission of katte disease of cardamom. Indian Phytopath 15, 1–10.

Varma, P.M., 1962. Control of Katte or mosaic disease of cardamom in North Kanara. Arecanut J 13, 79–88.

Varma, P.M., Capoor, S.P., 1964. "Foorkey" disease of large cardamom. Indian J. Agric. Sci. 34, 56–62.

Varma, S.K., 1987. Preliminary studies on the effect of honey bees on the yield of greater cardamom. Indian Bee J 49, 25–26.

Vasanthakumar, K., Mohanakumaran, N., 1998. Synthesis and translocation of photosynthates in cardamom. J. Plantation Crops 17, 96–100.

Vasanthkumar, K., Sheela, V.C., 1970. Sprinkler irrigation in cardamom plantations. Cardamom 23 (3), 5–9.

Vasanthakumar, K., Mohanakumaran, N., Narayanan, C.S., 1989. Quality evaluation of three selected cardamom genotypes at different seed maturity stages. Spice India 2, 25.

Vasanthakumar, K., Mohanakumaran, N., Wahid, P.A., 1989. Rate of photosynthesis and translocation of photosynthates in cardamom. J. Plantation Crops 17 (2), 96–100.

Vatsya, B., Duiesh, K., Kundapurkar, A.R., Bhaskaran, S., 1987. Large scale plant formation of cardamom (*Elettaria cardamomum*) by shoot tip cultures. Plant Physiol. Biochem. 14, 14–19.

Venugopal, M.N., 1995. Viral diseases of cardamom (*Elettaria cardamomum* Maton.) and their management. J. Spices Aromatic Crops 4, 32–39.

Venugopal, M.N., 1999. Natural disease escapes as a source of resistance against cardamom mosaic virus causing katte disease of cardamom (*Elettaria cardamomum* Maton.) J. Spices Aromatic Crops 8, 145–151.

Venugopal, M.N., Govidaraju, C., 1993. Cardamom vein clearing virus: a new threat to cardamom. Spice India 5 (2), 6–7.

Venugopal, M.N., Naidu, R., 1981. Geographical distribution of Katte disease of small cardamom in India (abstract). In: Third International Symposium on Plant Pathology, New Delhi, India, December 14–18, pp. 221–222.

Venugopal, M.N., Naidu, R., 1987. Effect of natural infection of Katte mosaic disease on yield of cardamom—a case study. In: Proc. PLACROSYM - V1. Oxford and IBH, New Delhi, India, pp. 115–119.

Venugopal, M.N., Saju, K.A., Mathew, M.J., 1997a. Primary spread of cardamom mosaic virus under different field situations. In: National Symposium on Resurgence of Vector Borne Diseases. Gujarat Agricultural University, Anand, India, August 1–3, 1997, p. 15.

Venugopal, M.N., Saju, K.A., Mathew, M.J., 1997b. Transmission and serological relationship of cardamom vein clearing disease (Kokke Kandu) (abstract). In: Symposium on Economically Important Plant Diseases, Bangalore, India, December 18–20, 1997, p. 60.

Venugopal, M.N., Mathew, M.J., Anandaraj, M., 1999. Breeding potential of *Pentalonia nigronervosa f. Caladii* Van der Groot (Horopotera: Aplindidae) on katte escaped cardamom (*Elettartia cardamomum* Maton.). J. Spices Aromatic Crops 8, 189–191.

Venugopal, R., Parameswar, N.S., 1974. Pollen studies in cardamom (*Elettaria cardamomum* Maton.). Mysore J. Agric. Sci. 8, 203–205.
Verghese, J., 1985. On the husk and seed oils of *Elettartia cardamomum* Maton. Cardamom 18 (10), 9.
Verghese, J., 1996. The worlds of spices and herbs. Spice India 12, 20.
Viswanath, S., Siddaramaiah, A.L., 1974. *Alpinia neutans,* a new host of Katte disease of cardamom. Curr. Res. 3, 96.
Wallis, T.E., 1967. Textbook of Pharmacognosy, Fourth ed. J&A Churchill, London, UK.
Warrier, P.K., 1989. Spices in Ayurveda. In: Strategies of Export Development of Spices. Spices Board, Cochin, Kerala, India, pp. 28–29.
Watt, G., 1872. Dictionary of Economic Products of India vol. 3, 227–236.
Wijesekera, R.O.B., Nethsingha, C., 1975. Compendium on Spices of Sri Lanka. 1: Cardamom. Ceylon Institute of Scientific and Industrial Research and National Science Council of Sri Lanka, Kularatne & Co. Ltd., Colombo, Sri Lanka, p. 10.
Willis, J.C., 1967. A Dictionary of the Flowering Plants and Ferns, 7th ed. Cambridge University Press,
Wilson, K.I., Rahim, M.A., Luke, P.L., 1974. *In vitro* evaluation of fungicides against azhukal disease of cardamom. Agric. Res. J. Kerala 12, 94–95.
Wilson, K.I., Sasi, P.S., James Mathew, J., 1979. *Fusarium* capsule disease of cardamom. Curr. Sci. 48, 1005.
Wilson, K.I., Sasi, P.S., Rajagopalan, B., 1997. Damping off of cardamom caused by *Rhizoctonia solani* Kuhn. Curr. Sci. 48, 364.
Wistreich, H.E., Schafer, W.F., 1962. Free-grinding ups product quality. Food Eng 34 (5), 62.
Wood, L.K., De Turk, E.E., 1941. The adsorption of potassium in soils in replaceable forms. Soil Sci. Soc. Am. Proc. 5, 152–161.
Xu, M.L., Liu, Z.Y., 1983. The nutrient status of soil–root interface. II. Potassium accumulation and depletion in rhizosphere soils. Acta Pedol. Sinica 20, 295–302.
Yamada, C.K., 1992. Mitogenic and complement activating activities of the herbal components. Planta Med. 58, 166–170.
Yamahara, J., Kashiwa, H., Kishi, K., Fujimura, H., 1989. Chem. Pharm. Bull. (Japan) 37, 855–856. (quoted from Hirasa and Takemasa, 1998)
Yaraguntaiah, R.C., 1979. *Curcuma neilgherensis* WT: a new host of Katte disease of cardamom. In: Proc. PLACROSYM II. Central Plantation Crops Research Institute, Kasaragod, Kerala, India, pp. 313–315.
Yaw Bin, H., JiaYou, F., ChenHsun, H., PawChu, W., Yittung, T., 1999. Cyclic monoterpene extract from cardamom oil as a skin permeation enhancer for indometacin; *in vitro* and *in vivo* studies. Biol. Pharm. Bull. 22, 642–646.
Yuri, Y., Izumi, K., 1994. Aromatopica 3, 65. (quoted from Hirasa and Takemasa, 1998).
Zachariah, P.K., 1976. Mulching in cardamom plantations. Cardamom News 8 (1), 3–9.
Zachariah, P.K., 1978. Fertilizer management for cardamom. In: Proceedings of the 1st Annual Symposium on Plantation Crops, Kottayam, pp. 141–156.
Zachariah, T.J., Lukose, R., 1992. Aroma profile of selected germplasm accessions in cardamom (*Elettaria cardamomum* Maton.). J. Plantation Crops 20 (Suppl), 310–312.
Zachariah, T.J., Mulge, R., Venugopal, M.N., 1998. Quality of cardamom from different accessions. In: Mathew, N.M., Jacob, C.Kuruvilla (Eds.), Developments in Plantation Crops Research (pp. 337–340). Allied Publishers Ltd., Mumbai.

Printed and bound by CPI Group (UK) Ltd, Croydon, CR0 4YY
08/06/2025
01896872-0005